# THE FORMATION OF THE GERMAN CHEMICAL COMMUNITY
(1720–1795)

# The Formation of the German Chemical Community (1720–1795)

## Karl Hufbauer

UNIVERSITY OF CALIFORNIA PRESS
*Berkeley • Los Angeles • London*

University of California Press
Berkeley and Los Angeles, California
University of California Press, Ltd.
London, England
© 1982 by
The Regents of the University of California
Printed in the United States of America

1  2  3  4  5  6  7  8  9

**Library of Congress Cataloging in Publication Data**

Hufbauer, Karl.
 The formation of the German chemical community.

 Includes index.
 1. Chemistry—Germany—History.  I. Title.
QD18.G3H83      540'.943      81-2988
ISBN 0-520-04318-9            AACR2
ISBN 0-520-04415-0 (pbk.)

# CONTENTS

Preface / vii

1 INTRODUCTION / 1

2 MORAL SUPPORT / 13

3 MATERIAL SUPPORT / 30

4 MANPOWER SUPPORT / 50

5 LORENZ CRELL: CHEMICAL JOURNALIST / 62

6 THE FORMATION OF THE GERMAN CHEMICAL COMMUNITY / 83

7 THE "FRENCH CHEMISTRY" / 96

8 THE NOTORIOUS REDUCTION EXPERIMENT / 118

9 CONCLUSION / 145

Appendix I  BIOGRAPHICAL PROFILES / 153

Appendix II  INSTITUTIONAL HISTORIES / 225

Appendix III  CRELL'S SUBSCRIBERS (1784–1791) / 271

Subject Index / 301

Person and Place Index / 305

# ILLUSTRATIONS

| | | |
|---|---|---:|
| Map 1. | Salaried Chemical Positions in German Institutions of Learning, 1780 | 31 |
| Map 2. | Residences of Selected Chemists on the Eve of the Formation of the German Chemical Community, 1780 | 52 |
| Figure 1. | G. E. Stahl | 9 |
| Figure 2. | J. C. Wiegleb | 27 |
| Figure 3. | L. Crell | 63 |
| Figure 4. | Title page, *Chemische Annalen* | 81 |
| Figure 5. | S. F. Hermbstaedt | 113 |
| Figure 6. | M. H. Klaproth | 116 |
| Figure 7. | F. A. C. Gren | 121 |
| Figure 8. | J. F. Westrumb | 124 |

# PREFACE

In 1964, when I began research on the social history of German chemistry in Hans Rosenberg's seminar, I planned to use a brief chapter on eighteenth-century developments as an introduction to a full study of the science's professionalization in the nineteenth century. Over the next two years, however, I abandoned this plan. I was intrigued by indications that Germany's chemists coalesced into a national discipline-oriented community during the 1780s and that their new social solidarity affected their reception of Lavoisier's revolutionary theory during the 1790s. I was also impressed by the multiplicity and richness of materials bearing on the background and early history of the German chemical community. Accordingly, encouraged by Paul Forman, I decided to concentreate on chemistry in eighteenth-century Germany. My investigation resulted first in a dissertation, then in various articles, and finally in the monograph that follows.

I can acknowledge here but a few of the many people who have generously assisted me as my work has progressed. Roger Hahn, Hans Rosenberg, Charles Susskind, Paul Forman, Arnold Thackray, Bob Multhauf, Mary Ellen Bowden, Steve Turner, and Eric Robinson gave my dissertation thoughtful critiques upon its completion, inspiring me to take the project further. Everett Mendelsohn, Jon Jacobson, and Joe Slavin played key roles in assuring the continuation of my academic career, and hence my research. Tom Saine, Mike Johnson, Jon Wiener, Ed Todd, Larry Holmes, and anonymous readers made many substantive suggestions for improving the drafts preceding the present text. Gunter Mende taught me how to read Crell's handwriting; David Heifetz aided me in assembling data regarding professorial careers; Ted Brunner, Richard Frank, Ray Oliver, and Craig Longuevan guided me through Latin and Italian passages; Cheryl Danieri assisted with the final editing of Appendices I and II; Henry Lowood, Paul Forman, and John Neu helped me locate and obtain the portraits; and Karin Fouts prepared the maps. Sally Hufbauer, Anne Rogers, Ellen Bork, and Jane Hedges have, in turn, served as my instructors in style. And the University of California Press has displayed virtuosity in transforming a complicated manuscript into the ensuing book.

Finally, I owe special thanks to those whose steadfast encouragement has enabled me to bring the project to completion—Paul Forman, Sally Hufbauer, Alan and Anne Rogers, Spece Olin, Jon Jacobson, Cathy Smith,

**Preface / viii**

Georg Harig, Mike Johnson, my children Sarah Beth, Ben, and Ruth Hufbauer, and my parents Clyde and Arabelle Hufbauer.

Irvine, California  K. H.
January 1982

# 1 INTRODUCTION

Late in the eighteenth century, German chemists came to constitute a community. They began, that is, to regard one another as important peers, as primary arbiters of truth and merit. In Helmstedt in 1790, L. Crell expressed this new sense of community when he wrote across Germany to K. G. Hagen in Königsberg that a recent discovery made in Hameln by their countryman J. F. Westrumb provided "a new buttress for our phlogiston which the French want to steal from us."[1] The following year in Berlin, F. Wolff voiced this same collective consciousness when he called upon "the most respected chemists of Germany" to serve as terminological "legislators for their German chemical *Mit bruder*."[2] Even before Germany was unified or chemistry was professionalized, German chemists had coalesced into one of the first national discipline-oriented communities.

The formation of the German chemical community, though hitherto unexamined by historians of science, warrants attention for three reasons. First, analyzing how support for chemistry rose to the level that German chemists could form into a scientific community contributes to a more complete understanding of the growth of science in eighteenth-century Germany. One cause of this growth, as contemporaries and historians have often remarked, was the ambition of a few German rulers to enhance their glory by imitating the French crown's *Académie*.[3] Another, as historians of German universities have recently shown, was the readiness of some German administrators and professors to promote university attendance, and revenues, by modernizing curricula and encouraging professorial publication.[4] Yet another, as historians of German literature have established, was the eagerness of many writers and readers to reflect on the wonders of God's creation.[5] However, it was not

---

1. A. Hagen, "K. G. Hagen's Leben und Wirken," *Neue preussische Provinzial-Blätter*, 9 (1850), 84. For biographical profiles of Crell, Hagen, and Westrumb, see Appendix I.

2. J. A. Chaptal, *Anfangsgründe der Chemie*, trans. F. Wolff, vol. I (Königsberg: Nicolovius, 1791), p. 4.

3. J. Ben-David has emphasized the role of French-style academies in supporting science in eighteenth-century Germany. See his *The Scientist's Role in Society: A Comparative Study* (Englewood Cliffs, N.J.: Prentice-Hall, 1971), pp. 85, 111.

4. R. S. Turner, "University Reformers and Professorial Scholarship in Germany 1760–1806," in *The University in Society*, ed. L. Stone, vol. II (Princeton: Princeton University Press, 1974), pp. 495–531; C. E. McClelland, "The Aristocracy and University Reform in Eighteenth-Century Germany," in *Schooling and Society: Studies in the History of Education*, ed. L. Stone (Baltimore, Md.: Johns Hopkins University Press, 1976), pp. 146–173; and J. L. Heilbron, *Electricity in the 17th and 18th Centuries: A Study of Early Modern Physics* (Berkeley: University of California Press, 1979), pp. 137–140.

5. For orientation to these literary studies, see W. Schatzberg, "Scientific Themes in the Popular Literature and Poetry of the German Enlightenment, 1720–1760," *German Studies in*

only to emulate the French, attract students, and worship God that Germans supported science during the century of Enlightenment. As the growth of moral, material, and manpower support for chemistry reveals, Germans also did so because they came to believe that natural knowledge could be *used* to improve health and increase production.[6]

Second, analyzing the way in which German chemists drew upon social support for chemistry to form into a national community illuminates an important phase in the genesis of the social organization of modern science. Since the fifth century B.C., Western intellectuals have divided natural knowledge into various subjects, each encompassing distinctive phenomena, techniques, and goals. Among the newer fields, as O. Hannaway has recently shown, was chemistry, which A. Libavius, in reaction to Paracelsian universalism, defined and delimited around 1600.[7] However, before the eighteenth century, the several branches of natural knowledge had more definitional integrity than social reality. While these "disciplines," as they were sometimes called, were used in book cataloguing and university teaching, they did not correspond, except in a tenuous way, to enduring social networks. For the most part, men of science lacked both incentives for sustained attention to a single discipline and reliable means for identifying and communicating with others, who, at any given time, happened to share their interests. Consequently, they oriented themselves to the "republic of letters" which, from the 1660s onwards, found its most visible manifestations in royal societies and academies and in comprehensive reviewing journals.[8] This state of affairs continued until the last third of the eighteenth century when, with the quickening of scientific activity and the appearance of specialized periodicals, the republic of letters began to break up along vernacular and disciplinary lines. Increasingly, men of science clustered into national discipline-oriented communities. In Germany such communities eventually had a significance extending far beyond their constituents. Between the 1790s and the 1840s they played an essential role in transforming the German universities into research centers, thereby making possible the professionalization of the sciences.[9]

---

*America*, 12 (1973); and T. P. Saine, "Natural Science and the Ideology of Nature in the German Enlightenment," *Lessing Yearbook*, 8 (1976), 61–88. Also see O. Sonntag, "The Motivations of the Scientist: The Self-Image of Albrecht von Haller," *Isis*, 65 (1974), 336–351.

6. The importance of eighteenth-century utilitarianism for the emergence of social support for chemistry in France and Britain has been emphasized by H. Guerlac, "Some French Antecedents of the Chemical Revolution," *Chymia*, 5 (1959), 73–112; and A. Donovan, "British Chemistry and the Concept of Science in the Eighteenth Century," *Albion*, 7 (1975), 131–144.

7. O. Hannaway, *The Chemists and the Word: The Didactic Origins of Chemistry* (Baltimore, Md.: Johns Hopkins University Press, 1975).

8. R. Hahn, *The Anatomy of a Scientific Institution: The Paris Academy of Sciences, 1666–1803* (Berkeley: University of California Press, 1971).

9. J. Ben-David's provocative analysis of the nineteenth-century development of German universities into scientific centers neglects the role of preexisting discipline-oriented communities in this process. See his *The Scientist's Role in Society*, pp. 108–138. R. S. Turner, by contrast, has correctly suggested that these communities began to influence the universities as early as the 1790s and were playing a prominent role in appointments and advancements by the mid-1830s. See his "University Reformers . . . ," in *The University and Society*, II: 510–511, 522–525, 531.

Hence, the formation of the German chemical community illustrates how scientists in one field, despite residing in a fragmented nation and working at various occupations, achieved sufficient solidarity in the late eighteenth century to transform their science into a profession during the nineteenth century.

Third, examining the response of the newly formed German chemical community to Lavoisier's antiphlogistic theory corroborates the importance of extrinsic factors in scientific revolutions. As early as the late eighteenth century, Lavoisier—evidently the first self-conscious revolutionary in science—noted that his allies tended to be physicists, mathematicians, and younger chemists and his foes, older and German chemists.[10] Since Lavoisier's day, scientists have often remarked that responses to competing theories have not depended solely on their empirical underpinnings, logical coherence, and heuristic value. An explanation for this seemingly aberrant behavior was offered by T. S. Kuhn in his provocative study of scientific revolutions. Because rival theories, he argued, are ultimately incomparable, or "incommensurable," theoretical allegiances are open to extraneous influences.[11] The late eighteenth-century debate between German phlogistonists and oxygenists both confirms and amplifies Kuhn's interpretation of scientific revolutions. The protagonists' stances and the revolution's timing, it seems clear, were profoundly influenced by the structure and values of the emergent German chemical community.

Evidence bearing on the formation of the German chemical community has proved to be abundant. Much relevant material reposes in the manuscripts and publications of the German chemists and their contemporaries.[12] More lies in scores of biographies[13] and institutional histories.[14] Still more is

---

10. I. B. Cohen, "The Eighteenth-Century Origins of the Concept of Scientific Revolution," *Journal of the History of Ideas*, 37 (1976), 257–288. Also see H. Guerlac, "The Chemical Revolution: A Word from Monsieur Fourcroy," *Ambix*, 23 (1976), 1–4.

11. T. S. Kuhn, *The Structure of Scientific Revolutions* (Chicago: University of Chicago Press, 1962). H. G. McCann has recently sought to test Kuhn's theory with a sociological study of the antiphlogistic revolution in France. See his *Chemistry Transformed: The Paradigmatic Shift from Phlogiston to Oxygen* (Norwood, N.J.: Ablex, 1978). Though McCann makes an admirable attempt at quantification, his analysis is flawed—see my "A Test of the Kuhnian Theory," *Science*, 204 (1979), 744–745.

12. For a fairly comprehensive yet rather slapdash guide to the publications of eighteenth-century German chemists, see the *Chemisch-Pharmazeutisches Bio- und Bibliographikon*, ed. F. Ferchl (Mittenwald: Nemayer, 1937).

13. Among the numerous biographies referenced in the chemists' profiles in Appendix I, I am particularly impressed by B. von Freyberg, "Johann Gottlob Lehmann (1719–1767): Ein Arzt, Chemiker, Metallurg, Bergmann, Mineraloge und grundlegender Geologe," *Erlanger Forschungen: Reihe B: Naturwissenschaften*, 1 (1955); G. E. Dann, *Martin Heinrich Klaproth (1743–1817): Ein deutscher Apotheker und Chemiker: Sein Weg und seine Leistung* (Berlin: Akademie Verlag, 1958); W. Herrmann, "Bergrat Henckel: Ein Wegbereiter der Bergakademie," *Freiberger Forschungshefte: Kultur und Technik*, D37 (1962); I. Mieck, "Sigismund Friedrich Hermbstaedt (1760 bis 1833): Chemiker und Technologe in Berlin," *Technik-Geschichte*, 32 (1965), 325–382; and W. Götz, "Zu Leben und Werk von Johann Bartholomäus Trommsdorff (1770–1837)," *Quellen und Studien zur Geschichte der Pharmazie*, 16 (1977).

14. Among the many studies bearing on chemistry's place in the schools listed in Appendix II, the most valuable are H. Lehmann, *Das Collegium medico-chirurgicum in Berlin als Lehrstätte der Botanik und der Pharmacie* (Diss., Berlin University; Berlin: Triltsch & Huther, 1936); G.-A. Ganss, *Geschichte der pharmazeutischen Chemie an der Universität Göttingen*,

scattered through thematic studies dealing with the diffusion of a rational-utilitarian concept of chemistry in Germany,[15] the growing activity of German pharmacists in chemical instruction and research,[16] the establishment of German periodicals devoted to chemistry and related subjects,[17] and the reception by German chemists of Lavoisier's theory.[18]

Using evidence from these various sources, I propose a cultural-institutional explanation of the formation of the German chemical community and a conflict interpretation of its members' full realization of their new social solidarity. In Chapters II, III, and IV, I maintain that educated and powerful Germans, as a consequence of their growing approval of the Enlightenment and enlightened governance, provided chemistry with increasing moral, material, and manpower support between 1720 and 1780. In Chapters V and VI, I go on to argue that Crell, by drawing upon chemistry's Enlightenment audience for support, developed his new chemical journal (founded 1778) into a forum for German chemists and thereby fostered their coalescence into the German chemical community. And in Chapters VII and VIII, I analyze the antiphlogistic revolution in Germany, showing that the new community's structure and values shaped this theoretical upheaval and suggesting that Ger-

---

*dargestellt in ihrem Zusammenhang mit der allgemeinen und der medizinischen Chemie* (Diss., Göttingen University; Marburg: Euker, 1937); G. Kallinich, *Das Vermächtnis Georg Ludwig Claudius Rousseaus an die Pharmazie: Zweihundert Jahre Pharmazie an der Universität Ingolstadt-Landshut-München 1760–1960* (Munich: Govi, 1960); W. Oberhummer, "Die Chemie an der Universität Wien in der Zeit von 1749 bis 1848 und die Inhaber des Lehrstuhls für Chemie und Botanik," *Studien zur Geschichte der Universität Wien*, 3 (1965), 126–202; E. Mayr, "Die Entwicklung der Chemie und der pharmazeutischen Chemie an der Universität Leipzig" (Diss., Leipzig University, 1965); A. Hampel, "Die beiden Lehrstühle für Chemie und für Pathologie an der Erfurter Medizinischen Fakultät während der ersten Hälfte des 18. Jahrhunderts" (Diss., Erfurt Medizinische Akademie, 1968); and C. Meinel, "Die Chemie an der Universität Marburg seit Beginn des 19. Jahrhunderts: Ein Beitrag zu ihrer Entwicklung als Hochschulfach," *Academia Marburgensis*, 3 (1978).

15. W. Strube, "Die Auswirkung der neuen Auffassung von der Chemie in Deutschland in der Zeit von 1745 bis 1785" (Diss., Leipzig University, 1961); E. Schmauderer, "Chemiatriker, Scheidekünstler und Chemisten der Barock- und der frühen Aufklärungszeit," in *Der Chemiker im Wandel der Zeiten*, ed. E. Schmauderer (Weinheim: Verlag Chemie, 1973), pp. 101–205; and H. Schimank, "Der Chemiker im Zeitalter der Aufklärung und des Empire (1720–1820)," in ibid., pp. 207–258.

16. G. E. Dann, "Berlin als ein Zentrum chemischer und pharmazeutischer Forschung im 18. Jahrhundert," *Pharmazeutische Zeitung*, 112 (1967), 189–196; D. Pohl, *Zur Geschichte der pharmazeutischen Privatinstitute in Deutschland von 1779 bis 1873* (Diss., Marburg University; Marburg: Mauersberger, 1972); B. H. Gustin, "The Emergence of the German Chemical Profession 1790–1867" (Diss., University of Chicago, 1975); and E. Hickel, "Der Apothekerberuf als Keimzelle naturwissenschaftlicher Berufe in Deutschland," *Pharmazie in unserer Zeit*, 6 (1977), 15–22.

17. H. Harff, *Die Entwicklung der deutschen chemischen Fachzeitschrift: Ein Beitrag zur Wesensbestimmung der wissenschaftliche Fachzeitschrift* (Berlin: Verlag Chemie, 1941); and D. von Engelhardt, "Die chemischen Zeitschriften des Lorenz von Crell," in *Indices naturwissenschaftlich-medizinischer Periodica bis 1850*, ed. A. Geus, vol. II (Stuttgart: Hiersemann, 1974).

18. G. W. A. Kahlbaum and A. Hoffmann, "Die Einführung der Lavoisier'schen Theorie im besonderen in Deutschland," *Monographieen aus der Geschichte der Chemie*, 1 (1897); and H. Vopel, "Die Auseinandersetzung mit dem chemischen System Lavoisiers in Deutschland am Ende des 18. Jahrhunderts" (Diss., Leipzig University, 1972).

man chemists came away from this conflict more fully aware of their collective responsibility and power as custodians of German chemistry.

The entire argument presupposes that the formation of the German chemical community was essentially a social process. So far as I can discern, German chemists did not form into a national discipline-oriented community as a result of embracing a common paradigm for investigating and interpreting chemical phenomena. The views that they came to share during the eighteenth century—confidence in chemistry's usefulness and profundity, enthusiasm for chemical experimentation, pride in their identity as German chemists—were more akin to an ideology than a Kuhnian paradigm. However, this shared viewpoint, which derived from their shared experiences of trying to promote, advance, and harness chemistry within the context of German culture and institutions, did not in itself engender social solidarity. So long as they lacked a reliable means of communicating with one another, German chemists oriented themselves to local or cosmopolitan audiences. Only after Crell's journal provided them with a forum did they begin thinking of the collectivity of German chemists as their most important reference group. Notwithstanding its social origins, the newly formed community soon proved capable of influencing the cognitive development of German chemistry by conditioning, perhaps in some instances dictating, its members' responses to Lavoisier's chemical system. The community's formation intensified consensual pressures among German chemists, thereby introducing a significant new factor into the discipline's development in Germany.

Three appendices follow the text. Appendix I presents sixty-five profiles of chemists who were active in Germany during the eighteenth century. Appendix II presents fifty-five institutional histories dealing with chemistry's place in German schools and academies during the eighteenth century. And Appendix III presents a master list of the subscribers to Crell's *Chemische Annalen* between 1784 and 1791. These appendices are intended to give readers an opportunity to check and, perhaps, to refine or refute my observations about the growth of social support for chemistry in eighteenth-century Germany. They are also intended to provide scholars who wish to look at related scientific, medical, and technological developments in Germany between roughly 1675 and 1825 with useful information and tools for their inquiries.

By way of setting the stage for all that follows, I shall now characterize cultural conditions in early eighteenth-century Germany and delineate the varied aims of self-styled chemists working within that milieu. Around 1700 "Germany" was a quarrelsome congeries of some three hundred principalities and imperial cities linked together not only by a common mother tongue but also by various feudal obligations, military alliances, and commercial dealings. The most pronounced cultural cleavage within this ramshackle nation was still between Catholics and Protestants. In the Catholic territories literacy was low, the lay intelligentsia was inconsequential, and the universities were moribund. In the Protestant lands, by contrast, conditions favored a comparatively spirited and open intellectual life. Thanks to the many schools founded during and since the Reformation to promote lay Bible reading, literacy was fairly high, especially in the towns. Thanks to opportunities for employment in towns and some state bureaucracies, the lay intelligentsia was sizable and, in cultural matters, increasingly influential. Thanks to the growing importance attached to religious and secular learning, the universities, especially those in Leipzig, Jena, and Halle, were moderately cosmopolitan and progressive. Extolling the fecundity of this culture, one Leipzig graduate, Hamburg school rector J. H. Hübner, exclaimed in 1712 that

> during the last fifty or so years, there has been such a marked increase in the number of learned sciences that it would be necessary to double the number of university chairs for every discipline to be taught separately. Moreover, so much has been added to each science that the old natural philosophers, mathematicians, and historians, if they could rise again with all their knowledge, would only pass for poor beginners. Furthermore, the population of the empire of learning has increased so greatly that now, wherever one turns, there are swarms of learned people and many lowly sciences which used to be left to mechanics are cultivated by literati. Finally, the present century is imbued with so much curiosity that everyone wants to know everything, or at least something about everything. Because these eager learners could not reach their goal so long as Latin had a monopoly in all learned matters, the Germans have followed the example of other nations and translated almost all the sciences into their mother tongue.[19]

To judge from his enthusiasm, Hübner's Germany—i.e., Protestant, urban Germany—must have been, by the standards of the early eighteenth century, a relatively congenial setting for scholarly inquiry, scientific investigation, and learned charlantry.[20]

In fact, self-styled chemists were fairly numerous in Protestant Germany. The alchemists, at least the honest ones, were pursuing noble metals, won-

---

19. Hübner's remarks appeared in the foreword to his popular medical-technological-scientific lexicon, the baroque title of which further reveals the interests of progressive Germans of the early eighteenth century. See his *Curieuses und Reales Natur-Kunst-Berg-Gewerck- und Handlungs-Lexicon darinnen nicht nur die in der Philosophie, Physic, Medicin, Botanic, Chymie, Anatomie, Chirurgie und Apothecker-Kunst, wie auch in der Mathematic, Astronomie, Mechanic, Bürgerlichen und Kriegs-Baukunst, Schifffahrten, etc., Ferner bey den galanten und Ritterlichen Exercitien; bey Bergwercken, Jägerey, Fischerey, Gärtnerey; wie auch in der Kauffmannschafft, bey Buchhalten und in Wechsel-Sachen, bey Künstlern und Handwerckern gebräuchliche Termini technici oder Kunst-Wörter, nach Alphabetischer Ordnung ausführlich beschrieben werden*, 3d ed. (Leipzig: Gleditsch, 1717), pp. [iii–iv].

20. Of course, even in the most progressive towns relatively few of the inhabitants had a taste for reading. See, for instance, R. Engelsing, *Analphabetentum und Lektüre: Zur Sozialgeschichte des Lesens in Deutschland zwischen feudaler und industrieller Gesellschaft* (Stuttgart: Metzler, 1973).

drous panaceas, and transcendent knowledge. They worked in secrecy and, if they reported their endeavors at all, did so in cryptic writings so that unworthy adepts could not debase the art. Such precautions were evidently necessary, for one rhymester complained that

> Everyone wants to be an alchemist;
> Young men and old, the village idiots,
> The soldiers, priests, and merry politicians,
> Shorn monks, old wives, and broken-down physicians.[21]

Characteristic of the serious alchemist in early eighteenth-century Protestant Germany was the wandering pietistic physician, J. C. Dippel. We find him as a young man, delving into the secrets of metals at a country estate near Darmstadt around 1703 and evaluating Count Cajetano's gold recipes for Prussia's Frederick I in 1707. In later years, he was extolling his medicaments in 1728, selling a "particular *arcanum chemicum*" to Hesse-Darmstadt's Ernst Ludwig in 1732, and prophesying seventy-five more years of life for himself shortly before dying in 1733.[22]

More prosaic than the alchemists, the iatrochemists of early eighteenth-century Protestant Germany regarded chemistry as a handmaiden to medicine. Though some still propounded the bold chemical interpretations of health and sickness advanced in the preceding two centuries, these men were no longer pursuing physiological insights in their laboratories. Instead, they were content to seek novel drugs or cheaper methods of preparing existing remedies. J. H. Jüngken, town physician in Frankfurt am Main, was a representative figure. Certain that he, as a physician, knew best how to prepare potent chemical remedies, he engaged in a sideline drug business. He was so successful in selling his products that a Nuremburg barber charged him with infringing an imperial patent in 1702 and five Frankfurt apothecaries denounced him for dispensing medicines in 1726.[23]

Shunning the alchemists and iatrochemists, a growing number of chemists in early eighteenth-century Protestant Germany insisted that chemistry be approached as a rational science with wide-ranging applications. A few, inspired by the mechanical philosophy which prominent French and English chemists had espoused since the 1660s, believed that chemical phenomena could be explained in terms of atomic shapes and motions.[24] Most, however,

---

21. For this doggerel, the essence of which was caught for me by Raymond Oliver, see Pantaleon [F. Gassmann], *Examen alchymisticum* . . . (1676), quoted in *Deutsches Theatrum Chemicum, auf welchem der berühmtesten Philosophen und Alchymisten Schrifften . . . vorgestellet werden*, ed. F. Roth-Scholtz, 3 vols. (Nuremberg: Felssecker, 1728–1732), II:290. My brief characterization of alchemy in early eighteenth-century Germany is, for want of adequate secondary studies, based on a wide variety of sources, including Roth-Scholtz's collection of alchemical writings.
22. For a biographical profile of Dippel, see Appendix I.
23. For a biographical profile of Jüngken, see Appendix I.
24. The chief advocates of a mechanistic approach to chemistry in early eighteenth-century Germany were E. W. von Tschirnhaus, G. W. Leibniz, and F. Hoffmann. In 1700 Tschirnhaus, who had arranged for a German translation of N. Lemery's text, recommended this work on account of its "methodical" approach. See his *Gründliche Anleitung zu nützlichen Wissenschaf-*

rejected this reductionist program. Working within the framework established by Libavius, they regarded chemistry as an independent science with its own methods, concepts, and domain.[25] Chief among the champions of the antireductionist viewpoint was G. E. Stahl, professor of medicine at Halle University from 1694 to 1714, then Royal Physician in Berlin until his death in 1734 (Figure 1).[26] In Berlin, he waged such a vigorous campaign for his version of the rational-utilitarian approach to chemistry that J. F. Henckel hailed him as a "clear-sighted Columbus" in 1722 and another disciple credited him with revealing chemistry's "true essentialness, effectiveness, attractiveness, and usefulness" in 1734.[27] Ultimately the approach to chemistry advocated by Stahl and his disciples became the rallying platform of the German chemical community. Stahlian rhetoric warrants, therefore, a full exposition.

According to Stahl, "true chemistry" was "a rational, deliberate, and comprehensible investigation and processing" of natural substances that led to "fundamental knowledge." The true chemist, consequently, was inspired by a "truly rational enthusiasm for research," a desire to find the "true knowledge of the material composition of natural substances," and an eagerness to illuminate "the truth of natural composition for its own sake." As he made his inquiries, he could expect "intellectual pleasure, clear knowledge, and moderate advantages and benefits." He could, for example, use "rational chemical processes" to make appropriate changes in the strength of medicines. Or, more importantly, he could use chemistry to understand and improve mineral processing, distilling, brewing, and "many other generally useful things."[28]

*ten*, facsimile of the 1729 ed. (Stuttgart: Frommann, 1967), p. 52. Leibniz and G. E. Stahl crossed pens over the value of mechanics in medicine and chemistry in 1708. For a general analysis of their exchange, see L. J. Rather and J. B. Frerichs, "The Leibniz-Stahl Controversy," *Clio Medica*, 3 (1968), 21–40; 5 (1970), 56–67. The origins of Hoffmann's corpuscularianism are illuminated in K. E. Rothschuh, "Studien zu Friedrich Hoffmann (1660–1742), Zweiter Teil: Hoffmann, Descartes und Leibniz," *Sudhoffs Archiv*, 60 (1976), 235–270. Hoffmann's unflagging loyalty to the mechanical philosophy was manifest in his recommendation of 1738 that medical students use the texts of Lemery and H. Boerhaave to learn chemistry. See his "Medicus Politicus," in his *Opuscula Pathologico-Practica* (Venice: Balleon, 1739), p. 278. For a biographical profile of Hoffmann, see Appendix I.

25. The transmission and transformation of Libavius's view of chemistry in the German universities during the seventeenth century is, I understand, being investigated by Hannaway. Recently A. G. Debus had made a case for the influence of the Paracelsian tradition on eighteenth-century chemists. See his *The Chemical Philosophy: Paracelsian Science and Medicine in the Sixteenth and Seventeenth Centuries*, vol. II (New York: Science History, 1977). Stahl's overt hostility to Paracelsianism (see below) suggests that, in his case at least, such influence was not very strong.

26. For a biographical profile of Stahl, see Appendix I. The best account of Stahl's life and mature views on chemistry is I. Strube, "Der Beitrag Georg Ernst Stahls (1659–1734) zur Entwicklung der Chemie" (Diss., Leipzig University, 1960).

27. Henckel, *Flora Saturnizans, die Verwandschafft des Pflanzen mit dem Mineral-Reich, nach der Natural-Historie und Chymie* (Leipzig: Martini, 1722), p. 419; and the anonymous translator's foreword to Stahl, *Zymotechnia fundamentalis, oder Allgemeine grund-erkanntniss der Gährungs-Kunst* (Frankfurt a. M.: Montag, 1734), p. [vii].

28. For these and subsequent quotations from Stahl, see his *Zufällige Gedanken und nützliche Bedenken über den Streit, von dem so genannten Sulphure, und zwar sowol dem*

Fig. 1. G. E. Stahl, chief spokesman for the rational-utilitarian approach to chemistry in early eighteenth-century Germany.
Source: *Georgii Ernesti Stahlii Opusculum chymico-physico-medicum* (Halle: Orphanotrophei, 1715)

Likewise Stahl's disciple Henckel, physician and private chemistry teacher in Freiberg, maintained that chemistry held the key to nature's secrets, for it

---

*gemeinen verbrennlichen oder flüchtigen, als unverbrennlichen, oder fixen* (Halle: Wäysenhaus, 1718), pp. 1–19, 55, 195; his *Ausführliche Betrachtung und zulänglicher Beweiss von den Saltzen, dass dieselbe aus einer Zarten Erde, mit Wasser innig verbunden, bestehen* (Halle: Wäysenhaus, 1723), pp. 2–3, 59, 332–335; and his "Bedencken von der Gold-Macherey" (1726), in J. J. Becher, *Chymischer Glücks-Hafen*, 3d ed. (Leipzig: Kraus, 1755), pp. i–xx.

penetrated to the "core and essence of bodies." Although the chemist could not, for the time being, reasonably expect to construct a satisfactory "system" of natural knowledge, he could provide "suitable building materials" for "posterity" to construct such a system by patiently analyzing natural substances. In recompense for his efforts, the chemist could expect the joy of wresting clear concepts from experience, the pleasure of discovering unexpected truths, and the satisfaction of announcing his findings to the world. While the promotion of truth should be the chemist's first aim, Henckel saw no need to renounce "personal gain as a goal." After all, God regarded "the care of one's self and one's own in any honorable occupation as a legitimate aim." If the chemist proceeded cautiously, he could expect results with "actual advantages and usefulness in common life." Success was certain if, rather than seeking directly to make gold or some other specific product, the chemist were always alert to possible applications as he systematically investigated one substance after another.[29]

Similarly, Stahl's protégé C. Neumann, Court Apothecary and professor of chemistry in Berlin, maintained that the goal of "pure" chemistry was to reveal "the true innermost nature of the components of all bodies created by God." This goal could only be reached by the chemist who was willing to "take charcoal into his hands," for chemistry was, "praise God, a science of demonstration." Not only was "the science and practice of chemistry the proper workshop for the true investigation of the natural creation," but it was also "useful and indispensable for all the world's professions"—"pharmaceutical or medical chemistry" dealt with the preparation of medicines; "docimastic chemistry," with assaying and smelting; "mechanical chemistry," with arts and crafts such as glassmaking, dyeing, and painting; and "economical chemistry," with industries using salts or fermentation processes.[30]

Believing that "true chemistry" deserved respect, patronage, and cultivation, the Stahlians expressed concern about the reputation that alchemists and iatrochemists had given the subject. Ever since Paracelsus had brought chemistry into "general view," Stahl maintained, "shameless" claims had been made on its behalf. These claims had aroused "almost completely false, delusionary hopes of making gold" and "equally futile" aspirations of producing "panaceas or powerful and infallible arcana for specific diseases." In pur-

29. For Henckel's advocacy of the Stahlian program for chemistry, see his *Flora Saturnizans*, pp. 13–15, 347, 463; his *Pyritologia; oder: Kiess-Historie, als des vornehmsten Minerals, nach dessen Nahmen, Arten, Lagerstätten, Ursprung, Eisen, Kupfer, unmetallischer Erde, Schwefel, Arsenic, Silber, Gold, einfachen Teiligen, Vitriol, und Schmeltz-Nutzung* (Leipzig: Martini, 1725), pp. 12–13, 25–31, 890–891; and the posthumous edition of his lectures, *Unterricht von der Mineralogie oder Wissenschaft von Wassern, Erdsäften, Salzen, Erden, Steinen, und Erzen nebst angefügtem Unterrichte von der Chymia Metallurgica*, ed. J. E. Stephani, 2d ed. (Dresden: Gerlach, 1759), pp. 132, 135, 137. For a biographical profile of Henckel, see Appendix I.

30. For Neumann's advocacy of the Stahlian program, see his *Lectiones chymicæ von Salibus Alkalino-Fixis und von Camphora* (Berlin: Schlechtiger, 1727), pp. 2, 35; his *Lectiones Publicæ von vier subjectis Pharmaceuticis* (Berlin: Haude, 1730), p. 3; his *Disquisitio de Ambra grysea* (Dresden: Hilscher, 1736), p. 5; and the posthumous edition of his lectures, *Prælectiones Chemicæ*, ed. J. C. Zimmermann (Berlin: Rüdiger, 1740), pp. [i, iii]. For a biographical profile of Neumann, see Appendix I.

suit of "gold recipes and medical miracles," an "unruly clique" composed mostly of pharmacists and mining men had lost "their wealth, time, honorable livelihoods, health, good reputations, and credit." The "bungling," "irrelevant snooping," and misfortunes of this "host of uninformed gullible people" had brought chemistry into such ill-repute that "respectable people" were rightfully reluctant to have anything to do with the science. Eager to change this state of affairs, to set themselves off from their rivals, and to win support for "true chemistry," Stahl and his allies stigmatized the alchemists as "gold-cooks" and "swindlers," the iatrochemists as "medicasters" and "recipe-collectors," and the mechanists as "mere cerebral chemists" and "unchemical corpuscle rummagers."

The Stahlians were particularly anxious to distinguish themselves from the alchemists. Admitting that "mere word and name mongers" still regarded *Alchymie* and *Chymie* as equivalent, Stahl insisted that the two words had come, "although not so many years ago," to denote two completely different enterprises. Alchemy was "the mostly confused" and "largely futile and vain undertaking . . . to make gold." Chemistry, by contrast, was devoted to rational experimentation as a means of expanding fundamental knowledge of natural substances. Alchemy had had its chance to accomplish something; now it was chemistry's turn. This was all the more true because alchemy had four grievous faults. First, it was a "civil" nuisance because it had driven many to "beggary" and nourished "swindling." Second, alchemy was "morally" objectionable because it encouraged hankering after gold, silver, and fantastic medicines and distracted its enthusiasts from their obligations to God. Third, there were neither reliable historical grounds nor sound rational-empirical arguments for believing in the present feasibility of alchemy. And fourth, there were no teachers capable of giving "rational or even reliable instruction" in alchemy. Consequently, it was foolish to incur "the numerous certain inconveniences, dangers, and losses" that resulted from the pursuit of alchemy's "so poorly grounded and attested hopes." Rather, one should apply his "spare time and extra money" to chemical investigations because such inquiries accorded with "nature, reason, and true skill and dexterity." It might, Stahl conceded, someday prove possible to make gold, but the time was not yet ripe. One should "learn the alphabet and how to spell" before trying to comprehend "not only unknown handwriting but also abbreviations, even tangled codes." In a similar spirit, Henckel maintained that the efforts of "the gold-thirsty subjects of the alchemical empire," of "the alchemical nitpickers," were doomed to failure. Though making gold was surely possible, the method for doing so would only be discovered after chemistry had made much more progress.

Thus the Stahlians wanted their countrymen to view chemistry as an independent and profound natural science which could be applied not only in pharmacy but also in metallurgy, brewing, dyeing, glassmaking, and many other crafts. They also wanted their countrymen to appreciate, finance, even participate in their endeavors to advance chemical knowledge. However, the Stahlians' rhetoric on behalf of chemistry could only succeed insofar as their message, or at least parts of it, struck responsive chords in their audiences. Fortunately for them, they lived in more propitious times than their like-minded forerunners. As the Enlightenment unfolded, Germans, especially educated Germans in Protestant lands, became increasingly receptive and responsive to claims that chemistry could contribute to public health and material progress. By 1780, as we shall see in the next three chapters, chemistry enjoyed considerable moral, material, and manpower support in Germany.

# 2 MORAL SUPPORT

Chemistry, J. F. Gmelin reminded the reader of his *Geschichte der Chemie* in 1797, used to be misunderstood, disdained, and ridiculed. In his day, however, the science had become "the idol before which all peoples and all orders, princes, and subjects, clergy and laymen, the educated and uneducated, high and low, bend their knees."[1] Though Gmelin exaggerated, the moral support or esteem for chemistry did increase during the eighteenth century. In the early 1700s, chemists in Germany enjoyed scant encouragement. They were, as the Stahlians complained, likely to be associated with sooty laborants, venal charlatans, rabid gold-bugs, and brazen swindlers. Between the 1720s and 1770s, however, these negative associations gave way to much more favorable ones, slowly at first and then with increasing rapidity. By 1780 many educated and powerful Germans esteemed chemistry as a basic natural science of broad utility. The present chapter seeks to illuminate this emergence of moral support for chemistry.[2]

Before the widespread establishment of effective institutional patronage for research during the nineteenth and twentieth centuries, the most important way in which social groups promoted scientific endeavors was by giving these endeavors their moral support. When educated men esteemed a science, they were likely to follow its developments, embrace its doctrines, acclaim its participants, and respect its patrons. And when powerful and wealthy men prized a science, they were apt to applaud its achievements, honor its leaders, reward its participants, and finance its instruction. Naturally such expressions of esteem by significant social groups reinforced the morale and dedication of the science's participants, thereby promoting its cultivation and development.

The moral support that a social group accords a science and its devotees

1. J. F. Gmelin, *Geschichte der Chemie seit dem Wiederaufleben der Wissenschaften bis an das Ende des 18. Jahrhunderts*, facsimile of the 1797–1799 ed., vol. I (Hildesheim: Olms, 1965), p. 2. Gmelin's testimony warrants credence because, as the son of a chemist and a chemist himself, he would have been sensitive to changes in attitudes toward the science since the 1740s. For a profile of Gmelin, see Appendix I.
2. In writing this chapter, I have drawn heavily on my "Social Support for Chemistry in Germany during the Eighteenth Century: How and Why Did It Change?" *Historical Studies in the Physical Sciences*, 3 (1971), 205–231. In recent years the emergence of moral support for chemistry in eighteenth-century Germany has received considerable attention. See W. Strube, "Die Auswirkung der neuen Auffassung von der Chemie in Deutschland in der Zeit von 1745 bis 1785" (Diss., Leipzig University, 1961); E. Schmauderer, "Chemiatriker, Scheidekünstler und Chemisten des Barock und der frühen Aufklärungszeit," in *Der Chemiker im Wandel der Zeiten*, ed. E. Schmauderer (Weinheim: Verlag Chemie, 1973), pp. 101–205; and H. Schimank, "Der Chemiker im Zeitalter der Aufklärung und des Empire (1720–1820)," in ibid., pp. 207–258.

depends, I believe, on the relation between the group's image of the science and the group's values. A group that regards a science as irrelevant to or incompatible with its priorities will be indifferent or hostile to the science's representatives. By contrast, a group that sees a science as connected with or contributing to its aspirations will bestow its moral support on those associated with the science. A group's evaluation of a science may change solely because its perception of that science has altered. However, the process is usually more complex. A group's moral support for a science may change as the result of concurrent and interacting changes in both its values and its image of the science. Indeed, my thesis is that moral support for chemistry emerged in eighteenth-century Germany because educated and powerful Germans embraced new values that made them more receptive to new information being disseminated about that science. In developing this thesis, I first examine German attitudes toward chemistry in the early eighteenth century. Then I consider the spread of new values and the dissemination of new information about the science between 1720 and 1780. Finally, I present evidence of the consequent increase in moral support for chemistry.

**DIVIDED OPINIONS**

The testimony of Stahl and his disciples indicates that early eighteenth-century Germans, when they had any opinion at all, generally viewed chemistry as an art pursued for alchemical or pharmaceutical purposes. As J. F. Henckel observed in 1736, his contemporaries usually associated chemistry with "gold-making or . . . medical preparations."[3] The Stahlian assessment of chemistry's public image was not the product of paranoia.

For instance, the main article on chemistry in J. H. Hübner's *Curieuses und reales Natur-Kunst-Berg-Gewerck- und Handlungs-Lexicon* (1712) appeared under the heading of *Alchymia*. Referring readers who looked up either *Goldmacherey* or *Chemia* to this entry, the lexicon described alchemy as

> an art by virtue of which the pure is separated from the impure, or an art [for acquiring] an active and working knowledge of natural things. It may be called practical natural philosophy because it resolves and dissolves all sublunary bodies into their first seeds or prime matter and recoagulates these into their former bodies so that the medicines which can be made from these [bodies] will be safer and healthier. . . . One can divide the chemical art into vulgar or common [chemistry], known by apothecaries and physicians, and secret [chemistry], properly called alchemy, which concerns itself with the preparation of the

---

3. J. F. Henckel, letter to A. von Korff (1 February 1736), in W. Herrmann, "Bergrat Henckel: Ein Wegbereiter der Bergakademie," *Freiberger Forschungshefte: Kultur und Technik*, D37 (1962), 161.

stone of wisdom upon which the transmutation of metals depends.

Other entries in the lexicon reinforced the associations of chemistry with the preparation of drugs and the making of gold. The article on medicine identified *Chymiam* as one of five branches of medicine, defining it as "the art that teaches how the best force can be drawn from natural things by means of fire." And the article on *Lapis Philosophorum* implied that most of those who devoted themselves to *Chymie* were fools for believing that they could create the philosopher's stone despite the unreliability of existing recipes, which were the products of either "vain hope" or "plain fraudulence."[4]

Chemistry was portrayed in much the same way by J. T. Jablonsky in his *Allgemeines Lexicon der Künste und Wissenschaften* (1721). He described *Chymie, Alchymie, Chymia*, which he regarded as one of the six branches of medicine, as

> an art that teaches how one should dissolve natural substances, separate them from one another, combine them, and prepare wholesome medicines out of them; or [how one should] analyze mixed, compound, or aggregate bodies into their fundamental parts or synthesize the same bodies out of such fundamental parts. . . . Chemistry, insofar as it investigates the powers and properties of nature and teaches the preparation of drugs, is very useful and necessary to the doctor. However, one must not allow oneself to be seduced by it [chemistry] into trying to make the notorious philosopher's stone and thereby gold and silver.

Anyone who did so, Jablonsky warned his readers, would soon be just as impoverished as "many poor *Chymisten*."[5]

The lexicographers' treatment of chemistry attests to the credibility of the Stahlian assessment of the subject's public image. When early eighteenth-century Germans thought of chemistry, they were likely to associate it with alchemy and/or pharmacy. Their particular responses to these two associations depended on their specific contacts with chemistry and their perception of its relevance to their aspirations. Consequently, the ruling aristocracy, the physicians, and the educated public had different opinions about chemistry's worth because their links to the subject and their goals were different.[6]

---

4. J. Hübner, *Curieuses . . . Lexicon*, 3d ed. (Leipzig: Gleditsch, 1717), pp. 47–49, 383, 931, 1039.

5. J. T. Jablonsky, *Allgemeines Lexicon* (Leipzig: Frisch, 1721), pp. 50, 137–138. A few years later another lexicographer, who drew upon Jablonsky's work, indicated that *Alchymie* and *Chymie* were often used as synonyms, but suggested that it was more precise to use the first term for gold making and the second for drug making. See Sperander [F. Gladow], *A la Mode-Sprach der Teutschen, oder Compendieuses Hand-Lexicon* (Nuremberg: Buggenl & Seitz, 1727), pp. 24–25, 114.

6. In addition to the aristocrats, physicians, and educated public, the political economists and natural philosophers also had opinions on chemistry. My reason for ignoring their positions is that in the early eighteenth century they did not yet constitute significant social groups in

German rulers and nobles received most of their information about chemistry from importunate alchemists who played upon their fantasies of unlimited wealth. Like their principal informants, aristocrats tended to equate chemistry with gold-making. Many, in fact, patronized chemistry-alchemy as an intriguing gamble for immense riches. Even as they did so, however, they were haunted by apprehensions of being swindled. Inevitably confronted with failure or dishonesty, they often put their adepts under house arrest, sometimes threw them into dungeons, and occasionally sent them to the gallows. Thus, to judge from their actions, powerful Germans regarded chemistry-alchemy with considerable avarice and even more displeasure.[7]

While the German aristocracy of the early eighteenth century associated chemistry with alchemy, the physicians associated it with pharmacy. They acquired most of their knowledge about chemistry from proponents of chemical medicines who appealed to their desire for effective treatments. The tendency of physicians to identify chemistry with the compounding of drugs led the Galenists, a vanishing breed who regarded elaborated medicines as universally harmful, to despise the subject.[8] But for every hidebound Galenist, there was at least one enthusiast who, still championing the ambitious claims of the seventeenth-century iatrochemists, praised chemistry as "the principal and most noble part of medical study."[9] Most physicians eschewed these

---

Germany. Generally speaking, the political economists valued chemistry for its broad potential usefulness. See, for instance, the views of J. B. von Rohr (1688–1742) as summarized in U. Troitzsch, "Ansätze technologischen Denkens bei den Kameralisten des 17. und 18. Jahrhunderts," *Schriften zur Wirtschafts- und Sozialgeschichte*, 5 (1966), 44–48, 57–58. The natural philosophers valued chemistry more as a source of data than of fundamental insights. See, for instance, C. Wolff, *Vernünfftige Gedancken von den Würckungen der Natur*, 3d ed. (Halle: Renger, 1734), pp. 48, 443, 566–577, 596–597.

7. There has been no systematic empirical study of aristocratic patronage for alchemy in Germany during the eighteenth century. For a suggestive survey of the secondary literature, see W. Strube, "Ueber die Rolle der Alchimie in der Zeit vom 16. bis zum 18. Jahrhundert in Deutschland," Leuna-Merseburg Technische Hochschule für Chemie: *Wissenschaftliche Zeitschrift*, 5 (1963), 109–117. For the life of an aristocratic patron of alchemy who eventually became an alchemist, see Baron de Bournet, letters to J. F. Henckel (1732–1740), in *Mineralogische, Chemische und Alchymistische Briefe von reisenden und anderen Gelehrten an den ehemaligen Chursächsischen Bergrath J. F. Henkel*, vol. I (Dresden: Walther, 1794), pp. 70–94. For the "enlightened" Frederick II's patronage of alchemy, see his letters to M. G. Fredersdorf (1753), in *Die Briefe Friedrichs des Grossen an seinen vormaligen Kammerdiener Fredersdorf*, ed. J. Richter (Berlin-Grunewald: Klemm, 1923), pp. 211–213, 218, 220, 224, 228, 238, 262–263. Frederick soon withdrew from the venture, angry with himself for being such a fool and sympathetic with his brother-in-law, the Duke of Brunswick, who had more than ten alchemists in prison.

8. Though I have not succeeded in identifying any strict Galenists among eighteenth-century German physicians, their existence is attested to in the *Grosses vollständiges Universal-Lexicon*, ed. J. H. Zedler, vol. II (Leipzig: Zedler, 1732), p. 1743.

9. This claim for chemistry's central place in medicine was made in 1712 by a committee inspecting Tübingen University. See K. Klüpfel, *Geschichte und Beschreibung der Universität Tübingen* (Tübingen: Fues, 1849), p. 163. The claim probably reflected the attitudes of the medical professors, for Haller characterized their approach to therapy as "Ettmüllerian," i.e., as strictly iatrochemical. See "Albrecht Hallers Tagebücher seiner Reisen nach Deutschland, Holland und England 1723–1727," ed. E. Hintzsche, *Berner Beiträge zur Geschichte der Medizin und der Naturwissenschaften*, N.F. 4 (1971), 27.

extreme positions. They appreciated chemistry's contribution to the pharmacopoeia, without thinking of it as the only source of efficacious drugs. Most physicians, that is, valued chemistry, but only as a useful auxiliary discipline.[10]

Unlike their rulers and their doctors, educated Germans associated chemistry equally with alchemy and pharmacy. Lacking close contacts with self-proclaimed generators of gold and concocters of cures, they had a more balanced view of the subject. To the degree that they associated the subject with alchemy, educated Germans were outraged. As good burghers, they were offended by the flagrant hankering after riches of both adepts and patrons, by the ruinous consequences for those who got involved in gold-making, and by the waste of funds that might better be used for religious or social purposes.[11] Their association of chemistry with pharmacy did little to counteract this sense of outrage. Of course, educated Germans approved of the remarkable healings that chemical drugs sometimes effected. But they knew that these drugs were unreliable, often exacerbating the patient's misery. Furthermore, they were disconcerted by the doctors' rodomontades for "their fountains of youth, their incombustible oils, their hermetical antidotes, their elixirs of gold, their snake-powders and precious stones, their remedies for snake bites, their tinctures and panaceas, and their six hundred other remedies out of Arabia."[12] All in all, therefore, the educated public regarded chemistry with disdain and distrust.

## A BASIS FOR BROAD ESTEEM

From the 1720s through the 1770s, German attitudes toward chemistry both metamorphosed and converged. By 1780 there was extensive, though certainly not universal, esteem for chemistry. This emergence of moral support for the discipline resulted from two concurrent developments. As educated and powerful Germans gave their allegiance to the beliefs, values, and aspi-

---

10. For an interesting portrayal of chemistry's relation to medicine in the 1720s, see F. Börner, *Nachrichten von den vornehmsten Lebensumständen und Schriften jeztlebender berühmter Aerzte und Naturforscher in und um Deutschland*, 3 vols. (Wolfenbüttel: Meissner, 1749–1756), III: 451–454, 465. For abundant evidence that physicians saw chemistry as an auxiliary branch of medicine, see the histories of chemistry's place in German medical schools in Appendix II.
11. The educated public was already taking a dim view of alchemy by the late seventeenth century. See F. H. Wagman, *Magic and Natural Science in German Baroque Literature: A Study in the Prose Forms of the Later Seventeenth Century*, facsimile of the 1942 ed. (New York: AMS, 1966), pp. 43, 58–59, 88. In 1725 an important moral weekly pointed out that success in gold making would have disastrous social-economic consequences. See *Der Patriot*, 2 (13 December 1725), 450–458.
12. J. B. Mencke, *The Charlatanry of the Learned*, trans. from the 1715 ed. by F. E. Litz with notes by H. L. Mencken (New York and London: Knopf, 1937), p. 59. Mencke, a distant ancestor of the American pundit, also remarked that "although they [the doctors] are profoundly ignorant about the real effect of their medicines, they administer pills, syrups, drops and I cannot tell what other panaceas to their patients with an assurance so overwhelming that sometimes they even promise to restore the dead to life" (p. 165).

rations of the Enlightenment, they became increasingly receptive to the ongoing efforts of reform-minded chemists to disseminate a rational-utilitarian conception of chemistry.[13]

## Embracing the Enlightenment

By the early eighteenth century, educated Germans were ready for new ideas. Memories and reminders of the severe ravages of the Thirty Years' War were rapidly disappearing. Many towns were flourishing as never before. And several universities enjoyed unprecedented attendance. Small wonder that some Germans, influenced by the moral weeklies and by the secular wing of Pietism, abandoned otherworldly Christianity and its pessimistic passivity. They sought instead to improve private and public morality by living as productive and socially useful Christians.[14] Somewhat later in the century, many Germans were attracted to the rationalism expounded by C. Wolff and his disciples. In the 1740s these two currents converged in the *Aufklärung*, a movement dedicated to achieving social betterment through economic growth, medical improvement, educational reform, and intellectual progress. During the 1750s and 1760s, heartened by increasing aristocratic approval and, after the Seven Years' War (1756–1763), invigorating prosperity, most educated Germans embraced at least some tenets of the Enlightenment.[15]

The rate at which this secular movement gained adherents was reflected in the books displayed at the annual fair in Leipzig. For instance, after remaining fairly constant for decades, the number and percentage of books concerning theology fell rapidly between the 1740s and 1760s. A good part of the-

13. In explaining why reform-minded chemists succeeded in transforming chemistry's image and repute, W. Strube argues that the growth of capitalistic production, especially in industries using chemical processes and/or making chemical products, made the German bourgeoisie and officialdom increasingly receptive to the chemists' propaganda. See his "Zur Annäherung von Wissenschaft und Produktion im 18. Jahrhundert—Dargestellt am Beispiel der Chemie," *Jahrbuch für Wirtschaftsgeschichte*, no. 3 (1974), 141–165. There is little evidence that the connection between production and moral support for chemistry was as direct as Strube maintains. Nonetheless, as I suggest below, it is likely that prosperity, especially after the Seven Years' War, helped generate interest in chemistry and other sciences that promised to promote technological progress.

14. For the influence of the moral weeklies, see W. Martens, *Die Botschaft der Tugend: Die Aufklärung im Spiegel der deutschen moralischen Wochenschriften* (Stuttgart: Metzler, 1968). For the influence of Pietism, see K. Deppermann, *Der hallesche Pietismus und der preussische Staat unter Friedrich III. (1.)* (Göttingen: Vandenhoeck & Ruprecht, 1961); G. Kaiser, *Pietismus und Patriotismus im literarischen Deutschland: Ein Beitrag zum Problem der Säkularisation*, 2d ed. (Frankfurt am Main: Athenäum, 1973); and I. Tönnies, "Die Arbeitswelt von Pietismus, Erweckungsbewegung und Brüdergemeinde: Ideen und Institutionen; Zur religiös-sozialen Vorgeschichte des Industriezeitalters in Berlin und Mitteldeutschland," *Jahrbuch für die Geschichte Mittel- und Ostdeutschlands*, 20 (1971), 89–133; 21 (1972), 140–183.

15. The accounts of the *Aufklärung* that I have found most helpful—because they deal with Enlightenment utilitarianism—are H. Brunschwig, *La Crise de l'état prussien à la fin du xviii$^e$ siècle et la genèse de la mentalité romantique* (Paris: PUF, 1947); H. M. Wolff, *Die Weltanschauung der deutschen Aufklärung in geschichtlicher Entwicklung*, 2d ed. (Bern: Francke, 1963); W. Martens, "Von Thomasius bis Lichtenberg: Zur Gelehrtensatire der Aufklärung," *Lessing Yearbook*, 10 (1978), 7–34; and for Catholic Germany, F. G. Dreyfus, *Sociétés et mentalités à Mayence dans la seconde moitié du xviii$^e$ siècle* (Paris: Colin, 1968), pp. 403–441.

TABLE 1: Theology's Decline at the Leipzig Book Fair, 1740–1770

|  | 1741–1745 | 1761–1765 |
|---|---|---|
| Average number (%) of *all* books in theology | 432 (37%) | 273 (20%) |
|  | 1740 | 1770 |
| Number (%) of *new* books in theology | 291 (39%) | 280 (24%) |
| Number (%) of *new* books in science, medicine, and technology | 83 (11%) | 222 (19%) |

Source: Based on F. Kapp, *Geschichte des Deutschen Buchhandels bis in das siebzehnte Jahrhundert* (Leipzig: Börsenverein der Deutschen Buchhändler, 1886), pp. 791–804; and R. Jentzsch, "Der deutschlateinische Büchermarkt nach den Leipziger Ostermess-Katalogen von 1740, 1770 und 1800 in seiner Gliederung und Wandlung," *Beiträge zur Kultur- und Universalgeschichte*, 22 (1912), 316 and tables in the unpaginated appendix.

TABLE 2: New Periodicals in Germany for the Natural, Medical, and Productive Sciences, 1741–1780

|  | 1741–1750 | 1751–1760 | 1761–1770 | 1771–1780 |
|---|---|---|---|---|
| Number for natural sciences, including mathematics | 11 | 22 | 19 | 23 |
| Number for medical sciences | 7 | 9 | 13 | 28 |
| Number for productive sciences | 4 | 8 | 36 | 35 |
| Total (% all new periodicals) | 22 (8%) | 39 (12%) | 68 (17%) | 86 (12%) |

Source: Based on J. Kirchner, *Die Grundlagen des deutschen Zeitschriftenwesens mit einer Gesamtbibliographie der deutschen Zeitschriften bis zum Jahre 1790*, vol. II (Leipzig: Hiersemann, 1931), p. 340.

ology's relative decline was offset by gains in science, medicine, and technology (see Table 1). Meanwhile, there was a dramatic increase in the number of periodicals established for the natural, medical, and productive sciences (see Table 2). As educated Germans became more worldly after 1740, they clearly became more concerned with scientific and material advancement.

They were particularly interested in material progress. As early as the 1760s, indeed, materialistic values were becoming so strong that some intellectuals began to complain. The philosopher M. Mendelsohn, for example, thought it necessary to remind his readers that the discoverer of a new truth was a far greater man than the inventor of a new flea powder. Likewise the scholar J. G. Hamann ridiculed the enlightened man who was so devoted to his belly that he only sought those truths that helped him satisfy his material needs.[16] And in 1775 the poet C. M. Wieland felt obliged to criticize those

---

16. For the reactions of Mendelsohn and Hamann to the burgeoning of crude utilitarianism, see H. M. Wolff, *Die Weltanschauung der deutschen Aufklärung*, pp. 219, 252. Though Hamann's reaction to the crass side of secularization was typical of many elite intellectuals, his pervasive hostility to natural studies was unusual. For his belief that science was a source of

"despisers of beauty" who were blind to the usefulness of endeavors that did not fill their money bags.[17] The German Enlightenment was clearly a utilitarian as well as philosophical and aesthetic movement.

While this movement was capturing the imagination of the educated public, the idea of enlightened governance was gaining ground with ruling elites in Germany. Rulers and their close advisors had long been striving to consolidate and strengthen state power. Prussia's Frederick II and Austria's Empress Maria Theresa, both crowned in 1740, were among the first to recognize that the growing commitment of educated Germans to the Enlightenment could be used to this end. Though the ultimate goal of these rulers and a growing number of emulators was greater power rather than social meliorism, they too began to promote material and intellectual progress. Initially they proceeded at a dilatory pace. But after the Seven Years' War, with its legacy of immense debts and uneasy peace, ruling groups approached the task of promoting progress with greater urgency.[18]

At an accelerating rate from the 1720s through the 1770s, therefore, both educated and powerful Germans came to value material and intellectual progress. Their growing commitment to these goals created an increasingly favorable climate of opinion for all the natural sciences. By the late 1770s, as the educator F. K. Bahrdt observed, the person who lacked "natural knowledge" could not be "a complete agriculturalist, merchant, military man, or true man of learning."[19]

## Disseminating the Rational-Utilitarian Image of Chemistry

In the early decades of the eighteenth century, as we have seen, the Stahlians were busy agitating for their version of the rational-utilitarian approach to chemistry. Condemning alchemical and iatrochemical goals and methods, they portrayed chemistry as a fundamental and generally useful experimental science in their lectures and books, their conversations and correspondence. Gradually this image diffused outward from the Stahlians and their disciples, reaching many educated and powerful Germans by 1780.

---

hubris and disbelief as well as hedonism, see W. M. Alexander, *Johann Georg Hamann: Philosophy and Faith* (The Hague: Nijhoff, 1966), pp. 40, 183, 188.

17. C. M. Wieland, "Ueber das Verhältniss des Angenehmen und Schönen zum Nützlichen" (1775), in *Wieland's Werke*, vol. XXXII (Berlin: Hempel, 1879), pp. 33–42.

18. My interpretation of enlightened governance derives largely from H. Rosenberg, *Bureaucracy, Aristocracy, and Autocracy: The Prussian Experience, 1660–1815* (Cambridge, Mass.: Harvard University Press, 1958). For the change in pace after the Seven Years' War, see, for instance, H. Baumgärtel, "Bergbau und Absolutismus: Der sächsische Bergbau in der zweiten Hälfte des 18. Jahrhunderts und Massnahmen zu seiner Verbesserung nach dem Siebenjährigen Kriege," *Freiberger Forschungshefte: Kultur und Technik*, D44 (1963); and K. Witzel, "Friedrich Carl von Moser: Ein Beitrag zur hessen-darmstädtischen Finanz- und Wirtschaftsgeschichte am Ausgang des 18. Jahrhunderts," Historische Kommission für den Volksstaat Hessen: *Quellen und Forschungen zur hessischen Geschichte*, 10 (1929).

19. W. Schöler, *Geschichte des naturwissenschaftlichen Unterrichts im 17. bis 19. Jahrhundert: Erziehungstheoretische Grundlegung und schulgeschichtliche Entwicklung* (Berlin: de Gruyter, 1970), p. 57.

Of the many networks that spread the rational-utilitarian conception of chemistry, the one originating in medical education was the most important because it was the most extensive. From the 1720s onwards, a growing number of Germany's many medical schools offered chemistry courses once or twice a year (see Chapter III). These courses were based increasingly on texts that, though still oriented toward chemistry's pharmaceutical uses, opened with enthusiastic remarks about the broad possibilities of this science. Students attending these courses—chiefly prospective physicians, but also increasing numbers of would-be pharmacists and administrators—sometimes found the ideas about chemistry's promise so congenial that they passed them on later in their careers. Some physicians passed favorable ideas on to powerful and influential patients who were in a position to support chemistry. Some passed such ideas on to the pharmacists whose shops they inspected and patronized. In turn, some apothecaries passed them on to their apprentices and journeymen. And some of these young men, in their turn, passed them on again as they moved from shop to shop preparing for permanent positions.[20]

The mining network, though considerably less extensive, also played a role in disseminating the new image of chemistry. Since the sixteenth century, a few thoughtful mining administrators had been aware of chemistry's potential relevance to their industry. Until Stahl's disciple Henckel began his course on metallurgical chemistry at Freiberg, however, systematic instruction in this science was lacking in mining towns. Thanks to his and similar courses (see Chapter III), a growing number of prospective mining administrators came to view chemistry as a penetrating science with numerous uses in their own and other fields. Later in their careers, some of these men—most notably Henckel's famous pupil F. A. von Heynitz, who directed major reforms of the Saxon and Prussian mining industries—propounded this view to their rulers and subordinates.[21]

The effectiveness of these and more transitory networks in spreading the new image of chemistry was manifest in the subscription list appearing in C. F. Wenzel's book on chemical affinities published in 1777. In addition to several foreign subscribers, the list included 247 persons residing in some 85

---

20. For information concerning the medical network, see most of the profiles in Appendix I; most of the institutional histories in Appendix II; and the cumulative subscription list to Crell's *Chemische Annalen* in Appendix III. There is abundant additional evidence scattered through F. Börner, *Nachrichten von . . . Aerzte und Naturforscher*; E. G. Baldinger, *Biographien jetztlebender Aerzte und Naturforscher in und ausser Deutschland* (Jena: Hartung, 1768–1772); and J. K. P. Elwert, *Nachrichten von dem Leben und den Schriften jeztlebender teutscher Aerzte, Wundärzte, Thierärzte, Apotheker und Naturforscher* (Hildesheim: Gerstenberg, 1799).

21. For information concerning the mining network, see the profiles of Henckel, J. A. Cramer, C. E. Gellert, J. G. Lehmann, N. J. Jacquin, G. A. Scopoli, C. A. Gerhard, and I. Born in Appendix I and the institutional histories of the Schemnitz, Freiberg, and Berlin mining schools in Appendix II. For Heynitz, see W. Weber, "Innovationen im frühindustriellen deutschen Bergbau und Hüttenwesen: Friedrich Anton von Heynitz," *Studien zu Naturwissenschaft, Technik und Wirtschaft im Neunzehnten Jahrhundert*, 6 (1976).

German towns that were scattered, mostly in Protestant lands, from Königsberg in the northeast to Bern in the southwest. Of these 247 subscribers, 16 were nobles and 231 commoners. And of the 214 subscribers whose occupations were given, 90 were in medicine and surgery, 85 in pharmacy, 22 in government (5 in mining, 4 in the military), 8 in science-teaching and science, and the remaining 9 distributed among a medley of occupations ranging from hospital pastor to silk-dyer. In short, Wenzel's audience consisted primarily of Protestant commoners in medicine and pharmacy; but it had considerable geographical, social, and occupational diversity. That a book on such a specialized chemical topic should attract so many and such varied subscribers indicates that chemistry's new image as a basic and broadly useful science was widely accepted by the late 1770s.[22]

## NEW ESTEEM FOR CHEMISTRY

As values favorable to the natural sciences spread and as chemistry's new image diffused, positive attitudes about chemistry emerged in Germany. In some Protestant cultural centers, chemists became aware of the growing esteem as early as the 1740s. For example, the Berlin chemist J. H. Pott proclaimed in 1746 that

> rational chemistry [and its] further investigation presently
> awaken such general approval from most learned and reasonable
> men, not only within the cultured nations but probably also
> among the barbaric peoples, that its experiments and endeavors
> are always received by them with gratitude.[23]

Though hyperbolic, Pott's appraisal of public opinion was an accurate harbinger of the prevailing respect for chemistry some three decades later.

Thanks to frequent contacts with reform-minded spokesmen for chemistry, physicians were among the first to acclaim the science. In 1752 F. Börner, a Leipzig physician who was editing a bio-bibliographical series on "famous living doctors and natural scientists," began his account of Pott's life by observing that

> it cannot be denied that there has never been a lack of persons
> who conduct themselves as the declared and sworn enemies of
> chemistry and its admirers and who have banned it as

---

22. For the list of Wenzel's subscribers, see his *Lehre von der Verwandschaft der Körper* (Dresden: Gerlach, 1777), pp. 485–491. Those persons who later subscribed to Crell's *Chemische Annalen* (1784–1791) are indicated in Appendix III. For a biographical profile of Wenzel, see Appendix I.

23. J. H. Pott, *Chymische Untersuchungen welche fürnehmlich von der Lithogeognosia oder Erkäntniss und Bearbeitung der gemeinen einfacheren Steine und Erden ingleichen von Feuer und Licht handeln*, vol. I (Potsdam: Voss, 1746), p. [1]. The book is dedicated to G. D. von Arnim, Privy State and War Minister, President of the Justice Council . . . and "a great connoisseur and true promoter of *reeler* sciences." For a biographical profile of Pott, see Appendix I.

completely useless or unnecessary from the republic of letters. On the other hand, there has never been a lack of men who, as experienced judges of useful knowledge, have not counted it among the sciences that are indispensable for a true doctor and perform the most beneficial services for people of nearly all ranks. For it is chemistry alone that unlocks the secrets of nature and leads us to the very heart of things, discovering most wonderful contrivances that otherwise would be eternally hidden from us. It alone teaches us how to imitate nature, how to discern the essential components of things, how to create new substances. It alone explains the different characteristics of animal substances, an exact knowledge of which is necessary for understanding them. It alone teaches us how, starting with natural substances, to extract and compound numerous drugs, some of which are very salutary for the well-being of mankind. It alone shows us the reasons why we must avoid this or that thing as extremely harmful to our bodies while we can enjoy other natural substances as a means of sustaining health and prolonged life. It alone is the key to most human contrivances, arts, and crafts. . . . It furnishes painters, goldsmiths, dyers, glassmakers with countless useful things for reaching their goals. It teaches us how to smelt metals and minerals and to purify and combine them. It prepares useful liquids through fermentation such as wine, beer, mead, alcohol, and vinegar. It contributes much to the commonweal. It delights, entertains, and benefits. It is indispensable.[24]

Even though Börner admitted that some intellectuals continued to have serious doubts about chemistry, neither he nor his readers, at least that was his assumption, shared in this prejudiced view of the science.

Nine years after Börner's panegyric, the Berlin chemist J. G. Lehmann prefaced a volume of A. S. Marggraf's chemical treatises with remarks that further illuminate the growing esteem for chemistry. "There has," he claimed,

probably never been a century richer in chemical authors than the present and it goes with chemistry as with medicine, *Fingunt se Chymicos omnes*. The state official, the financier, the barber and surgeon, the brewer and distiller, the dyer, the tanner, the old woman, the charcoal carrier and woodcutter, and, yes, the project maker (oh, what a deplorable name), all are clever enough to count themselves among the chemists. Thousands, who have been ruined by such people, are the sad witnesses to this fact.

24. F. Börner, *Nachrichten von . . . Aerzte und Naturforscher*, II:485–487.

Lehmann went on to insist that his colleague had nothing in common with such frauds. Then, after summarizing the contents of Marggraf's papers, he condescendingly gave some practical hints to those "mechanical" men who were only interested in chemistry's applications. Finally, he declared that he was proud to serve as editor for "such useful and important treatises" because he was sure that they needed no recommendation for true men of learning.[25] Lehmann, though he was distressed that so many mountebanks could masquerade as chemists, clearly assumed that chemistry's audience included many discerning men. Indeed, both his haughty treatment of narrow utilitarians and his confidence in a favorable reception from true intellectuals suggest a new self-assurance on the part of chemists.

Lehmann and his fellow chemists had reason to be more confident than their predecessors. Besides physicians like Börner, influential proponents of economic development were portraying chemistry as an important subject. In 1763 D. G. Schreber, a cameralist in Bützow, gave chemistry a prominent place in an article recommending the establishment of scientific-technical schools for prospective administrators. He thought such schools should have five professors, one of whom would be responsible for teaching mineralogy and chemistry. This professor would teach

> physical chemistry in its entirety, demonstrating the requisite experiments in the laboratory; economic chemistry, which rests on the former, explaining the theory and practice of its various parts, namely, dyeing, salt and saltpeter works, glassmaking, lime and brick firing, ceramics, porcelain making, economic metallurgy (steel making, gold and silver work, wire making); mining science, namely mining, metallurgy, smelting, assaying, etc., by exhibiting the necessary models of ovens, etc. in the model collection.

This professor would also be expected to investigate the uses of local earths, stones, and minerals and to manage any of the school's factories that depended upon chemical operations.[26]

Again, in 1776 the Berlin encyclopedist J. G. Krünitz defined *Chemie* in his successful *Oeconomische Encyclopädie* as

---

25. J. G. Lehmann, "Vorrede" (25 March 1761), in A. S. Marggraf, *Chymischer Schriften*, vol. I (Berlin: Wever, 1761). "Everyone imagines himself a physician" was apparently an old saying. See J. B. Mencke, *The Charlatanry of the Learned*, pp. 37–38, 168. For biographies of Lehmann and Marggraf, see Appendix I.

26. D. G. Schreber, "Entwurf von einer zum Nutzen eines Staats zu errichtenden Academie der öconomischen Wissenschaften," in his *Sammlung verschiedener Schriften, welche in die öconomischen, Policey- und Cameral- auch andere Wissenschaften einschlagen*, vol. X (Halle: Curts, 1763), pp. 417–436. For evidence of Schreber's influence, see H. Webler, "Die Entwicklung der Kaiserslauterer Textilindustrie seit dem 18. Jahrhundert," Institut für Landeskunde des Saarlandes, Saarbrücken: *Veröffentlichungen*, 8 (1963), 13; and W. Stieda, "Das Projekt zur Errichtung einer 'Kameral-Hohenschule' in München im Jahre 1777," *Forschungen zur Geschichte Bayerns*, 16 (1908), 91–93.

that science which has as its subject the study of the nature and properties of all substances by decomposing and combining them. . . . Chemistry not only acquaints us with the nature and properties of substances, but also teaches us the correct handling of substances in order to make them useful in the world.

In support of this contention he proceeded to discuss chemistry's many applications in agriculture and the crafts.[27]

Two years later the *Bergmännisches Wörterbuch*, which appeared anonymously in the Saxon mining town of Chemnitz, defined *Chymie* as "a science that investigates, purifies, transforms, decomposes, combines, and determines the natural character and effect of natural substances and makes these useful to various sciences and crafts." Significantly, it was only after giving this general definition that the article mentioned assaying and smelting, the two branches of chemistry most relevant to the intended reader.[28]

Even men whose interests were distant from chemistry were beginning to have favorable things to say about the subject. In 1768 J. D. Michaelis, an orientalist at Göttingen, described natural history, chemistry, and physics as the three main branches of natural science. He believed that students, especially wellborn students destined for high positions, should give special attention to each of these subjects. Chemistry, Michaelis implied, was indispensable for the study of nature, especially for the study of minerals. It was also essential for understanding mining and smelting, dyeing, minting, and other practical activities that a man of quality might supervise later in life. And it was the best prophylactic against alchemical delusions.[29]

The following year a lessor of two Pomeranian estates gave chemistry a central place in his response to the Berlin Academy's prize question on how natural science might be brought into closer connection with agriculture. Extensive education, he argued, would achieve the desired end. Virtually every science had something to offer the agriculturalist, but especially *Chemie*, which would teach him

> how he should dye and otherwise treat flax; press oil from linseed, hemp seed, rape seed; make malt; mill and bake meal, malt, and groats; treat wine; brew beer and vinegar; distill brandy; preserve cucumbers, cabbage, beets, walnuts, and all kinds of fruit; make all kinds of jams; soften honey and wax; dip candles; make soap; render scraps and fat; make wurst; pickle and smoke meat; iron trousers and linen; make butter, cheese,

27. J. G. Krünitz, *Oeconomische Encyclopädie*, vol. VIII (Berlin: Pauli, 1776), p. 53.
28. *Bergmännische Wörterbuch* (Chemnitz: Stössel, 1778), p. 123.
29. [J. D. Michaelis], *Raisonnement über die protestantischen Universitäten in Deutschland*, vol. I (Frankfurt a. M. and Leipzig: Andreä, 1768), p. 243.
30. H.-H. Müller, "Akademie und Wirtschaft im 18. Jahrhundert: Agrarökonomische Preisaufgaben und Preisschriften der Preussischen Akademie der Wissenschaften," *Studien zur Geschichte der Akademie der Wissenschaften der DDR*, 3 (1975), 132–134.

and meat loaf; also the art of fertilizing and gypsum, lime, and brick firing.

Besides chemistry proper (as he understood it), the essayist also thought that *Cymotechemie*, *Halotechemie*, *Metallurgie*, and *Pyrotechemie* could prove of value to the enterprising land manager.[30]

In 1774 the poet F. G. Klopstock alluded to chemistry in his *Die deutsche Gelehrtenrepublik*. He divided his imaginary republic's inhabitants into various orders and guilds, placing the *Chymiker* into the "great and venerable guild" of the natural scientists. He might just as well, Klopstock observed, have assigned the chemists to a separate guild. In contrast to the natural scientists who merely described nature, they used synthesis and analysis to act upon it.[31]

A few years later, advisors to the Elector of Mainz mentioned chemistry several times in their reform proposals for the ailing universities under his charge. In a proposal bearing on Mainz University, the jurist J. B. von Horix recommended that the science be taught by two professors. Besides urging that a new chair of chemistry and pharmacy be established for medical instruction, he suggested that the professor of natural philosophy give chemistry courses for artisans whose crafts involved chemical operations. In a proposal for Erfurt University, the poet Wieland, whom we have already seen criticizing crass materialism, urged that chemical instruction be offered every year. He insisted that chemistry was not only essential for physicians but also valuable for "philosophers, cameralists, and many other scholars."[32]

In 1780, attesting to the new interest in and esteem for chemistry, F. Nicolai graced the frontispiece of the forty-second volume of his *Allgemeine deutsche Bibliothek* with a portrait of J. C. Wiegleb (see Figure 2). This periodical was Germany's premier reviewing journal. Wiegleb was a small-town apothecary whose only claim to fame was as a chemist.[33]

That Michaelis, a Pomeranian estate manager, Klopstock, Horix, Wieland, and Nicolai, men whose central interests were far from chemistry, should have dealt so favorably with the science indicates that it enjoyed considerable

---

31. F. G. Klopstock, *Die deutsche Gelehrtenrepublik* (Hamburg: Bode, 1774), p. 5. His innovative use of the suffix *er* on the term for chemist was probably motivated by his cultural nationalism. His chief informant regarding chemistry seems to have been the metallurgical chemist Cramer. See F. Cramer, "Johann Andreas Cramer," *Harzboten* (1828), 278.

32. L. Just and H. Mathy, *Die Universität Mainz: Grundzüge ihrer Geschichte* (Trautheim and Mainz: Mushake & Franzmathes, 1965), pp. 104–105, 107; and W. Stieda, "Erfurter Universitätsreformpläne im 18. Jahrhundert," Akademie gemeinnütziger Wissenschaften zu Erfurt: *Sonderschriften*, 5 (1934), 178, 186, 211–212.

33. Wiegleb was the first chemist to appear on the frontispiece of Nicolai's journal. L. Crell followed in 1784 and J. G. R. Andreae in 1787. See *Allgemeine deutsche Bibliothek*, 59 (1784); 77 (1787). For biographies of Wiegleb and Andreae, see Appendix I. For a biography of Crell, see Chapter V. For Nicolai and his journal, see H. Möller, "Aufklärung in Preussen: Der Verleger, Publizist und Geschichtsschreiber Friedrich Nicolai," Historische Kommission zu Berlin: *Einzelveröffentlichungen*, 15 (1974); and T. Bauman, "The Music Reviews in the Allgemeine Deutsche Bibliothek," *Acta Musicologica*, 49 (1977), 69–85.

Fig. 2. J. C. Wiegleb, the first chemist to appear on a frontispiece of F. Nicolai's prestigious reviewing journal.
Source: *Allgemeine deutsche Bibliothek*, 42 (1780)

respect among intellectuals and officials during the 1770s. Interest in chemistry became so widespread, in fact, that authors of chemistry texts no longer needed to orient their books to the science's pharmaceutical applications. Only by taking a general approach, J. C. P. Erxleben observed in his pioneering text of 1775, could the chemistry teacher satisfy the needs of the many students who wanted to study the science but had no desire to become doctors.[34] At long last, Erxleben's pupil and emulator, C. E. Weigel, exclaimed in his text of 1777, chemistry had been "recognized" as "a generally useful [science] because of its weighty influence on so many branches of learning and on the productive estate."[35]

Since the 1720s, educated and powerful Germans had come to see chemistry not as a menial, pharmaceutical, and alchemical art, but rather as a useful and fundamental science. During this same period, such men had come to prize activities that promoted material and intellectual progress. The combined result of these two developments was a tremendous improvement in chemistry's prestige by the 1770s.[36] During this decade and the next, the prevailing image of chemistry harmonized closely with prevailing values and aspirations. In consequence, chemistry appealed to "the friends of natural science, medicine, domestic economy, and manufacturing,"[37] to "scientists, metallurgists, cameralists, and agricultural and financial officials as well as to doctors,"[38] and in Berlin to "persons of all orders [including] distinguished

---

34. J. C. P. Erxleben, *Anfangsgründe der Chemie* (Göttingen: Dieterich, 1775), pp. [i–ii]. For Erxleben, see M. Speter, "Joh. Christ. Polykarp Erxleben, der erfolgreiche Bekämpfer des 'Acidum pingue'-Interregnums in der Chemie (1775–1776)," *Isis*, 27 (1937), 11–19.

35. C. E. Weigel, *Grundriss der reinen und angewandten Chemie*, vol. I (Greifswald: Röse, 1777), p. vii. For a biographical profile of Weigel, see Appendix I.

36. Of course, respect for chemistry was not universal in the 1770s. But to my knowledge, it was only in Catholic Germany that chemists were still on the defensive. See L. Rousseau, *Rede von dem wechselweisen Einfluss der Naturkunde und Chymie auf die Wohlfahrt eines Staats, in Erweiterung der Künste und Wissenschaften*, 2d ed. (Nuremberg: Schwarzkopf, 1771), p. 7; G. A. Suckow, *Von dem Nuzzen der Chymie zum Behuf des bürgerlichen Lebens, und der Oekonomie* (Mannheim and Kaiserslautern, 1775), pp. [ii–iii]; and J. G. Menn, *Rede von der Nothwendigkeit der Chemie* (Cologne: Universitätsdruckerey, 1777), p. 3. In Protestant Germany, spokesmen for chemistry were able to concentrate on expounding the science's virtues. See, for instance, C. E. Weigel, *Einladungs-Schrift vom Vorträge der Chemie auf Academien* (Greifswald: Röse, 1775). Failing to distinguish the addresses given in Catholic Germany from those given in Protestant Germany, B. H. Gustin mistakenly construes all such speeches as evidence of chemistry's low standing in the 1770s. See his "The Emergence of the German Chemical Profession 1790–1867" (Diss., University of Chicago, 1975), pp. 33–37.

37. This phrase is from the title of Crell's chemical journal, the *Chemisches Journal für die Freunde der Naturlehre, Arzneygelahrtheit, Haushaltungskunst und Manufakturen*, 1–6 (1778–1781). This journal and its successors will be considered at length in Chapter V.

38. The minor chemist P. L. Wittwer believed that, except for the doctors, this audience for chemistry emerged between 1763 and 1783. See his "Lebensgeschichte Dr. Jac. Reinbold Spielmann, der Arzneygelahrtheit Prof. in Strassburg," *Chemische Annalen*, no. 1 (1784), 563.

persons of the fair sex."[39] It enjoyed, that is, the status of a *Lieblingswissenschaft*.[40]

---

39. H—— in Berlin, letter to Crell, *Chemische Annalen*, no. 1 (1784), 342. While chemistry's audience in late eighteenth-century Germany was as diverse as the chemists claimed, most of its members were still in medical occupations. See the analysis of Crell's subscribers in Chapter VI.

40. In 1760 J. H. G. von Justi described *Naturkunde* as the "*Modewissenschaft* of our century." See his *Gesammlete Chymische Schriften*, vol. I (Berlin and Leipzig: Real-Schule, 1760), p. 127. In 1775 G. A. Suckow claimed that "the economic sciences" had become the "*Leiblings Wissenschaft* of our century." See his *Von dem Nuzzen der Chymie*, p. [i]. While I am convinced that chemistry began to be regarded as a *Lieblingswissenschaft* around this time, I have not seen the term applied to chemistry proper before the mid-1790s. See J. C. Fabricius, *Ueber Academien, insonderheit in Dännemark* (Copenhagen: Proft & Storch, 1796), p. 105; J. F. Gmelin, *Geschichte der Chemie*, I:2; and W. von Humboldt, letter to J. H. Campe (15 October 1798), in J. Leyser, *Joachim Heinrich Campe*, 2d ed., vol. II (Brunswick: Vieweg, 1896), p. 318.

# 3 MATERIAL SUPPORT

No one lamented when Marburg University's long-idle chemical laboratory was demolished in 1731 so the stables could be expanded. But five decades later, the medical professors in Freiburg im Breisgau University did not sit idly by when authorities denied their request for a laboratory, proposing instead that the professor of chemistry teach the science in "an apothecary shop in town or a kitchen in his residence." They successfully campaigned for construction funds by arguing that the Medical Faculty's single most pressing need was for a laboratory.[1] These incidents are indicative of the improvement in material support for chemistry in German institutions of learning between 1720 and 1780. The number of salaried chairs involving some responsibility for this science climbed from six to thirty-six and the number of usable laboratories associated with these chairs, from one to eleven (see Map 1). In the present chapter, I characterize, explain, and explore the consequences of this growth in institutional patronage for chemistry in Germany.[2]

The emergence of this patronage required a willingness to innovate. Then, as today, this was no small requirement. Those who manage learned institutions are usually imbued with a reverence for tradition and a distaste for debates about resource allocation. Nonetheless, as patrons and their officers perceive new opportunities and needs, they sometimes move in new directions. They are particularly likely to do so when new monies are filling their coffers. Since revenues did increase in many German states between 1720

---

1. Dedicated in 1685, Marburg University's second laboratory was last used circa 1707. The first complaint about the laboratory's destruction was evidently made in 1748 by J. G. Duising. See R. Schmitz, *Die Naturwissenschaften an der Philipps-Universität Marburg 1527–1977* (Marburg: Elwert, 1978), pp. 205–208. For the negotiations concerning the laboratory in Freiburg im Breisgau, see A. Lüttringhaus and C. Baumfelder, "Die Chemie an der Universität Freiburg i. Br. von den Anfängen bis 1920," *Beiträge zur Freiburger Wissenschafts- und Universitätsgeschichte*, 18 (1957), 30–31. For further information about chemistry's place in Marburg and Freiburg im Breisgau universities, see Appendix II.

2. This chapter, which is based on the institutional histories in Appendix II, builds upon prior studies of the development of patronage for chemistry in German institutions of learning during the eighteenth century. See G. Lockemann, "Der chemische Unterricht an den deutschen Universitäten im ersten Viertel des neunzehnten Jahrhunderts," in *Studien zur Geschichte der Chemie, Festgabe Edmund O. v. Lippmann*, ed. J. Ruska (Berlin: Springer, 1927), pp. 138–158; A. Schleebach, *Die Entwicklung der chemischen Forschung und Lehre an der Universität Erlangen von ihrer Gründung (1743) bis zum Jahre 1820* (Bayreuth: Seuffer, 1937), pp. 1–5; and R. Schmitz, *Die deutschen pharmazeutisch-chemischen Hochschulinstitute: Ihre Entstehung und Entwicklung in Vergangenheit und Gegenwart* (Ingelheim am Rhein: Boehringen, 1969).

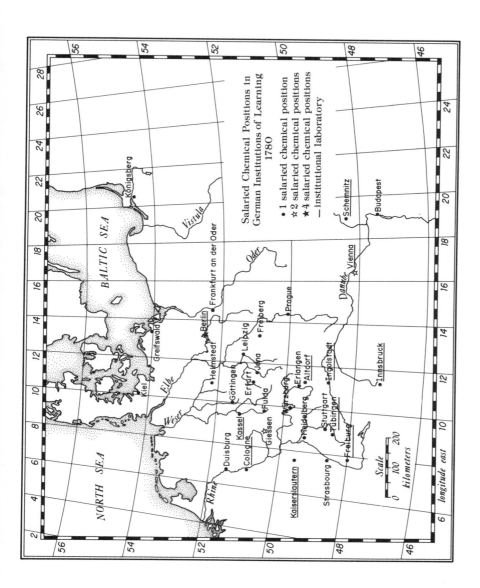

Map 1.

and 1780,[3] German authorities could easily translate their growing respect for chemistry into institutionalized patronage for the science. They took advantage of this opportunity in medical schools, administrative schools, and the Berlin Academy.

## MEDICAL EDUCATION

Of the many occupations dealing with chemical phenomena, medicine was the first in Europe in which familial and master-apprentice systems of training were supplanted by academic instruction. Formal medical education began with the rise of Salerno's school of medicine in the eleventh century.[4] It spread into Germany with the founding of university medical faculties there in the fourteenth century. Eventually, certainly by the eighteenth century in Germany, elites regarded formal training as a necessary prerequisite for beginning a medical practice.[5]

Once academic instruction was viewed as essential for the physician, medical schools stood to benefit from any substantial increase in concern over health. German medical schools reaped such benefits during the eighteenth century. Thanks to economic growth, urbanization, and secularization, Germans became more ardent in their quest for physical well-being, the demand for physicians intensified, and attendance at medical schools multiplied.[6] Meanwhile, influenced by military needs and melioristic ideals, German rulers evinced greater concern for the health of their soldiers and subjects. One

---

3. For a useful discussion of state finances in Germany during the eighteenth century, see H. Hassinger, "Politische Kräfte und Wirtschaft 1350–1800," in *Handbuch der deutschen Wirtschafts- und Sozialgeschichte*, ed. H. Aubin and W. Zorn, vol. I (Stuttgart: Union, 1971), pp. 646–657. In addition to growing tax revenues, Catholic authorities also had new revenues from confiscated Jesuit properties beginning in the mid-1770s. Of course, the availability of new revenues does not, in itself, explain disbursements for chemistry.

4. G. Baader, "Die Schule von Salerno," *Medizinhistorisches Journal*, 13 (1978), 124–128.

5. German elites regarded persons who attended the sick without a formal medical education as "quacks." Physicians, apparently motivated by concern about income as well as public health, played on this view in campaigning against quackery from the late seventeenth century on. See M. Stürzbecher, "Beiträge zur Berliner Medizingeschichte," Historische Kommission zu Berlin: *Veröffentlichungen*, 18 (1966), 1–66, 134–138.

6. Medical matriculations at Duisburg, Freiburg im Breisgau, Halle, Heidelberg, Strasbourg, and Würzburg universities increased from 1,209 (6 percent) for the period 1711 to 1730 to 3,432 (18 percent) for the period 1771 to 1790. Combining these data with information about total matriculations at other institutions, I estimate that the number of medical students in Germany climbed from around 500 in 1720 to around 1,750 in 1780. The data and my estimate are based primarily on E. T. Nauck, "Die Zahl der Medizinstudenten deutscher Hochschulen im 14.-18. Jahrhundert," *Sudhoffs Archiv*, 38 (1954), 175–186; 39 (1955), 378–379; and F. Eulenburg, "Die Frequenz der deutschen Universitäten von ihrer Gründung bis zur Gegenwart," Sächsische Gesellschaft der Wissenschaften: Philologisch-Historische Klasse: *Abhandlungen*, 24, no. 2 (1904). The estimate is also based, however, on the following works not utilized by Nauck: F. Nicolai, *Beschreibung einer Reise durch Deutschland und die Schweiz im Jahre 1781*, vol. IV: supp. (Berlin and Stettin: Nicolai, 1784), p. 59; Ingolstadt University, *Matrikelbuch*, ed. F. X. Freninger, vol. I (Munich: Eichleiter, 1872); "Die Matrikel des preussischen Collegium medico-chirurgicum in Berlin 1730 bis 1797," ed. A. von Lyncker, *Archiv für Sippenforschung*, 12 (1935), 97–135; and M. Braubach, *Die erste Bonner Hochschule: Maxische Akademie und kurfürstliche Universität, 1774/77 bis 1798* (Bonn: Bouvier & Röhrscheid, 1966).

approach that authorities took to the problem of public health was to improve medical education. They increased the availability of such education by founding medical schools, by adding professors to existing medical schools, and by increasing salaries so that professors could devote more time to teaching. As a result of this patronage, there were forty-eight medical schools in Germany by 1780. These schools employed some 160 teachers in all and some 80 teachers at salaries sufficient to permit concentration on academic activities.[7] Besides increasing the availability of medical education, German authorities took steps to improve its quality. To ensure that prospective doctors would receive instruction in essential medical subjects, they sometimes assigned these subjects to specific chairs. And to enable students to go beyond mere book learning, they sometimes funded anatomical theaters, botanical gardens, chemical laboratories, and teaching hospitals. Such reforms and improvements involved decisions about how existing or new resources should be allocated to the various medical subjects. One result of these decisions was that material support for chemistry grew markedly in German medical schools between 1720 and 1780.

## Pattern of Growth

By 1720, thanks to sixteenth- and seventeenth-century campaigns for the use of chemical drugs and ideas in medicine, chemistry was no longer an exotic subject in German medical schools. At most of these schools a professor or private lecturer would offer chemical instruction at least every other year. At six schools—the medical faculties of the Protestant universities in Altdorf, Jena, Königsberg, Rostock, Strasbourg, and Tübingen—one of the salaried professors was formally responsible for teaching chemistry. And at one school—that in Altdorf—the salaried chemical representative had access to an institutional laboratory in which to demonstrate processes and products.[8]

---

7. Twenty of these medical schools were established between 1720 and 1780. Eight were university medical faculties—Trier (1722), Fulda (1732), Göttingen (1734), Erlangen (1742), Bützow (1760), Bamberg (ca. 1770), Budapest (founded at Tyrnau in 1770, moved to Budapest in 1777), and Stuttgart (1776). For details, see Appendix II. Twelve were medical-surgical colleges—Berlin (1723), Kassel (1738), Dresden (1748), Brunswick (1750), Mannheim (1754), Frankfurt am Main (1763), Karlsruhe (1763), Düsseldorf (1765), Meinberg (1770), Munich (1772), Klagenfurt (1775), Linz (1775), Bruchsal (1777), and Brünn (1778). See K. Goldmann, *Verzeichnis der Hochschulen* (Neustadt an der Aisch: Degener, 1967); F. Ekkard, *Litterarisches Handbuch der bekannten höhern Lehranstalten in und ausser Teutschland*, vol. I (Erlangen: Schleich, 1780), pp. 84, 95, 269; vol. II (Erlangen: Palm, 1782), pp. 211, 236, 279; I. Fischer, *Medizinische Lyzeen: Ein Beitrag zur Geschichte des medizinische Unterrichtes in Oesterreich* (Vienna and Leipzig: Braumüller, 1915), pp. 24–27, 32–35; and the institutional histories of the Berlin, Brunswick, and Kassel Colleges in Appendix II. The only medical-surgical colleges to achieve university status between 1720 and 1780 were those in Berlin and, for a short time, Kassel. By 1780 there were fifteen medical schools with five or more salaried professors—Berlin (7), Budapest (5), Erlangen (5), Freiburg im Breisgau (5), Göttingen (6), Halle (5), Heidelberg (5), Ingolstadt (5), Innsbruck (5), Kassel (5), Kiel (5), Leipzig (5), Prague (5), Vienna (10), and Würzburg (5). The reason for speaking of "salaried" professors is that the teaching staffs of some medical schools also included nonsalaried professors and private lecturers who, while awaiting a chance at a salaried position, were recompensed solely by student fees.

8. For information about chemical instruction in German medical schools before 1720, see Appendix II. In speaking of men as "salaried chemical representatives," I do not mean to imply

TABLE 3: Material Support for Chemistry in German Medical Schools, 1720–1780

|  | 1720 | 1740 | 1760 | 1780 |
|---|---|---|---|---|
| In Protestant medical schools |  |  |  |  |
| Number of salaried chairs | 6 | 12 | 11[a] | 17 |
| Number of associated laboratories | 1 | 1 | 2 | 3 |
| Median salary[b] | 150 Rtl | 300 Rtl | 350 Rtl | 400 Rtl |
| In Catholic medical schools |  |  |  |  |
| Number of salaried chairs | 0 | 3 | 8 | 11 |
| Number of associated laboratories | 0 | 0 | 1 | 5 |
| Median salary[b] | — | 300 Rtl | 350 Rtl | 650 Rtl |

Source: Based on Appendix II.
Note: The Erfurt and Heidelberg medical faculties have been counted among the Catholic schools because they were under the jurisdiction of Catholic authorities.

[a] This number includes the temporarily vacant chair of medical theory, chemistry, anatomy, and botany in the Frankfurt an der Oder Medical Faculty.

[b] To facilitate comparison, all salaries have been converted to Reichstaler. The figures have been rounded to the nearest 50 Rtl.

Between 1720 and 1780, thanks to the action of Protestant and especially Catholic authorities, the number of salaried chemical positions in German medical schools increased manyfold. This growth entailed considerable disbursements because these authorities went beyond adjusting the teaching assignments of existing professors to promote chemical instruction. They established new salaried positions carrying responsibility for the subject. They increased the salaries of the chemistry teachers in their employ. And in some instances they provided the chemistry teachers with laboratories in which to prepare for classes (see Table 3).

As a result of all these developments, chemistry had a solid foothold in German medical education by 1780. To be precise, twenty-eight salaried professors were responsible for teaching the subject in German medical schools. Eighteen were occupants of chairs that had been established at least partly for this purpose. Some fifteen were receiving adequate salaries. And eight had the use of institutional laboratories. A substantial gain had been registered since 1720.[9]

## Reasons for Growth

The primary rationale for patronage of chemistry in medical schools was the science's close connection with pharmacy. In 1727, for instance, Heidel-

---

that chemistry was their sole responsibility. The occupants of "salaried chemical positions" were usually responsible for other subjects as well.

9. Chemistry was, to be sure, not the only subject to benefit from the eighteenth-century campaign to improve medical education. Still, I believe, it fared relatively well in the competition for support. The claim that "some fifteen" salaried chemical representatives were receiving "adequate salaries" in 1780 is based on the assumption that faculty in the Protestant medical schools could supplement their income with student fees (see below). In 1777 J. B. von Horix advised the Elector of Mainz that professors would fritter away their time on outside activities unless their total academic incomes were above 600 Rtl. See L. Just and H. Mathy, *Die Universität Mainz: Grundzüge ihrer Geschichte* (Trautheim and Mainz: Mushake & Franzmathes, 1965), p. 96.

TABLE 4: Orientation of Salaried Chemistry Chairs in German Medical Schools, 1720–1780

| | Number of Chairs | | | |
|---|---|---|---|---|
| | 1720 | 1740 | 1760 | 1780 |
| Chairs also responsible for therapeutic subjects[a] | 2 | 6 | 8 | 6 |
| Chairs also responsible for pharmaceutical subjects[b] | 3 | 6 | 9 | 17 |
| Chairs without other specified responsibilities[c] | 1 | 3 | 6 | 7 |

Source: Based on Appendix II.
Note: The totals for each year exceed the totals indicated in Table 3 because chairs that carried both therapeutic and pharmaceutical subjects have been counted twice.
[a]By therapeutic subjects, I mean therapy, pathology, physiology, surgery, and anatomy.
[b]By pharmaceutical subjects, I mean pharmacy, materia medica, and botany.
[c]The occupants of these chairs were, to be sure, often expected to teach other subjects. However, they were free to choose which ones.

berg's Medical Faculty successfully recommended that one of the professors be charged with chemistry so that students would learn to "recognize and distinguish simple and compound medicines."[10] When R. A. Vogel was called to Göttingen for chemistry in 1753, he was told that, if he decided to establish a laboratory in the University Apothecary Shop, he would have to meet the operating costs by selling the drugs that he and his students prepared.[11] In 1777 the Berlin Academy advised that someone skilled in pharmacy be appointed to the recently vacated chair of chemistry in the Berlin Medical-Surgical College.[12] The pharmaceutical justification was also manifest in the subjects that were assigned along with chemistry to salaried chemical positions. A growing number of these chairs combined responsibility for chemistry with responsibility for pharmacy, materia medica, botany, and/or natural history (see Table 4).

But why did chemistry's old connection with pharmacy prompt new material support for the science in German medical schools between 1720 and 1780? Why was there a new concern that the physician, whose job was to attend sick people rather than prepare drugs, know pharmaceutical chemistry? Two developments underlay this new response. As Germans became

10. *Urkundenbuch der Universität Heidelberg*, ed. E. Winkelmann, vol. I (Heidelberg: Winter, 1886), p. 409. For subsequent developments at Heidelberg, see Appendix II.
11. Vogel rejected the prospect of funding his experimentation in this way, setting up a laboratory in his residence instead. See G.–A. Ganss, *Geschichte der pharmazeutischen Chemie an der Universität Göttingen, dargestellt in ihrem Zusammenhang mit der allgemeinen und der medizinischen Chemie* (Diss., Göttingen University; Marburg: Euker, 1937), p. 22. For subsequent developments at Göttingen, see Appendix II and the profiles of Vogel and J. F. Gmelin in Appendix I.
12. H. Lehmann, *Das Collegium medico-chirurgicum in Berlin als Lehrstätte der Botanik und der Pharmacie* (Diss., Berlin University; Berlin: Triltsch & Huther, 1936), p. 59. There is no reason to infer from the Academy's recommendation that its members could not distinguish chemistry from pharmaceutical chemistry. Rather, they were apparently hoping that the appointment of a pharmacist as Pott's successor in the Medical-Surgical College would enable the Academy to reclaim that part of his salary which had been diverted from its treasury decades earlier. The strategy succeeded. See the institutional histories of the Berlin Academy and Berlin Medical-Surgical College in Appendix II.

more enamored of rationalism, elite physicians sought to enhance the legitimacy of their profession by requiring all doctors to have a good grasp of theory, including a scientific knowledge of the preparation and composition of drugs.[13] More importantly, with the spread of the gospel of meliorism and the practice of enlightened governance, elite physicians and political theorists urged rulers to regulate the pharmaceutical profession as a means of promoting public health. With the establishment of tests for prospective apothecaries and inspections of apothecary shops, it became increasingly evident that regulation could only be effective if doctors knew enough pharmaceutical chemistry to enforce the new edicts.[14]

If, however, the spread of Enlightenment values and beliefs underlay the new patronage for chemistry in medical schools, why were Protestant authorities comparatively frugal when it came to funding the science? The Enlightenment, after all, attracted a much broader following in Protestant than in Catholic lands. It would seem, therefore, that Protestant authorities ought to have been more, not less, generous than their Catholic counterparts. The reason for their relative parsimony was that the universities in the Protestant states were generally quick to respond to new concerns and interests. Most courses in these universities were given on a fee basis. Hence, whenever demand was considerable, a teacher interested in chemistry was likely to offer to teach the science. When demand was high, the potential earnings might even engender competitive offerings. To win the day, teachers would sometimes accompany their lectures with demonstrations and opportunities for practical experience. Thus, at a thriving Protestant university, good chemical instruction was generally forthcoming if the students were willing to pay for it. The only drawback was that this spontaneous response to student demand tended to undercut the case for state patronage.[15]

By contrast, most universities under Catholic jurisdiction were so tightly regulated that their professors rarely responded to new student interests by offering new courses. Forbidden to charge for instruction, these professors had little incentive to teach any subject for which they were not responsible.

13. Physicians who neglected chemistry came to be made the objects of ridicule. See E. G. Kurella, *Entdeckung der Maximen ohne Zeitverlust und Mühe ein berühmter und reicher Arzt zu Werden* (Berlin and Potsdam: Voss, 1750), pp. 14–15; and J. G. Zimmermann, *Das Leben des Herrn von Haller* (Zürich: Heidegger, 1755), pp. 103–105. E. G. Baldinger observed in 1779 that "it is self-evident that a [medical] practice without fundamental theory is mere quackery." See his "Vorrede," *Neues Magazin für Aerzte*, 1 (1779), [i].

14. For the tightening regulation of pharmacists and pharmaceutical practice, see A. Fischer, *Geschichte des deutschen Gesundheitswesens*, vol. II (Berlin: Rothacker, 1933); and A. Adlung and G. Urdang, *Grundriss der Geschichte der deutschen Pharmazie* (Berlin: Springer, 1935). For an actual inspection report and Prussian edicts directing physicians how to inspect apothecary shops and test pharmacists and surgeons, see the *Magazin für die gerichtliche Arzneikunde und medizinische Polizei*, 1 (1783), 943–958; 2 (1784), 322–328.

15. For an exceptionally thoughtful analysis of the Protestant universities, see [J. D. Michaelis], *Raisonnement über die protestantischen Universitäten in Deutschland*, 2 vols. (Frankfurt a. M. and Leipzig: Andreä, 1768–1770). That the availability of private courses could detract from arguments for patronage of chemical instruction is particularly evident from the relative neglect of chemistry at Halle, Jena, and Leipzig universities. See Appendix II.

When they thought a new subject should be introduced into the curriculum, they were content to recommend that their schools be provided with the appropriate resources for teaching it. As they came to recognize chemistry's importance and to envy leading Protestant universities for their chemistry courses, they made such recommendations. In 1718, for instance, the Medical Faculty in Vienna petitioned for "a well-equipped course on chemistry where young doctors and medical students could observe the operations daily and have the opportunity of learning them . . . as at other leading universities."[16] At first such proposals did not elicit much response. But as Catholic authorities saw the advantages in enlightened governance, they paid increasing heed to their petitioners. In 1749, for example, the Medical Faculty in Vienna again complained about its inability to teach chemistry, blaming low enrollments on this and other gaps in the curriculum. This time action was taken, because Empress Maria Theresa's personal physician, G. van Swieten, also believed that the science should be taught. Soon the government announced the establishment of a new chair for botany and chemistry, expressing the hope that this measure would augment the university's "splendor and eminence" as like measures had done "at other universities."[17] The same spirit of emulation was manifested in the labors of the Bavarian Elector's personal physician, J. A. von Wolter, to improve medical education at Ingolstadt. In 1763 he urged Maximillian Joseph to continue supporting a new chemistry course there even though the experiments were costing 200 Rtl a year. He prevailed, perhaps because he argued that "such courses had been proving their worth for physicians and even jurists at the chief universities, especially those of the Protestants, for the last thirty years."[18] Thus, once Catholic authorities wanted medical students to learn as much about chemistry as could be learned in the better Protestant schools, they were likely, given the organization of the Catholic universities, to be relatively generous in their patronage of chemical instruction.[19]

16. *Acta Facultatis Medicae Universitatis Vindobonensis 1399–1724*, ed. K. Schrauf and L. Senfelder, vol. VI (Vienna: Wiener medizinischen Doktorenkollegiums, 1912), p. 399.
17. W. Oberhummer, "Die Chemie an der Universität Wien in der Zeit von 1749 bis 1848 und die Inhaber des Lehrstuhls für Chemie und Botanik," *Studien zur Geschichte der Universität Wein*, 3 (1965), 128. The establishment of this chair exemplified Swieten's genius for arranging patronage by, as one contemporary put it, "making princes aware of the useful side of the sciences." See E. G. Baldinger, *Biographien jetztlebender Aerzte und Naturforscher in und ausser Deutschland*, pt. 1 (Jena: Hartung, 1768), p. 6.
18. G. Kallinich, *Das Vermächtnis Georg Ludwig Claudius Rousseaus und die Pharmazie: Zweihundert Jahre Pharmazie an der Universität Ingolstadt-Landshut-München 1760–1960* (Munich: Govi, 1960), p. 379.
19. This interpretation of Catholic patronage for chemical instruction receives ample support from contemporary comparisons of Catholic and Protestant universities. See F. Nicolai, *Beschreibung einer Reise durch Deutschland*, IV:682–687; and C. Meiners, *Geschichte der Entstehung und Entwicklung der hohen Schulen unsers Erdtheils*, vol. I (Göttingen: Röwer, 1802), pp. 310–323. For concurrent Catholic reforms oriented toward the subjects of law and history, see N. Hammerstein, "Aufklärung und katholisches Reich: Untersuchungen zur Universitätsreform und Politik katholischer Territorien des Heiligen Römischen Reichs Deutscher Nation im 18. Jahrhundert," *Historische Forschungen*, 12 (1977).

### Effects of Patronage

Primarily intended for the improvement of pharmaceutical practice and thereby public health, material support for chemistry in German medical schools can have been only moderately successful in achieving its goals. The purity of drugs available to the public may have improved as physicians became better able to test prospective pharmacists and inspect apothecary shops. But, given the state of medicine in the eighteenth century, it is doubtful that standards of pharmaceutical performance had a very direct bearing on mortality or even comfort.

Besides its necessarily modest impact on public health, medical school patronage for chemistry had four unintended benefits for the science. First, such patronage increased chemistry's visibility. Any student who attended a university with a salaried chemical representative would at least learn that the science enjoyed official approbation. And any pharmacist who lived nearby would know that his knowledge of chemistry's fundamentals and his proficiency in elaborating drugs was likely to undergo periodic scrutiny.

Second, as indicated in the preceding chapter, salaried chemical representatives participated in the transformation of chemistry's public image. Although they oriented their instruction to pharmaceutical applications, they usually devoted one or more lectures to extolling this science's potential depth and scope. In doing so, they helped circulate an image of chemistry that transcended pharmacy.[20]

Third, salaried chemical representatives participated in the education of men who subsequently achieved recognition as chemists. Most notably Stahl's protégé, J. H. Pott, who lectured at Berlin's Medical-Surgical College from 1724 to 1770, taught A. S. Marggraf, J. R. Spielmann, J. G. R. Andreae, R. A. Vogel, F. A. Cartheuser, C. A. Gerhard, and J. C. F. Meyer. Although no other chemistry professor came close to matching this record, several did give instruction to one or two persons who later contributed to the science.[21]

---

20. That the salaried chemical representatives in medical schools gave at least passing attention to chemistry's depth and scope between 1720 and 1780 can be seen by looking at the textbooks that a few wrote while in their chairs. See C. Neumann, *Prælectiones chemicæ; seu, Chemia medico-pharmaceutica experimentalis & rationalis*, ed. J. C. Zimmermann (Berlin: Rüdiger, 1740); P. Gericke, *Fundamenta chymiae rationalis* (Berlin: Rüdiger, 1740); R. A. Vogel, *Institutiones chemiae* (Göttingen: Luzac, 1755); J. R. Spielmann, *Institutiones chemiae* (Strasbourg: Bauer, 1763); C. E. Weigel, *Grundriss der reinen und angewandten Chemie*, 2 vols. (Greifswald: Röse, 1777); and J. F. Gmelin, *Einleitung in die Chemie* (Nuremberg: Raspe, 1780). For profiles of Neumann, Gericke, Vogel, Spielmann, Weigel, and Gmelin, see Appendix I.

21. For biographical profiles of Pott and his students, see Appendix I. In addition to Pott, the following salaried chemical teachers helped train chemists with profiles in Appendix I: P. Gericke (professor of anatomy, pharmacy, and chemistry in Helmstedt's Medical Faculty) taught J. A. Cramer; J. F. Cartheuser (professor of chemistry, pharmacy, materia medica, anatomy, and botany in Frankfurt an der Oder's Medical Faculty) taught his son F. A. Cartheuser and C. A. Gerhard; P. F. Gmelin (professor of chemistry and botany in Tübingen's Medical Faculty) and N. J. Jacquin (professor of chemistry and botany in Vienna's Medical Faculty) taught J. F. Gmelin; G. C. Beireis (professor of theoretical medicine, materia medica, and chemistry in Helmstedt's Medical Faculty) taught L. Crell; and R. A. Vogel (professor of

TABLE 5: Salaried Chemistry Professors in German Medical Schools Publishing on Chemistry, 1720–1780

|                     | 1720 |       | 1740 |         | 1760 |        | 1780 |        |
|---------------------|------|-------|------|---------|------|--------|------|--------|
| In Protestant schools | 3    | (50%) | 8    | (70%)   | 8    | (70%)  | 14   | (80%)  |
| In Catholic schools   | —    |       | 1    | (30%)[a]| 4    | (50%)[b]| 6   | (50%)[b]|
| In all schools        | 3    | (50%) | 9    | (60%)   | 12   | (60%)  | 20   | (70%)  |

Source: Based on Appendix II.
Note: Percentages are based on the number of salaried chemistry professors in the respective categories and are rounded to the nearest 10 percent.
[a] A Protestant.
[b] Includes two Protestants.

Last, salaried chemical representatives, especially the Protestants, often went beyond teaching the science to publish their own findings and interpretations (see Table 5). Indeed, several—C. Neumann and J. H. Pott in Berlin, R. A. Vogel and J. F. Gmelin in Göttingen, J. R. Spielmann in Strasbourg, and C. E. Weigel in Greifswald—published enough to achieve considerable renown as chemists during the eighteenth century.[22]

Such publication was, it must be reiterated, an unintended consequence of medical school patronage for chemistry. Some authorities still regarded teaching as a professor's sole obligation. Erfurt's Governor, C. von Dalberg, expressed this common attitude in 1777, when he wrote that

> a professor need not be a Leibniz or a Bacon. A teacher does not have to be a discoverer, an enlarger of the sciences. He teaches youth what has long been known and how to apply it. This does not require transcendental abilities. Nothing more is needed than clear concepts, adequate knowledge of one's branch of science, industry, and good will.[23]

Going beyond this traditional view, other officials perceived a connection between professorial fame and student enrollments. They wanted professors to make themselves known, as a means of attracting students, especially wealthy students. Advisors with this mercantilistic view of higher education were sometimes willing to reward professors whose texts and tracts attracted the public eye. But as late as the 1770s, such rewards were still too rare to

---

chemistry in Göttingen's Medical Faculty) taught C. E. Weigel. Salaried chemistry teachers also participated in the training of several chemists of lesser renown.

22. For the reputations of Neumann, Pott, Vogel, Gmelin, Spielmann, and Weigel, see Appendix I.

23. Dalberg also believed that it was an "incontestable truth" that "the purpose of a university was to form useful tools (*Werkzeuge*) for the good of the state." For Dalberg's views on universities, see W. Stieda, "Erfurter Universitätsreformpläne im 18. Jahrhundert," Akademie gemeinnütziger Wissenschaften zu Erfurt: *Sonderschriften*, 5 (1934), 100–127. There is ample evidence that other "enlightened" rulers had the same narrow view of universities. See, for instance, C. Bornhak, *Geschichte der preussischen Universitätsverwaltung bis 1810* (Berlin: Reimer, 1900), p. 147; and R. Kink, *Geschichte der kaiserlichen Universität zu Wien*, vol. I (Vienna: Gerold, 1854), pp. 546–547.

constitute an important material incentive for professorial publication in chemistry.[24]

While salaried chemical representatives in medical schools rarely encountered material incentives for publication, they were, especially in the Protestant institutions, increasingly likely to receive collegial encouragement for scholarly productivity and originality. The emerging enthusiasm for professorial research found expression in J. D. Michaelis's influential work on the Protestant universities. Recognizing that professors had no obligation to publish, he nevertheless thought it desirable for those who had something new to say to be authors. If the professor proved to be one of those rare men who was genuinely original, he could greatly enhance a university's reputation. Indeed, "the author whom the learned world honors as its teacher can, despite shortcomings in or even neglect of his lecturing, be a professor whose loss would be detrimental to his university."[25]

Related to this new appreciation for originality was a new respect for specialization. The rising flood of worthwhile books and periodicals, German intellectuals realized, confirmed Pope's observation that

> One science only will one genius fit;
> So vast is art, so narrow wit.[26]

Polymathy was degenerating into dilettantism. As a German encyclopedist put it in 1741, the scholar who "scatters his energy and work over too many things cannot achieve real mastery in his discipline and consequently cannot be as useful to others as he would if he devoted all his energy to one subject."[27] Recognition of the need for specialization led university authorities to abandon the common practice of distributing medical subjects among the professors on the basis of seniority. Eager that students receive up-to-date instruction, they assigned the same subjects to a given professor for the duration of his career. This reform in turn spurred on the advance of speciali-

---

24. For the mercantilistic approach to professorial publication, see L. Just and H. Mathy, *Die Universität Mainz: Grundzüge ihrer Geschichte*, pp. 95–97; R. S. Turner, "University Reformers and Professorial Scholarship in Germany 1760–1806," in *The University and Society*, ed. L. Stone, vol. II (Princeton: Princeton University Press, 1974), pp. 495–531; C. E. McClelland, "The Aristocracy and University Reform in Eighteenth-Century Germany," in *Schooling and Society: Studies in the History of Education*, ed. L. Stone (Baltimore, Md.: Johns Hopkins University Press, 1976), pp. 146–173. McClelland suggests, correctly I believe, that mercantilistic advocacy of professorial publication facilitated the emergence of a research ethos in the Protestant universities during the second half of the eighteenth century.

25. [J. D. Michaelis], *Raisonnement über die protestantischen Universitäten in Deutschland*, I: 18, 92; II: 123–151, 225–230.

26. A. Pope, *An Essay on Criticism*, facsimile of the 1711 ed. (Menston, Engl. Scolar Press, 1970), p. 6, lines 60–61.

27. C. Wiedemann, "Polyhistors Glück und Ende: Von Daniel George Morhof zum jungen Lessing," in *Festschrift Gottfried Weber zu seinem 70. Geburtstag überreicht*, ed. H. O. Burger and K. von See (Bad Homberg: Gehlen, 1967), p. 233. B. Fabian suggests that traditional *Gelehrsamkeit*, with its ideal of polymathy, had completed its dissolution into dilettantism and specialization by the late eighteenth century. See his "Der Gelehrte als Leser," Wolfenbütteler Arbeitskreis für Geschichte des Buchwesens: *Schriften*, 1 (1977), 52–54.

zation, since a professor who taught the same subjects throughout his career was apt to focus his scholarly efforts on them.[28]

At Göttingen, the library's development also reflected and reinforced the rise of specialized originality. The purchasing program placed ever more emphasis on the acquisition of the latest works within the fields covered by the university. And the catalogues, especially the subject catalogues, facilitated the use of the rich holdings by advanced students, lecturers, and professors in their work.[29]

Although relentless, the advance of specialization in the universities was slow. For instance, in 1775, after helping Gmelin get appointed to the vacant chair of chemistry at Göttingen, A. von Haller felt called upon, as he informed a friend, to advise the young professor that

> I wish for his own and the university's good that he would limit his view and choose a narrower horizon where he can concentrate his powers. The general mistake of the Germans is that they undertake too many sciences, handling them in an elementary way and often just copying. The great utility of the universities is that one divides the sciences into small parts and gives each man a small and limited responsibility.[30]

To judge from their increasing involvement in research, many German professors, certainly more than Haller imagined, accepted his vision of the ideal university.[31] By the 1770s most of the chemical representatives who were

---

28. In 1734 *Aufrücken* (moving from one chair to the next as one gained in seniority) was explicitly forbidden at Würzburg in an attempt to improve instruction. See F. X. von Wegele, *Geschichte der Universität Würzburg*, vol. II (Würzburg: Stahel, 1882), p. 373. Three years later the practice was forbidden at Königsberg on the grounds that it was "not possible for the learned physician to have equal proficiency in all parts of medicine." See D. H. Arnoldt, *Ausführliche, und mit Urkunden versehene Historie der Königsbergischen Universität*, vol. II: supp. (Königsberg: Hartung, 1746), pp. 67–68. For other examples, see E. T. Nauck, "Zur Geschichte des medizinische Lehrplans und Unterrichts der Universität Freiburg i. Br.," *Beiträge zur Freiburger Wissenschaft- und Universitätsgeschichte*, 2 (1952), 43; and A. Burckhardt, *Geschichte der medizinischen Fakultät zu Basel* (Basel: Rheinhard, 1917), pp. 226, 324. However, at Jena, despite occasional complaints about the practice, *Aufrücken* was not abandoned until the early nineteenth century. See I. Jahn, "Geschichte der Botanik in Jena von der Gründung der Universität bis zur Berufung Pringsheims (1558–1864)" (Diss., Jena University, 1963), pp. 121, 154.

29. B. Fabian, "Göttingen als Forschungsbibliothek im achtzehnten Jahrhundert: Plädoyer für eine neue Bibliotheksgeschichte," *Wolfenbütteler Forschungen*, 2 (1977), 209–239. The size of the university library did not dissuade most of Göttingen's professors from acquiring fairly large personal libraries (i.e., 1,000 or more volumes). The diversity of their collections probably reflected not only, as G. Streich suggests, the breadth of their teaching (by today's standards) but also their satisfaction with the university library's coverage of their more esoteric research needs. See his "Die Büchersammlungen Göttinger Professoren im 18. Jahrhundert," ibid. 2 (1977), 241–299.

30. Haller, letter to E. F. von Gemmingen (20 April 1775), in "Briefwechsel zwischen Albrecht von Haller und Eberhard Friedrich von Gemmingen," ed. H. Fischer, Litterarischer Verein in Stuttgart: *Bibliothek*, 219 (1899), 84. Haller had been urging the advantages of specialization since the 1740s. See O. Sonntag, "The Idea of Natural Science in the Thought of Albrecht von Haller" (Diss., New York University, 1971), pp. 23–24.

31. J. L. Heilbron, *Electricity in the 17th and 18th Centuries: A Study of Early Modern*

publishing were concentrating on their assigned subjects. In fact, thanks to the rising esteem for chemistry, they usually allotted a generous share of their scholarly energies to this single science.[32]

## ADMINISTRATIVE EDUCATION

While formal education was already regarded as indispensable for physicians by the early eighteenth century, it still played but a modest role in the training of those Germans destined to make decisions affecting large-scale production. Most of these men—royal and aristocratic owners as well as the lessors, agents, and minor officials actually supervising production—learned about their jobs by observing their fathers and other close relatives or by serving an informal apprenticeship to a successful man, usually a family connection. Only the men who mediated between owners and supervisors—wealthy lessors and middle-level officials—were at all likely to supplement familial and master-apprentice systems of training with higher education. Those that did so attended either the universities, where they studied in the philosophical and law faculties, or the *Ritterakademien*, a new kind of school that offered French, geography, diplomacy, and other "modern" subjects. Despite such formal education, they rarely acquired any more knowledge of the natural phenomena underlying production than their superiors or subordinates. This is hardly surprising. In the early eighteenth century, few Germans were acquainted with, much less gave credence to, the idea that natural knowledge could serve as a fecund source of technological innovations. In ensuing decades, however, a growing number of intellectuals argued that prospective administrators should be given scientific training to enable them to understand the productive activities they would be supervising. From the 1760s on, their arguments redounded to chemistry's advantage.[33]

---

*Physics* (Berkeley: University of California Press, 1979), pp. 137–140. By 1802 the research ethos had made such headway at Halle that officials were complaining that the professors had forgotten that the purpose of universities was to teach the sciences, not widen them. See Bornhak, *Geschichte der preussischen Universitätsverwaltung bis 1810*, p. 147.

32. My impressions of publication patterns are based on the bibliographies of the salaried professors of chemistry in German medical schools, all of whom are listed in Appendix II.

33. For administrative education in Germany during the seventeenth and eighteenth centuries, see W. Stieda, "Die Nationalökonomie als Universitätswissenschaft," Sächsische Gesellschaft der Wissenschaften: Philologisch-Historische Klasse: *Abhandlungen*, 25, no. 2 (1906); U. Troitzsch, "Ansätze technologischen Denkens bei den Kameralisten des 17. und 18. Jahrhunderts," *Schriften zur Wirtschafts- und Sozialgeschichte*, 5 (1966); W. Schöler, *Geschichte des naturwissenschaftlichen Unterrichts im 17. bis 19. Jahrhundert: Erziehungstheoretische Grundlegung und schulgeschichtliche Entwicklung* (Berlin: de Gruyter, 1970); W. Bleek, "Von der Kameralausbildung zum Juristenprivileg: Studium, Prüfung und Ausbildung der höheren Beamten des allgemeinen Verwaltungsdienstes in Deutschland im 18. und 19. Jahrhundert," *Historische und Pädagogische Studien*, 3 (1972). For the *Ritterakademien*, see K. Bleeck, "Adelserziehung auf deutschen Ritterakademien: Die Lüneburger Adelsschulen 1566–1850," *Europäische Hochschulschriften*, Reihe III, 89 (1977). For the actual training of high-level administrators, see scattered remarks in H. Rosenberg, *Bureaucracy, Aristocracy, and Autocracy: The Prussian Experience, 1660–1815* (Cambridge, Mass.: Harvard University Press, 1958); and J. Lampe, "Aristokratie, Hofadel und Staatspatriziat in Kurhannover: Die Lebens-

Harbingers of patronage for chemistry started appearing in German administrative education in the 1690s. From time to time, professors at well-attended Protestant universities offered courses on the uses of chemistry outside medicine.[34] Then in the 1730s, a few mining officials in Electoral Saxony and the Duchy of Brunswick began using their facilities for teaching chemistry to would-be administrators.[35] These professors or mining officials did not, however, receive any part of their salaries specifically for teaching about the applications of chemistry to productive technology. Saxon authorities, indeed, saw no reason to fund such instruction. In 1729 they denied Wittenberg's request for a new chair of chemistry and mining.[36] And in the late 1740s, they ignored C. F. Zimmermann's advice that a mining school, including a chair of chemistry and a laboratory, be founded in Freiberg.[37] Even the influential van Swieten in Austria, despite his commitment to educational reform and his awareness of chemistry's nonmedical uses, seems not to have promoted chemical instruction for administrators prior to the Seven Years' War.[38]

After the war, however, there was a new willingness to act. The costs of debts, reparations, and rearmament intensified the interest of rulers and their advisors in economic development. To those who entertained a broad view of chemistry's possibilities, it now seemed worthwhile to bear the costs of providing future subordinates with training in the science. Van Swieten led the way, armed with the French metallurgist A. G. Jars's recent report castigating Hapsburg mining officials for their ignorance of the operations under their direction. In 1763/64 he arranged for the appointment of chemistry teachers in Schemnitz, Prague, and Idria and the construction of a laboratory

---

kreise der höheren Beamten an den kurhannoverschen Zentral- und Hofbehörden, 1714–1760," Historische Kommission für Niedersachsen: *Veröffentlichungen*, 24, pt. 2, no. 1–2 (1963).

34. At Halle University, for instance, both Stahl and Hoffmann taught metallurgical chemistry during the 1690s and J. J. Lange, professor of mathematics, taught chemistry courses with a technological orientation during the 1740s. See Halle University, lecture catalogues, Universitäts-Archiv, Halle, German Democratic Republic. At Leipzig University, both A. F. Petzold and S. T. Quellmalz offered courses on metallurgical chemistry during the 1730s. See *Grosses vollständiges Universal-Lexicon*, ed. J. H. Zedler, vol. XXVII (Leipzig: Zedler, 1741), pp. 1156–1162; vol. XXX, (1741), pp. 188–189.

35. For the early teaching of chemistry by mining officials, see the profiles of Henckel, Cramer, and Gellert in Appendix I and the information on H. M. Kaulitz in the institutional history of Brunswick's Collegium Carolinum in Appendix II.

36. Saxon officials denied the request on the grounds that the existing professors were already responsible for teaching chemistry. In fact, no professor had been formally charged with teaching the science. See W. Friedensburg, *Geschichte der Universität Wittenberg* (Halle: Niemeyer, 1917), p. 577; and Appendix II.

37. C. F. Zimmermann, "Von der Beschaffenheit, Einrichtung und Nutzen einer Academie derer Bergwercks-Wissenschaften," in his *Ober-Sächsische Berg-Academie, in welcher die Bergwercks-Wissenschaften nach ihren Grund-Wahrheiten untersuchet, und nach ihrem Zusammenhange entworffen werden* (Dresden and Leipzig: Hekel, 1746), pp. 9–56.

38. Swieten's assessment of the official climate in Vienna probably deterred him from proposing chemical instruction for prospective administrators prior to 1762. His subsequent advocacy of such instruction (see below) suggests that he was fully cognizant of chemistry's broad applicability. In fact, he had studied metallurgical chemistry with J. A. Cramer in Leiden in 1737/38. See *A Selection of the Correspondence of Linnaeus and other Naturalists*, ed. J. E. Smith, vol. II (London: Longmans, 1821), pp. 171–185.

TABLE 6: Salaried Chemistry Chairs and Laboratories oriented toward Administrative Education, 1763–1780

|  | Number established 1763–1780 | Surviving in 1780 | Median salaries 1780[a] |
|---|---|---|---|
| In Protestant schools |  |  |  |
| Number of salaried chairs | 5 | 3 | 100 Rtl[b] |
| Number of associated laboratories | 1 | 0 |  |
| In Catholic schools |  |  |  |
| Number of salaried chairs | 5 | 3 | 400 Rtl |
| Number of associated laboratories | 2 | 2 |  |
| In all schools |  |  |  |
| Number of salaried chairs | 10 | 6 | 200 Rtl |
| Number of associated laboratories | 3 | 2 |  |

Source: Based on Appendix II.
[a]To facilitate comparison, all salaries have been converted to Reichstaler.
[b]Achard and Baumer, the occupants of the chairs in the Berlin Mining School and Giessen University's Economics Faculty, held better salaried chemical positions in other institutions.

in Schemnitz.[39] A year later, F. A. von Heynitz persuaded his parsimonious Saxon colleagues to take advantage of the chemist C. E. Gellert's presence in Freiberg by developing a mining school there and appointing him to the chair of chemistry.[40] During the next decade or so, authorities established six more salaried chemical positions and two laboratories in administrative schools: Kassel's Collegium Carolinum (ca. 1766; lab. 1769), Berlin's Mining School (1770), Brunswick's Collegium Carolinum (1771), Kaiserslautern's Cameral High School (1774; lab. 1775), Vienna's Theresianum (ca. 1775), and Giessen's Economics Faculty (1777).[41]

Although four of these positions were abolished or transferred to medical schools because they failed to satisfy the expectations of their patrons or occupants, six positions and two laboratories still existed in 1780 (see Table 6). Five of the surviving positions—those in Schemnitz, Freiberg, Berlin, Vienna, and Giessen—were oriented toward youths preparing to direct state mines and smelteries. Only that in Kaiserslautern had a broader orientation, probably because the Palatine Elector, whose realm was poor in mineral resources, had opted for a more diversified entrepreneurial program than had many German rulers.[42]

---

39. For Swieten's role in creating the positions in Schemnitz, Prague, and Idria, see A. T. Hornoch, "Zu den Anfängen des höheren bergtechnischen Unterrichtes in Mitteleuropa," *Berg- und Hüttenmännische Monatshefte*, 89 (1941), 33; W. Oberhummer, "Die Chemie an der Universität Wien," *Studien zur Geschichte der Universität Wien*, 3 (1965), 145–150; and E. Lesky, *Arbeitsmedizin im 18. Jahrhundert: Werksarzt und Arbeiter im Quecksilberbergwerk Idria* (Vienna: Notrings, 1954), p. 30.
40. H. Baumgärtel, "Bergbau und Absolutismus: Der sächsische Bergbau in der zweiten Hälfte des 18. Jahrhunderts und Massnahmen zu seiner Verbesserung nach dem Siebenjährigen Kriege," *Freiberger Forschungshefte: Kultur und Technik*, D44 (1963), 80–81.
41. For details about these six schools, see Appendix II.
42. For the place of the mineral-products industry in mercantilistic doctrine, see H.

Primarily intended to promote more effective direction of chemical technology in the state sector of the economy, material support for chemistry in administrative education must have played some part in the dramatic growth of the German mining industry during the last third of the eighteenth century.[43] This patronage also benefited the science in three ways. Its very existence advertised chemistry's breadth. Moreover, its recipients helped circulate a favorable picture of the science by their teaching. And several—five of the six salaried chemical representatives of 1780—participated in the science by publishing texts or monographs during their tenure.[44]

## THE BERLIN ACADEMY

While universities and other schools were supposed to dispense existing knowledge, learned societies and academies were expected to enlarge understanding. Expectations were especially high for endowed academies. According to the rationale developed in the seventeenth century, such academies were ideally suited for expanding the frontiers of natural knowledge. Unlike universities or voluntary societies, they could afford to nurture the talents of those rare geniuses capable of discovery. In doing so, academies would contribute to progress and thereby enhance the glory and power of their patrons.[45]

During the seventeenth century, the Paris Academy was the only scientific association in Europe to employ several academicians as the above theory

---

Baumgärtel, "Bergbau und Absolutismus," *Freiberger Forschungshefte*; and F. Heynitz, *Mémoire sur les produits du regne minéral de la monarchie prussienne et sur les moyens de cultiver cette branche de l'économie politique* (Berlin: Decker, 1786). For economic doctrines and planning in the Palatinate, see M. J. Funk, "Der Kampf der merkantilistischen mit der physiokratischen Doktrin in der Kurpfalz," *Neue Heidelberger Jahrbücher*, 18 (1914), 103–200; and W. Freitag, "Die Entwicklung der Kaiserslauterer Textilindustrie seit dem 18. Jahrhundert," Institut für Landeskunde des Saarlandes, Saarbrücken: *Veröffentlichungen*, 8 (1963), 20–23.

43. Evidence of the growth of the mineral-products industry in Germany after the Seven Years' War is widely scattered. For a useful summary, see W. Zorn, "Gewerbe und Handel 1648–1800," in *Handbuch der deutschen Wirtschafts- und Sozialgeschichte*, I:541–555. For statistics, see A. M. Heron de Villefosse, *De la Richesse Minérale*, vol. I (Paris: Levrault, 1810), pp. 15–18, 153, 156, 221, 239–240, 432, and facing tables.

44. J. W. Baumer in Giessen and G. A. Suckow in Kaiserslautern received strong encouragement to publish from officials who were eager to boost the renown of and attendance at their respective schools. The encouragement of Baumer was rather heavy-handed—see K. Witzel, "Friedrich Carl Moser: Ein Beitrag zur hessen-darmstädtischen Finanz- und Wirtschaftsgeschichte am Ausgang des 18. Jahrhunderts," Historische Kommission für den Volksstaat Hessen: *Quellen und Forschungen zur hessischen Geschichte*, 10 (1929), enclosure 16.

45. For eighteenth-century German expressions of the rationale for academies, see C. Wolff, *Vernünfftige Gedancken von dem gesellschafftlichen Leben der Menschen und insonderheit dem gemeinen Wesen zu Beförderung der Glückseeligkeit des menschlichen Geschlechts*, 6th ed. (Frankfurt and Leipzig: Renger, 1747), pp. 241–254; *Grosses vollständiges Universal-Lexicon*, vol. LVII (1748), pp. 1428–1431; and A. von Haller's report on the first meeting of the Göttingen Society of Sciences, *Göttingische Zeitungen von gelehrten Sachen* (18 November 1751), 1129–1135. Toward the end of the eighteenth century, however, some intellectuals believed that the increase in professorial research made academies a needless luxury. See F. Nicolai, *Beschreibung einer Reise durch Deutschland*, IV:682, 701–703.

required.[46] It served Leibniz as a model in his many attempts to persuade German rulers to found academies. His efforts finally met with modest success in 1700, when Brandenburg Elector Frederick III, who was soon to become Frederick I, King in Prussia, established a Society of Sciences in Berlin. Initially this Society's income, which derived from a monopoly on calendars, sufficed to pay but one academician. During the reign of the founder's son, King Frederick William I, revenues increased as the market for calendars improved. However, acting on the advice of Stahl, this practical monarch diverted most of the new monies into medical education. His son, King Frederick II, broke with this policy. Eager for the glory that the Paris Academy reflected on its patron, he completely reorganized the Berlin Society in 1744. From then on, the new Academy of Sciences and Fine Literature employed several academicians whose primary responsibility was research.[47]

After 1750, intellectuals in several other German states tried to convince their rulers to create academies on the model of those in Paris, St. Petersburg, and Berlin. They succeeded in Bavaria and the Palatinate, where academies with a few salaried members were founded in Munich (1759) and Mannheim (1763) respectively. However, since neither academy patronized chemistry prior to 1780, the focus here is on the Berlin Academy.[48]

Chemistry figured prominently in Leibniz's proposals of 1700 for the creation of a scientific society in Berlin. He suggested that such a society should have a salaried member for chemistry and mineralogy and a laboratory. Elector Frederick III responded to these suggestions by assigning chemistry to the Society's "physical-mathematical section" and by promising to establish chemical facilities. Subsequently, however, the monarch turned a deaf ear to

---

46. R. Hahn, *The Anatomy of a Scientific Institution: The Paris Academy of Sciences, 1666–1803* (Berkeley: University of California Press, 1971).

47. A. Harnack, *Geschichte der Königlich Preussischen Akademie der Wissenschaften zu Berlin*, 3 vols. (Berlin: Reichsdruckerei, 1900).

48. For a general discussion of the development of academies in Germany during the eighteenth century, see A. Kraus, *Vernunft und Geschichte: Die Bedeutung der deutschen Akademien für die Entwicklung der Geschichtswissenschaft im späten 18. Jahrhundert* (Freiburg, Basel, and Vienna: Herder, 1963), pp. 206–227. For the Munich Academy, see L. Hammermayer, "Gründungs- und Frühgeschichte der Bayerischen Akademie der Wissenschaften," *Münchener Historische Studien: Abteilung Bayerische Geschichte*, 4 (1959). The Munich Academy eventually had a salaried chemical representative (see Appendix II). For the Mannheim Academy, see A. Kistner, *Die Pflege der Naturwissenschaften in Mannheim zur Zeit Karl Theodors* (Mannheim: Altertumsverein, 1930). The Mannheim Academy never did employ a chemical representative. There is, I must admit, a certain arbitrariness in passing over academies and societies that lacked salaried positions for chemistry. Several, in fact, sponsored prize questions on pure and applied chemistry. For the context and response to prize questions bearing on agricultural chemistry, see. H.-H. Müller, "Akademie und Wirtschaft im 18. Jahrhundert: Agrarökonomische Preisaufgaben und Preisschriften der Preussischen Akademie der Wissenschaften," *Studien zur Geschichte der Akademie der Wissenschaften der DDR*, 3 (1975). And in 1776 Dalberg established a laboratory for the Erfurt Academy of Useful Sciences. See D. Oergel, "Die Akademie nützlicher Wissenschaften zu Erfurt von ihrer Wiederbelebung durch Dalberg bis zu ihrer endgültigen Anerkennung durch die Krone Preussen (1776–1816)," Akademie gemeinnütziger Wissenschaften zu Erfurt: *Jahrbücher*, N.F. 30 (1904), 166, 207. However, prizes and even facilities are not of much enduring consequence unless accompanied by material support for personnel.

## Material Support / 47

the Society's proposals that a tax on brandy or a monopoly on *aqua fortis* be used to endow a laboratory.

During the reign of Frederick William I, three men who achieved renown as chemists were named to the Society—Neumann in 1721, Pott in 1722, and Marggraf, as Neumann's successor, in 1738. Starting in 1724, both Neumann and Pott drew salaries from the Society as chemistry professors in the Medical-Surgical College and in 1731 Neumann began drawing an additional salary as the Society's treasurer. But they did not draw salaries as the Society's chemical representatives. The only material advantages of membership were that the Society dispensed modest funds for chemical experimentation and bore the costs of publishing articles in its memoirs.

Frederick II's Academy provided chemistry with more ample patronage. At its inception in 1744, Pott was paid a salary of 150 Rtl as academician in addition to his salary of 400 Rtl as professor. Marggraf, his junior, started with a salary of 100 Rtl. In subsequent years, however, Marggraf received preferential treatment. His salary was raised to 350 Rtl in 1746, 600 Rtl in 1753, 800 Rtl in 1762, and 900 Rtl in 1770. Meanwhile, in 1753 he was put in charge of the Academy's new laboratory, a building which included a free residence. In 1766, after the laboratory was extensively remodeled, he obtained an annual budget of 250 Rtl for experiments. Finally, in 1779 he helped arrange for his disciple, F. C. Achard, to receive a salary of 400 Rtl from the Academy. Between 1744 and 1780, therefore, the Academy's direct support for its chemical representatives increased from 250 Rtl in salaries to 1,300 Rtl in salaries, 250 Rtl in budget, and a laboratory with a residence.[49]

Why did Frederick II, who played the dominant role in guiding the Berlin Academy, increase its patronage for chemistry so dramatically? He did not do so because he believed in the usefulness of the science. His original faith in chemistry's utility had evaporated during the first decade of his reign along with 6,000 Rtl he spent on Pott's unsuccessful experiments to develop porcelain.[50] Nor did he do so because he valued chemistry's insights into nature. As he confided to d'Alembert in 1768, he was disappointed that "all modern scientific investigations into electricity, gravity, and chemistry have not im-

---

49. For chemistry in the Berlin Society and Academy, see A. Harnack, *Geschichte der Königlich Preussischen Akademie der Wissenschaften zu Berlin*, I: 176–177, 185, 216, 231, 236–237, 244, 265–267, 300, 323, 344, 363, 378–383, 440–441, 487, 490–491; II: 59–60, 66–67, 69, 73, 79–80, 104, 107, 123, 173–179, 220–222, 237, 239, 245, 264, 276, 305; and the institutional history in Appendix II. I am indebted to Dr. C. Kirsten, Director of the Academy's Archives, for information concerning salaries and budgets.

50. For Frederick's patronage of Pott's attempt to invent porcelain, see *Friedrichs des Grossen Korrespondenz mit Aerzten*, ed. G. L. Mamlock, facsimile of the 1907 ed. (Wiesbaden: Sändig, 1966), pp. 45, 51–52; A. B. König, *Versuch einer Historische Schilderung der Hauptveränderungen, der Religion, Sitten, Gewohnheiten, Künste, Wissenschaften, etc. der Residenzstadt Berlin seit den ältesten Zeiten, bis zum Jahre 1786*, vol. V, pt. 2 (Berlin: Pauli, 1799), pp. 246–249; [S. Formey], "Éloge de M. Pott," Akademie der Wissenschaften, Berlin: *Historie* (1777), 59; A. D. Bensch, *Die Entwicklung der Berliner Porzellanindustrie unter Friedrich dem Grossen* (Berlin: Heymann, 1928), pp. 22–23; and J. Schulze, "Die ersten Versuche der Porzellanfabrikation in Brandenburg," *Forschungen zur Brandenburgischen und Preussischen Geschichte*, 4 (1935), 149–153.

TABLE 7: Material Support for Chemistry in German Schools and the Berlin Academy, 1720–1780

|  | 1720 | 1740 | 1760 | 1780 |
|---|---|---|---|---|
| Number of salaried chairs | 6 | 15 | 20 | 36 |
| Number of associated laboratories | 1 | 1 | 4 | 11 |
| Median salary | 150 Rtl | 300 Rtl | 350 Rtl | 400 Rtl |
| Number of chairs in |  |  |  |  |
| Medical schools | 6 | 15 | 19 | 28 |
| Administrative schools | 0 | 0 | 0 | 6 |
| Berlin Academy | 0 | 0 | 1 | 2 |
| Number of chairs in |  |  |  |  |
| Protestant institutions | 6 | 12 | 12 | 22 |
| Catholic institutions | 0 | 3 | 8 | 14 |

Source: Based on Appendix II.

proved men, nor changed their moral conditions."[51] Frederick did thirst, however, for glory. He wanted to be known and remembered not only as a Caesar, but also as a Maecenas. Chemistry provided an avenue for this ambition. Prompted by Maupertuis, who informed him in 1749 that "our chemists are superior to all the chemists of Europe," Frederick had good reason to think that patronage of chemistry would enhance his fame.[52] Hence, he was willing to use part of the Academy's growing income from calendars for the benefit of the science.

Intended to contribute to his glory, Frederick's generous patronage for the Academy's chemists achieved its goal. Pott, Marggraf, and Achard did important research, published their findings, attained European renown, and thereby advertised the splendor of their patron.[53]

In summary, material support for chemistry in German institutions of learning multiplied between 1720 and 1780. This growth was accompanied by diversification. Initially, Protestant medical schools housed all salaried chemical positions. During the ensuing decades, the Berlin Academy and administrative schools became important additional loci of patronage. Meanwhile, Protestant authorities were joined by their Catholic counterparts in providing funds for the science (see Table 7).

The basic reasons for the growth and diversification of material support for chemistry in German schools and the Berlin Academy are clear. As rulers and their advisors became committed to enlightened governance and concur-

---

51. Frederick II, letter to J. d'Alembert (7 January 1768), in A. Harnack, *Geschichte der Königlich Preussischen Akademie der Wissenschaften zu Berlin*, I:137.
52. P. L. de Maupertuis, letter to Frederick II (10 November 1749), in A. Harnack, *Geschichte der Königlich Preussischen Akademie der Wissenschaften zu Berlin*, II:276. It was Maupertuis who persuaded Frederick to build the laboratory in the early 1750s (I:487).
53. For the careers and reputations of Pott, Marggraf, and Achard, see their biographical profiles in Appendix I.

rently came to entertain broader views of chemistry, they became increasingly aware of its relevance to their goals. Accordingly, they patronized chemical education for physicians in order to promote pharmaceutical reform, public health and, ultimately, their power. They supported chemical education for mining and other administrators in order to promote technological progress, greater prosperity, and their power. And Frederick II patronized chemical research in order to promote scientific advancement and, thereby, his fame. The patrons' material support for chemistry depended not only on their growing awareness of its relevance to their goals, but also on their assessment of the need for patronage. Hence, Catholic authorities tended to be more generous than their Protestant counterparts because the need was greater in their more tightly regulated educational system.

Though primarily intended for other purposes, the new material support for chemistry in German schools and the Berlin Academy benefited the science in four main ways. First, the very existence of this patronage enhanced chemistry's visibility. Second, its recipients hastened the diffusion of chemistry's image as a useful and penetrating science. Third, its recipients helped train many men who subsequently engaged in chemical research. And fourth, the patronage placed its recipients, especially the Protestants, in situations where their colleagues increasingly expected them to publish on their specialty. A growing percentage of them did so, and several achieved contemporary renown as chemists, and one—Marggraf in Berlin—enduring fame.

# 4 MANPOWER SUPPORT

Moral and material support for chemistry multiplied in Germany during the eighteenth century. Did manpower keep pace? This question is crucial because, in the final analysis, a science's fate depends upon the persons who fill its ranks. To see that German manpower support for chemistry grew between 1720 and 1780, it suffices to examine changes in the number of chemists, their geographical distribution, and their rates of publication. But to obtain a full picture of this growth, it is necessary to look at concurrent changes in the chemists' educational backgrounds and occupational activities.[1] In carrying out these tasks, I focus on men who were born in Germany, did much of their chemical research there, and won recognition as contributors to chemistry from five or more attentive observers of the eighteenth and early nineteenth centuries.[2] Adhering to these criteria and counting those men who were twenty-five to sixty-five years old,[3] I obtain forty-two men—the "chemists" of this chapter—who were active in Germany between 1720 and 1780 (see Table 8). These German chemists, who at any given time probably comprised a fifth to a sixth of all Germans carrying out chemical investigations, were more resolute and original than their fellows. Then, as today, recognition tended to reflect quantity and quality of work. There is no reason, however, to suppose that the chemists selected were unrepresentative in other respects.[4]

  1. The use of prosopographic methods for investigating chemistry in eighteenth-century Germany was first suggested to me by H. Rosenberg. For orientation to this methodology and an exemplary study of early modern physicists, see L. Pyenson, "'Who the Guys Were': Prosopography in the History of Science," *History of Science*, 15 (1977), 155–188; and J. L. Heilbron, *Electricity in the 17th and 18th Centuries: A Study of Early Modern Physics* (Berkeley: University of California Press, 1979), pp. 98–166.
  2. For a general description of the evidence used to assess recognition, see the introduction to Appendix I. For a listing of the observers who identified a given German chemist as a meritorious contributor to chemistry, see that chemist's biographical profile in Appendix I. In this chapter I do not consider three fairly prominent foreign-born chemists who worked in Germany: G. A. Scopoli (1723, Cavalese–1788, Pavia), N. J. Jacquin (1727, Leiden–1817, Vienna), and J. Ingen-Housz (1730, Breda–1799, Bowood Park). (However, because they worked in Germany, I do include their profiles in Appendix I.) I also exclude from consideration three respected German-born chemists who emigrated before reaching the age of thirty: H. D. Gaubius (1705, Heidelberg–1789, Leiden), J. G. Model (1711, Rothenburg ob der Tauber–1775, St. Petersburg), and C. W. Scheele (1742, Stralsund–1786, Köping, Sweden).
  3. My rationale for using an age-range, rather than restricting attention to those who had begun publishing, is that before the last third of the eighteenth century German chemists often accumulated findings for years, even decades, before venturing into print. By the age of twenty-five (see below), most of those who earned reputations as chemists were moving beyond learning about chemistry to carry out their own investigations. By the age of sixty-five, most were behind-the-times, inactive, or dead—the only significant exception between 1720 and 1780 was G. E. Stahl.
  4. Comparison of visible German chemists of the late eighteenth century with German sub-

TABLE 8: Lifespans of Selected German Chemists

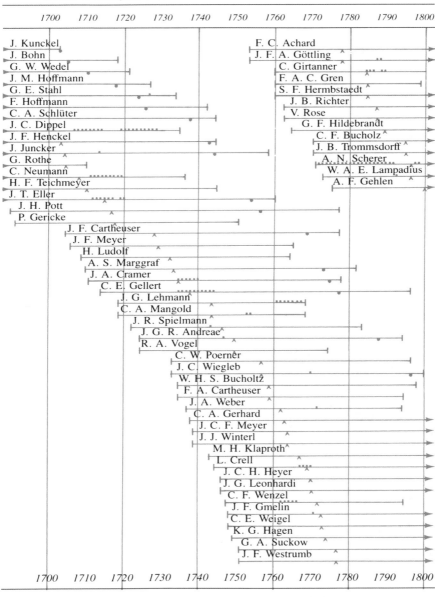

Note: The only chemists included are those listed in Table A1 in Appendix I.
^ = date of twenty-fifth birthday.
• = date of sixty-fifth birthday.
* = traveling or residing abroad.

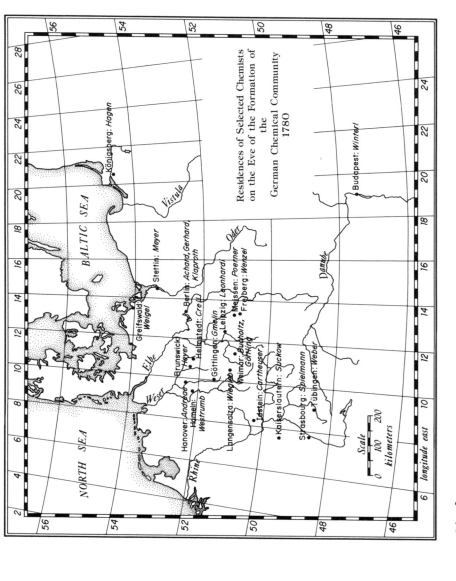

Map 2.

Between 1720 and 1780 the number of German chemists climbed steadily, registering the largest absolute gain after the Seven Years' War.[5] The importance of moral support in promoting this numerical growth is manifest in the pattern of their geographical dispersal. Despite the emergence of Catholic patronage for chemical instruction, very few of the selected chemists were educated or employed in Catholic lands.[6] Almost all studied and worked in Protestant towns where the Enlightenment became sufficiently deep-rooted to generate broad approval for the pursuit of chemistry. As chemists became more numerous and dispersed in Germany, they also became more regular in their contributions to the chemical literature. Besides reflecting the growth in public respect for chemistry, this increased productivity probably reflected, as we see in this chapter, shifts in the educational backgrounds and occupational activities of the chemists and, as we see in the next two chapters, the establishment of a chemical journal in the late 1770s. Thus Germany, especially urban Protestant Germany, provided chemistry with ever greater manpower support. By 1780 our chemical population numbered twenty-two men. They resided in thirteen Protestant educational, administrative, and economic centers, three Protestant burgs, and two Catholic educational centers (see Map 2). And they were typically publishing on chemistry one or more times a year (see Table 9).

## EDUCATIONAL BACKGROUNDS

Before one can participate in a science, one must acquire a familiarity with its current concepts, theories, data, and techniques. Who does so depends on the availability of opportunities for learning the science. In the eighteenth century, students at the better educational institutions could acquire the information and skills needed for contributing to most branches of natural knowledge. Moreover, young men in other walks of life were increasingly able to engage in autodidactic endeavors, share in learned conversations, and audit worthwhile courses. As a result of these opportunities, most men who made names for themselves in the natural sciences during the eighteenth century were, regardless of their status, introduced to the subjects they later cultivated while preparing for the world of work.

Our German chemists conformed closely, perhaps exceptionally closely, to

---

scribers to chemical publications of that era indicates that the visible chemists were similar to their audience in geographical, social, and educational backgrounds. See my "Chemistry's Enlightened Audience," *Studies on Voltaire and the Eighteenth Century*, 53 (1976), 1069–1086.

5. For an enthusiastic autobiographical account of the surge in chemical activity after the Seven Years' War, see J. C. Wiegleb, *Geschichte des Wachstums und der Erfindungen in der Chemie in der neuern Zeit*, vol. II (Berlin and Stettin: Nicolai, 1791), pp. [ii–iv].

6. Winterl was the sole Catholic among the men who joined the ranks of our sample chemists between 1720 and 1780. Besides Winterl, only four of our chemists spent more than a few weeks in Catholic Germany: Ludolf worked in Mainz from 1753 to 1764; Gmelin studied in Vienna during winter semester 1770/71; Suckow worked in Kaiserslautern from 1774 to 1784; and Cramer worked in the Hungarian mining district 1775/76. See their profiles in Appendix I.

TABLE 9: Number, Geographical Distribution, and Publication Rates of Selected German Chemists, 1720–1780

|  | 1720 | 1740 | 1760 | 1780 |
|---|---|---|---|---|
| Number of chemists aged 25–65, residing in Germany[a] | 9 | 11 | 15 | 22 |
| Number of Protestant towns with chemists[a] | 6 | 8 | 12 | 16 |
| Number of Catholic towns with chemists[a] | 0 | 0 | 1 | 2 |
| Number not yet publishing on chemistry[b] | 3 | 2 | 1 | 0 |
| Number with less than one article equivalent/year on chemistry[b] | 5 | 7 | 9 | 7 |
| Number with one or more article equivalent/year on chemistry[b] | 1 | 2 | 5 | 15 |
| Median rate of publication (in article equivalents/year)[b] | .3 | .7 | .7 | 1.4 |

Source: Based on Appendix I.
Note: This table is devoted to the forty-two German chemists who were recognized as contributors to chemistry by five or more attentive observers of the eighteenth and early nineteenth centuries and who were twenty-five to sixty-five years old between 1720 and 1780.
[a]Based on residence at the end of the year.
[b]Based on bibliographies for 1716–1724, 1736–1744, 1756–1764, and 1776–1784. In estimating the equivalents for books, I have been guided by the importance and originality of their contents. For bibliographies, I have relied on the works cited at the end of each profile in Appendix I; J. G. Meusel, *Lexikon der vom Jahr 1750 bis 1800 verstorbenen teutschen Schriftsteller*, 15 vols., facsimile of the 1802 ed. (Hildesheim: Olms, 1967–1968); G. C. Hamberger and J. G. Meusel, *Das gelehrte Teutschland oder Lexicon der jetzt lebenden teutschen Schriftsteller*, 23 vols., facsimile of the 1796–1834 ed. (Hildesheim: Olms, 1965–1966); and *Chemisch-Pharmazeutisches Bio- und Bibliographikon*, ed. F. Ferchl (Mittenwald: Nemayer, 1937). In determining medians, I have excluded those who had not yet begun publishing on chemistry.

TABLE 10: Educational Backgrounds of Selected German Chemists, 1720–1780

|  | 1720 | | 1740 | | 1760 | | 1780 | |
|---|---|---|---|---|---|---|---|---|
| Number who began studying chemistry | | | | | | | | |
| As prospective physicians | 7 | (5) | 8 | (2) | 7 | (5) | 10 | (8) |
| As prospective pharmacists | 1 | (1) | 2 | (2) | 6 | (4) | 10 | (7) |
| In other capacities | 1 | (0) | 1 | (1) | 2 | (1) | 2 | (1) |

Source: See Table 9.
Note: The numbers of chemists under forty-five years of age are indicated in parentheses in order to reveal the more recent recruits.

the general pattern. Almost all were introduced to chemistry as they were training for careers in therapy and pharmacy.[7] Around 1720 most chemists in our sample were men who had begun learning the science as prospective physicians. In the ensuing decades, however, those joining the ranks of our chemists were about as likely to have been introduced to chemistry as prospective apothecaries. By 1780, consequently, the educational backgrounds of our German chemists had undergone a pronounced shift from therapy toward pharmacy (see Table 10).

7. Only Schlüter, Cramer, Gellert, and Achard did not begin getting acquainted with chemistry as prospective physicians or pharmacists. Of these men, Cramer subsequently spent some time preparing for a medical career (see Appendix I).

About half of the thirty-one men entering our sample between 1720 and 1780 were introduced to chemistry as medical students.[8] Nearly all these men received considerable chemical instruction before reaching their mid-twenties. Some built upon foundations laid by their fathers. For instance, from 1752 to 1753, F. A. Cartheuser, son of J. F. Cartheuser, the chemistry professor in Frankfurt an der Oder, followed up his father's instruction with soujourns in Berlin, where he attended Pott's lectures and cultivated Marggraf's acquaintance, and the Saxon mining district, where he obtained practical training in assaying and smelting.[9] Likewise, from 1769 to 1771, J. F. Gmelin, son of Tübingen's chemistry professor, P. F. Gmelin, rounded out his education on a grand tour, enrolling in, among other things, the chemistry courses of Gaubius in Leiden and Jacquin in Vienna.[10] Others got their start in the science from enthusiastic instructors. At Leipzig in the early 1750s, for example, Poerner studied with A. Ridiger, an energetic rival of the aging chemistry professor A. F. Petzold. Fifteen years later, young Leonhardi was, in turn, initiated into chemistry by Poerner, who had assumed the role of energetic rival after Ridiger succeeded Petzold.[11]

Most of the remaining new chemists in our sample were men who began learning chemistry as pharmaceutical apprentices.[12] Their entry into the field was a remarkable development, for prospective apothecaries have rarely sought to transcend the realm of recipes. A rising number did so in eighteenth-century Germany because latent energies within the pharmaceutical profession were released by state intervention in recruitment. Starting with Prussia in 1725, more and more German states required journeymen who wanted to become masters of major apothecary shops to pass an examination in which, among other things, they were expected to demonstrate a knowledge of chemistry. Although there must have been some grumbling, pharmacists soon accepted this intrusion. Migrating journeymen, impressed by the outstanding pharmacist-chemists, Neumann and Marggraf, spread the view that all apothecaries owed it to their profession and themselves to be-

8. Fifteen of our new chemists began learning chemistry as prospective physicians: J. F. Cartheuser, Ludolf, Lehmann, Mangold, Vogel, Poerner, F. A. Cartheuser, Weber, Gerhard, Winterl, Crell, Leonhardi, Gmelin, Weigel, and Suckow (in order of birth). For their educations, see Appendix I.
9. An autobiographical account of Cartheuser's education appears in F. W. Strieder, *Grundlage zu einer Hessischen Gelehrten und Schriftsteller Geschichte*, vol. III (Kassel: Cramer, 1783), p. 537.
10. Gmelin, seven letters to A. von Haller (26 March 1770–10 March 1771), Bern Stadtbibliothek. Besides studying chemistry, Gmelin also pursued botany. In fact, to judge from his letters, botany was his favorite subject at this stage in his life. It was his subsequent appointment as Göttingen's professor of chemistry that turned him decisively toward this science.
11. For brief autobiographical accounts of Poerner's and Leonhardi's educations, see S. T. Quellmalz, *Panegyrin Medicam* (Leipzig: Langenheim, 1754), p. xv; and A. G. Plaz, *Panegyrin Medicam* (Leipzig: Langenheim, 1771), pp. xiii–xiv. For chemistry at Leipzig University during the 1750s and 1760s, see E. Mayr, "Die Entwicklung der Chemie und der pharmaceutischen Chemie an der Universität Leipzig" (Diss., Leipzig University, 1965), pp. 22–27.
12. Thirteen of our new chemists began learning the science as prospective pharmacists: J. F. Meyer, Marggraf, Spielmann, Andreae, Wiegleb, Bucholtz, J. C. F. Meyer, Klaproth, Heyer, Wenzel, Hagen, Westrumb, and Göttling. For their educations, see Appendix I. Spielmann, Bucholtz, and Hagen eventually obtained M.D.s.

come proficient chemists. The pharmacist who knew chemistry, it was argued, could not only achieve greater efficiency but also produce purer drugs. Moreover, he could rise above his status as a shopkeeper and gain entry into the relatively prestigious world of learning. Thanks to the spread of such views, prospective pharmacists had ever greater opportunities for bridging the gap between compounding medicines and pursuing chemistry.[13]

Nearly all the men who, beginning in apothecary shops, eventually became recognized chemists availed themselves of such opportunities. Several supplemented their pharmaceutical training with formal instruction in chemistry. Between 1726 and 1735, for example, Marggraf studied chemistry with Pott and Neumann in Berlin, F. Hoffmann and others in Halle, and Henckel in Freiberg.[14] Similarly, in the mid-1740s Andreae attended Pott's lectures in Berlin and Gaubius's in Leiden.[15] Again, about two decades later, Wenzel, after learning pharmacy from a Dutch pharmacist-surgeon, evidently studied the natural sciences, including chemistry, with Poerner in Leipzig.[16] In addition to those who obtained academic instruction, some got their grounding in chemistry almost or completely within the pharmaceutical ambit. In Hanover during the mid-1760s, for instance, the journeyman Klaproth set himself the task of learning chemistry as a science during his years in the town's well-appointed Hof-Apotheke, most likely drawing inspiration from the example and advice of the nearby apothecary Andreae, who was systematically analyzing local soils on government commission.[17] Klaproth's example and advice, in turn, inspired Westrumb, a young apprentice working under his supervision, to start studying chemistry.[18] Meanwhile in Langensalza,

13. For illuminating discussions of the growing participation of German pharmacists in chemical instruction and research, see G. E. Dann, "Berlin als ein Zentrum chemischer und pharmazeutischer Forschung im 18. Jahrhundert," *Pharmazeutische Zeitung*, 112 (1967), 189–196; D. Pohl, *Zur Geschichte der pharmazeutischen Privatinstitute in Deutschland von 1779 bis 1873* (Diss., Marburg University; Marburg: Mauersberger, 1972); B. H. Gustin, "The Emergence of the German Chemical Profession 1790–1867" (Diss., University of Chicago, 1975), pp. 55–71; and E. Hickel, "Der Apothekerberuf als Keimzelle naturwissenschaftlicher Berufe in Deutschland," *Pharmazie in unserer Zeit*, 6 (1977), 15–22. While Gustin and Hickel properly emphasize the importance of the influx of pharmaceutically trained men into German chemistry, they have not given sufficient attention to the role of the medical profession and medical schools in introducing pharmacists to chemistry as a science.

14. J. H. Pott, *Fortsetzung seiner physikalisch-chymischen Anmerkungen* (Berlin, 1756), p. 7; and E. G. Baldinger, *Biographien jetztlebender Aerzte und Naturforscher in und ausser Deutschland*, pt. 3 (Jena: Hartung, 1771), pp. 90–91.

15. F. Schlichtegroll, *Nekrolog auf das Jahr 1793*, vol. I (Gotha: Perthes, 1794), pp. 164–165. Andreae's letters to the apothecary J. A. Beurer in the Erlangen University Library would probably shed further light on his education.

16. For a brief account of Wenzel's colorful youth, see his obituary in the *Allgemeine Literatur-Zeitung: Intelligenzblatt* (1793), 706–707. My supposition that Poerner was Wenzel's chemistry teacher is supported by the fact that Poerner was later his supervisor at the Meissen porcelain works.

17. G. E. Dann, *Martin Henrich Klaproth (1743–1817): Ein deutscher Apotheker: Sein Weg und seine Leistung* (Berlin: Akademie Verlag, 1958), pp. 24–26. My suggestion that Andreae provided Klaproth with a role-model is based on their proximity and the similarity of their research.

18. For a vivid account of Westrumb's training and autodidactic endeavors, see A. Westrumb, "Johann Friedrich Westrumb," *Neues vaterländisches Archiv des Königreichs Hannover*, 7 (1825), 25–28.

Göttling, encouraged and assisted by his master, the pharmacist-chemist Wiegleb, was striving to learn the science.[19]

Notwithstanding the desire and efforts of all these youths to be more than pharmacists, most had somewhat circumscribed educations. Even those who spent time at one or another university usually concentrated on chemistry and other subjects related to pharmacy.[20] As they entered on their careers, therefore, they had fewer intellectual options than did those with university educations. For this reason, the shift toward pharmacy in our chemists' educational backgrounds probably helped promote dedication to and productivity in chemistry.

## OCCUPATIONAL ACTIVITIES

In order to participate in a science, one must not only know enough to make contributions but also have access to the requisite resources. In the eighteenth century, scientists solved the problem of resources in several ways. Some drew upon family wealth. Some obtained patronage. Some used earnings from jobs lacking any close relation to their scientific interests. Many depended on jobs that gave them the opportunity, even the responsibility, to apply or teach their scientific knowledge. Only a fortunate few—the salaried members of major academies—enjoyed direct material support for research comparable to that received by so many of today's scientists.

Like other eighteenth-century men of science, our German chemists got the wherewithal for their research in various ways. Some were born into fairly prosperous families.[21] A few found wealthy wives or aristocratic patrons.[22] But none was blessed with a legacy so large or a spouse or benefactor so open-handed that he could blithely devote himself to the pursuit of chemistry. Nor did any derive a substantial fraction of his income from a job that was completely unrelated to the science. Rather, it was by working within the medical-technical-scientific realm that our chemists supported their research. Around 1720 most were earning their living by practicing medicine and teaching others how to practice medicine. However, during the next six decades there was a sharp decline in our chemists' dependence on therapy. By 1780 they were much more likely to be working in pharmacy, technology, or science (see Table 11).

This shift away from therapy resulted, in part, from incipient trends within

---

19. R. Möller, "Chemiker und Pharmazeut der Goethezeit," *Pharmazie*, 17 (1962), 624.

20. For example in 1763, the year he took his M.D. in Jena, Bucholtz's knowledge of anatomy and surgery was found to be deficient by those examining his qualifications to begin practicing medicine. See R. Möller, "Ein Apotheker des klassischen Weimar," *Pharmazie*, 15 (1960), 182.

21. To judge from their parents, their travels, and their careers, at least ten of our new chemists came from wealthy families: Marggraf, Cramer, Lehmann, Spielmann, Andreae, F. A. Cartheuser, J. C. F. Meyer, Crell, Gmelin, and Achard.

22. Bucholtz and Klaproth acquired their apothecary shops with money from their wives' families. Pott, Mangold, Weber, and possibly Wenzel obtained patronage from rulers or nobles. Lehmann obtained a series of personal grants from the Berlin Academy. For details, see Appendix I.

TABLE 11: Primary Occupational Areas of Selected German Chemists, 1720–1780

|  | 1720 | 1740 | 1760 | 1780[a] |
|---|---|---|---|---|
| Therapy | 7 (5) | 6 (2) | 5 (3) | 3 (2) |
| Pharmacy | 1 (1) | 2 (2) | 4 (4) | 8 (6) |
| Technology | 1 (0) | 2 (1) | 2 (0) | 4 (3) |
| Science | — | 1 (0) | 4 (3) | 6 (5) |

Source: See Table 9.
Note: The numbers of chemists under forty-five years of age are indicated in parentheses in order to reveal the occupational areas of the more recent recruits to chemistry. Some guesswork is involved in assigning chemists to primary occupational areas because several worked in two or more areas simultaneously.
[a] F. A. Cartheuser, who retired from the Giessen chair of natural philosophy to an estate near Idstein in 1779, has not been included in this column.

the world of work. Thanks to the pharmaceutical profession's growing respect for chemistry, the knowledgeable journeyman had an edge in the competition for better jobs. For example, Klaproth, having gained the confidence of the pharmacist-chemist V. Rose, Sr. in Berlin in 1770, was entrusted with administering the thriving Apotheke zum weissen Schwan upon the latter's untimely death a year later.[23] Similarly, Göttling, having completed his apprenticeship with Wiegleb in 1775, easily won the job of managing the Hof-Apotheke in Weimar for the pharmacist-chemist Bucholtz.[24] A few years later Wiegleb capitalized on his profession's growing enthusiasm for chemistry by establishing Germany's first successful chemical-pharmaceutical boarding school in Langensalza.[25]

In the meantime, opportunities for men with chemical expertise were emerging in the technological and scientific realms. Thanks to increasing concern about the revenues yielded by state-owned enterprises, several of our chemists found good jobs in the bureaucracies directing metallurgical and porcelain works.[26] In the early 1740s, for instance, Cramer, who had begun learning chemistry at his father's ironworks and subsequently studied the science at Halle, Helmstedt, and Leiden, was charged with upgrading the mining industry in the environs of Blankenburg.[27] Again, in 1768 the physi-

23. Dann, *Martin Heinrich Klaproth*, pp. 28–30.
24. Möller, "Chemiker und Pharmazeut der Goethezeit," *Pharmazie*, p. 625.
25. For Wiegleb's school, which was established around 1779, see Pohl, *Zur Geschichte der pharmazeutischen Privatinstitute*, pp. 23–28. Over four decades earlier, Neumann established such a school in Berlin. But he died too soon afterwards for it to become well known. See H. Gossmann, "Das Collegium Pharmaceuticum Norimbergense und sein Auffluss auf das Nürnbergische Medizinalwesen," *Quellen und Studien zur Geschichte der Pharmazie*, 9 (1966), 191.
26. Among our new chemists, Cramer, Gellert, Gerhard, Poerner, and Wenzel obtained positions in the Brunswick, Saxon, and Prussian mining services. In Catholic Germany, Jacquin, Scopoli, and I. Born found employment in the Hapsburg mining service. For all these men, see Appendix I.
27. For Cramer's stormy career, see F. Cramer, "Johann Andreas Cramer," *Harzboten* (1828), 194–204, 277–287. A fresh, archivally based study of this influential technical chemist would illuminate the actual relations between science and technology in mid-eighteenth-century Germany.

cian-chemist Gerhard secured the Berlin Academy's backing in his successful bid for a post in the Prussian mining administration.[28] That same year, Leipzig's physician-chemist Poerner evidently used a job offer from St. Petersburg to get an important post at the Meissen porcelain works.[29] About this time, a few of our chemists also became involved in private manufacturing ventures.[30] In 1771 Weber began running a chemical works in Tübingen and advising others all over Germany on the manufacture of sal ammoniac, saltpeter, and other compounds.[31] Beginning around 1774, Suckow helped direct a factory for dyeing cloth that had recently been established in Kaiserslautern by a joint-stock company.[32] In 1780 J. C. F. Meyer founded a distillery in Stettin to produce French brandy and liqueurs according to a process that he had developed in his apothecary shop.[33] Meanwhile, as we saw in the last chapter, the improvement of material support for chemistry in learned institutions resulted in salaried positions for several of our chemists as chemistry teachers and for a few as academicians.[34]

Besides reflecting the emergence of new employment opportunities, the shift away from therapy in our chemists' occupational activities also reflected the emergence of more demanding standards in chemistry. Increasingly, experimental research, in contrast to erudite conjecturing, was recognized as the key to the science's progress. As Rose in Berlin observed in 1770, "thinking chemists who did their own work" contributed incomparably more to chemistry than "mere speculators."[35] But to do good experimental work, the chemist needed a well-equipped laboratory, a stock of reagents, an assistant

28. For Gerhard's hiring and career, see K. Wutke, "Aus der Vergangenheit des schlesischen Berg- und Hüttenlebens," in *Festschrift zum XII. Allgemeinen Deutschen Bergmannstage in Breslau 1913: Der Bergbau im Osten des Königreichs Preussen*, vol. V (Breslau: Nischkowsky, 1913), pp. 9, 11–12, 16, 35, 37, 51–52, 66, 211–214, 432, 437–440.
29. My reconstruction of Poerner's path to Meissen is based on the circumstantial evidence presented in his profile in Appendix I.
30. Despite the examples that follow, publishing chemists were not closely associated with the burgeoning of private chemical works in Germany during the last third of the eighteenth century. See, for instance, S. Jacob, "Chemische Vor- und Frühindustrie in Franken: Die vorindustrielle Produktion wichtiger Chemikalien und die Anfänge der chemischen Industrie in fränkischen Territorien des 17., 18. und frühen 19. Jahrhunderts," *Technikgeschichte in Einzeldarstellungen*, 9 (1968); and J. Kermann, "Die Manufakturen im Rheinland 1750–1833," *Rheinisches Archiv*, 82 (1972).
31. R. Multhauf, "A Premature Science Advisor: Jacob A. Weber (1737–1792)," *Isis*, 63 (1972), 356–369.
32. W. Freitag, "Die Entwicklung der Kaiserslauterer Textilindustrie seit dem 18. Jahrhundert," Institut für Landeskunde des Saarlandes, Saarbrücken: *Veröffentlichungen*, 8 (1963), 23–24, 30.
33. "Die Liköre des Hof-Apothekers Meyer in Stettin und Friedrich der Grosse," *Pharmazeutische Zeitung*, 80 (1935), 797.
34. Among our new chemists, J. F. Cartheuser, Vogel, Mangold, Ludolf, Spielmann, Gellert, Weigel, Suckow, Gerhard, Gmelin, and Achard became salaried chemistry teachers between 1720 and 1780; and Pott, Marggraf, and Achard became salaried academicians for chemistry. See their profiles in Appendix I and the corresponding institutional histories in Appendix II.
35. [Rose], review of L. J. D. Suckow's *Entwurf einer Physischen Scheidekunst* (1769), *Allgemeine deutsche Bibliothek*, 13 (1770), 430. That Rose was the reviewer is revealed in G. F. C. Parthey, *Die Mitarbeiter an Friedrich Nicolai's Allgemeiner deutscher Bibliothek nach ihren Namen und Zeichen* (Berlin: Nicolai, 1842).

or two, and plenty of time. This was no small order. Indeed, in 1774 Wiegleb argued that only states could employ and provision chemists in such a way that they could regard chemical research "as their sole and constant concern."[36]

Among those who knew enough chemistry to consider contributing to its advancement, physicians would have found it especially difficult to respond to the increasing emphasis on experimentation. Their calling gave them few, if any, incentives for maintaining a laboratory. They had little to gain by doing more than a perfunctory job of inspecting apothecary shops and spa waters, conducting inquests in suspected cases of poisoning, or apprehending food and beverage adulterators. Moreover, as apothecaries became better able to satisfy orders for exotic remedies, physicians had less and less to gain by manufacturing patent medicines.[37] Hence, doctors who wanted to cultivate chemistry generally had to devote a good deal of their time and wealth to this end.

Pharmacists, technical chemists, and salaried chemistry teachers were better situated to pursue the science. While carrying out their daily tasks and training their assistants, pharmacists and technical chemists could polish their experimental skills and improve their conceptual mastery. While preparing their lectures and demonstrating their assertions, salaried chemistry teachers could maintain their readiness to undertake research. Besides being in a better position than doctors to make efficient use of whatever spare time they chose to devote to chemistry, pharmacists, technical chemists, and salaried chemistry teachers could experiment without drawing so heavily on their discretionary income. Pharmacists could use their shops and subordinates as research facilities and staffs. Likewise, technical chemists could utilize their work-places and assistants for their investigations. And salaried chemistry teachers, though they often were not provided with laboratories, could usually depend on students to volunteer their services as experimental assistants. Thus their advantage in solving problems of time and cost helps explain why pharmacists, technical chemists, and salaried chemistry teachers became more numerous among our chemists as physicians became rarer.

Our chemists' growing economic dependence on applying and teaching chemistry, like the shift toward pharmacy in their educational backgrounds, probably helped foster a narrowing of their interests. Indeed, to judge from their scientific endeavors, the men in our sample were increasingly willing,

---

36. [Wiegleb], review of Poerner's *Chymische Versuche und Bemerkungen zum Nutzen der Färbekunst* (vol. II, 1772), *Allgemeine deutsche Bibliothek*, 21 (1774), 579.

37. Though regulations prohibiting physicians from competing with apothecaries usually did not cover patent medicines, the medical profession evidently looked on such remedies with increasing disapproval. Consequently, the prospect of financing a laboratory by producing them became less attractive. The last of our physician-chemists to sell patent remedies seems to have been Ludolf, who did so in Erfurt from the early 1740s to 1753. However, in the early 1770s, Weigel's father was still selling his special drops in Stralsund (see Appendix I).

even eager to seek identities solely as chemists.[38] The intensification of their occupational involvement with chemistry seems to have done more than encourage specialization. It evidently reinforced, as well as reflected, the rising ardor for experimentation. Men whose day-to-day work was more likely to take them into a laboratory than a study quite naturally valued experimental proficiency more than bibliographical thoroughness or theoretical ingenuity. In 1775 Wiegleb expressed the increasingly dominant viewpoint when, in criticizing a tract on saltpeter, he observed that "although the author . . . fails to announce that he is not a *Chymist von Profession*, that he does not perform his own experiments, that he builds only on other people's results, his every page betrays more than once that this is the case."[39] In fact, by the 1790s, as we shall see in Chapter VIII, our chemists placed such a high value on experiments that their reception of Lavoisier's theory, an issue of great theoretical complexity, was made to depend entirely on his experimental reliability.

38. Between 1720 and 1780, the fraction of our sample chemists devoting themselves primarily to chemistry in their intellectual work climbed from about 50 percent to about 90 percent. This estimate is based on bibliographies and biographies. See source note and note b, Table 9, and the profiles in Appendix I.
39. [Wiegleb], review of C. F. Selig's *Chymische Abhandlung vom Salpeter* (1774), *Allgemeine deutsche Bibliothek*, 27 (1775), 186.

# 5 LORENZ CRELL: CHEMICAL JOURNALIST

By the late 1770s, German chemists were ready to form into a national discipline-oriented community. Their dispersal was great enough for them to constitute a *national* community. Their outlook was similar enough for them to constitute a *discipline-oriented* community. They viewed chemistry as a fundamental natural science with a virtually inexhaustible range of applications. They regarded painstaking experimentation rather than agile theorizing as the key to chemistry's advancement. And, as we shall see in this and the following chapters, they thought of themselves as "German chemists."

Yet, for want of an adequate means of communicating with one another, German chemists had not yet coalesced into a community. Finding it wellnigh impossible to keep abreast of the work of chemists elsewhere in Germany, they rarely thought of other German chemists as their most important peers. Only a periodical that could serve as a forum would establish the regular communications necessary to bind them into a German chemical community. But then as now, scientific periodicals did not just spring into existence. An editor was needed who possessed the drive and talent both to rally a large enough audience to make the periodical financially viable and to mobilize enough good contributors to make it intellectually worthwhile. The man who stepped forward to fill this need for German chemists was Lorenz Crell (Figure 3).[1]

## THE MAKING OF AN EDITOR

Florens Lorenz Friedrich Crell was born in the Duchy of Brunswick's quiet university town of Helmstedt on 21 January 1745.[2] His family was rich in

---

1. For the problems confronting editors of scientific periodicals in the late eighteenth century, see D. A. Kronick, *A History of Scientific & Technical Periodicals: The Origins and Development of the Scientific and Technical Press 1665–1790*, 2d ed. (Metuchen, N.J.: Scarecrow, 1976); and J. E. McClellan, "The Scientific Press in Transition: Rozier's Journals and the Scientific Societies in the 1770s," *Annals of Science*, 36 (1979), 425–449. For Crell's role in the development of German chemical periodicals, see H. Harff, *Die Entwicklung der deutschen chemischen Fachzeitschrift: Ein Beitrag zur Wesensbestimmung der wissenschaftliche Fachzeitschrift* (Berlin: Verlag Chemie, 1941); and D. von Engelhardt, "Die chemischen Zeitschriften des Lorenz von Crell," in *Indices naturwissenschaftlich-medizinischer Periodica bis 1850*, ed. A. Geus, vol. II (Stuttgart: Hiersemann, 1974), pp. 11–16. While all these scholars acknowledge Crell's contribution in founding the first successful chemical journal, they do not explore the consequences of his success for the social organization of German chemistry.

2. Though Crell's services to chemistry were widely acclaimed from the mid-1780s to the early 1800s, his endeavors were already passing into oblivion by the time of his death in 1816.

Fig. 3. L. Crell, founder of the first successful chemical journal.
Source: *Allgemeine deutsche Bibliothek*, 59 (1784)

medical talent and learning. His maternal grandfather, Lorenz Heister (professor of therapy, surgery, and botany; court councillor; and ducal physician), was one of Germany's most famous surgeons. His father, Johann Friedrich Crell (professor of anatomy, physiology, and pharmacy), was a promising anatomist.[3] Despite this medical expertise, Crell's father succumbed to a chronic cough in 1747. Young Crell was raised, therefore, in his illustrious grandfather's household.

On 20 June 1759, a year after grandfather Heister died, Crell matriculated at Helmstedt University.[4] Still a youth, he spent the next few years completing his general education. Then around 1765 he embarked on professional training in medicine. He decided to stay at Helmstedt, most likely for such nonacademic reasons as a reluctance to leave his mother and a desire to save money. The Medical Faculty there was good but by no means first-rate. Its three chairs were held by P. C. Fabricius (anatomy, physiology, pharmacy), J. T. Adolph (surgery and botany), and G. C. Beireis (theoretical medicine, materia medica, and chemistry). Crell's favorite teacher was the polymath Beireis, an entertaining and indefatigable lecturer who in later years became one of Helmstedt's chief curiosities.[5]

On 22 June 1768, nine years after entering the University and one year after his mother's death, Crell took his M.D. In anticipation of a scholarly career, he wrote his own dissertation, dedicating it to his "teacher and patron" Beireis. His topic was "putrid" diseases. A. M. Plenciz had recently speculated that such diseases were caused by microscopic organisms. Attacking

---

The most complete biographies appearing after his death were F. Saalfeld, *Versuch einer academischen Gelehrten-Geschichte von der Georg-Augustus-Universität zu Göttingen*, vol. III (Hanover: Helwing, 1820), pp. 80–85; and J. F. Blumenbach, *Memoria Laurentii de Crell* (Göttingen: Dieterich, 1822). Preoccupied with originality, historians of science have neglected Crell. However, one local historian has written a brief biography. See W. Schrader, "Professor Johann Friedrich Crell und Lorenz v. Crell: Zwei Helmstedter Mediziner," *Alt-Helmstedt: Blätter der Heimatkunde für die Stadt und den Landkreis Helmstedt*, 19, no. 2 (March 1955), 1–2. Following Saalfeld or Blumenbach, most biographical dictionaries report Crell's birthdate as 21 January 1744. Using baptismal records in Helmstedt, Schrader ascertained his actual birthdate as 21 January 1745. Around 1772 Crell abandoned his full name and started signing "Lorenz Crell."

3. For L. Heister, see G. A. Will, *Nürnbergisches Gelehrten-Lexicon*, vol. II (Nuremberg and Altdorf: Schüpfel, 1757), pp. 66–69 and P. W. van der Pas, "Heister, Lorenz," in *Dictionary of Scientific Biography*, ed. C. C. Gillispie et al., vol. VI (New York: Scribner's, 1972), pp. 231–232. For J. F. Crell, see *Monumentum sylloge quibus memoriam . . . Ioh. Friderici Crellii* (Helmstedt: Schnorr, 1747). For documents bearing on Heister's and Crell's responsibilities, titles, and salaries at Helmstedt professors, see Niedersächsisches Staatsarchiv, Wolfenbüttel (37 Alt 257, 259, 439, 440, 444).

4. I am indebted to R. Volkmann, Director of Helmstedt's Ehemalige Universitäts-Bibliothek, for the date of Crell's matriculation.

5. For Helmstedt's Medical Faculty in the mid-1760s, see the lecture catalogues in the *Gelehrte Beyträge zu den Braunschweigischen Anzeigen*, 5 (1765), 170–174, 609–614; 6 (1766), 153–160, 553–560; 7 (1767), 129–136, 494–500; 8 (1768), 137–144; and Niedersächsisches Staatsarchiv, Wolfenbüttel (37 Alt 445–449). For G. C. Beireis, see *Sammlung von Briefen gewechselt zwischen J. F. Pfaff und . . . Anderen*, ed. C. Pfaff (Leipzig: Hinrichs, 1853), pp. 125, 134–137; G. Peacock, *Life of Thomas Young, M.D., F.R.S.* (London: Murray, 1855), pp. 98–100; and *Neue deutsche Biographie*, vol. II (Berlin: Duncker & Humblot, 1955), pp. 20–21.

this theory, Crell suggested a chemical etiology. He proposed that volatile alcali (ammonia) originating from a ferment in the humors caused the putrefaction. In presenting his case, Crell revealed that he, like so many physicians before him, had acquired a familiarity with chemistry during his medical training. At this point, however, therapy was still his primary interest.[6]

Soon after taking his degree, Crell set out on a European tour. He made his way through Göttingen, Frankfurt am Main, Stuttgart, and Strasbourg to Paris. From the French capital he sent a report about his travels to the Göttingen Society of Sciences. Although most of his remarks concerned medical subjects and institutions, he described the processes and products of a saltpeter works near Stuttgart.[7]

In mid-1769 Crell proceeded from Paris to Edinburgh. To his delight, the university there measured up to its high reputation as "one of the most outstanding schools for young doctors." Edinburgh University owed its supremacy, he later informed Beireis,

> to the advantageous arrangement of the hospital and particularly to the worth of the present teachers. A man of great accomplishment represents each subject. Cullen and Gregory alternately offer instruction in physiology and practical medicine, Black in chemistry, Hope in botany, and Home in materia medica.[8]

This was indeed a group of outstanding men. Though Crell probably studied under all of them, he was most inspired by J. Black and W. Cullen. Years later he expressed his gratitude to Black in stiff English:

> I own, I had already a very great passion for Chemistry, as I came to Edinburgh; but you have still increased it, by the clearness and preciseness of all your ideas, which you gave to your scholars, among whom, I recollect allways with pleasure to

---

6. Crell's dissertation, *Contagium vivum lustrans* (Helmstedt: Schnorr, 1768), was soon extracted in a new series designed to make theses more accessible to the medical-scientific public. See E. G. Baldinger, *Auszüge aus den neuesten Dissertationen über die Naturlehre, Arzneiwissenschaft, und alle Theile derselben*, vol. I, pt. 2 (Berlin and Stralsund: Lange, 1769), pp. 137–143. Crell learned chemistry from Beireis, who offered courses on "pharmaceutical," "metallurgical," and "physical chemistry." See Crell's autobiographical remarks in J. Black, *Vorlesungen über die Grundlehren der Chemie*, trans. L. Crell, vol. I (Hamburg: Hoffmann, 1804), p. ciii; and Helmstedt University's lecture catalogues.

7. For a summary of Crell's report to the Göttingen Society, see *Göttingische Anzeigen von gelehrten Sachen* (11 January 1770), 41–43. During his stay in Paris, Crell met C. R. Hopson, a recent M.D. from the University of Leiden, who told him about Black's theory of heat. See J. Black, *Vorlesungen über die Grundlehren der Chemie*, I:407. Crell and Hopson became such good friends that they journeyed to Edinburgh together. See Hopson, letter to Crell, *Beyträge zu den chemischen Annalen*, 4 (1790), 442. For Hopson, whose role in disseminating Black's views merits attention, see the *Dictionary of National Biography*, vol. IX (Oxford: Oxford University Press, 1917), p. 1238; and J. R. Partington, *A History of Chemistry*, vol. III (London: Macmillan, 1962), pp. 629–630.

8. Crell, "Briefe an den Herrn Hofrath Beireis über den jetziger Zustand der Arzneygelahrheit in Edinburgh," *Gelehrte Beyträge zu den Braunschweigischen Anzeigen*, 11 (1771), 325.

have been. . . . I do owe you so very great a deal of my chemical knowledge. . . .[9]

Crell's contact with Black extended beyond the lecture hall. In his mentor's laboratory he performed various experiments, including some purporting to buttress the possibility of alchemical transmutation.[10] If Crell's "passion for chemistry" was great in Edinburgh, his passion for therapy was greater. Cullen's ideas about fevers aroused Crell to enthusiastic comment and criticism during the next few years, a response that Black's more significant ideas on heat, phlogiston, and gases failed to elicit.[11]

After two semesters in Edinburgh, Crell proceeded to London, probably with the intention of pursuing his studies in the city's hospitals. While there he carried out a series of experiments on putrefaction, a subject that, as we have seen, began to interest him in Helmstedt. The account of these experiments, which he submitted to the Royal Society in the late fall of 1770, indicates both his competence in chemistry and his continued allegiance to therapy. The various tests that he used to determine the presence of volatile alcali and his explanation of odors as caused by combinations of saline matter and phlogiston suggest a good command of current chemical techniques, facts, and theories. Still, his desire to understand decay stemmed more from a concern to benefit "mankind [whose] health depends greatly upon [knowledge of] putrefaction" than from a concern to shed new light on "natural philosophy."[12]

In late 1770 Crell returned to the Duchy of Brunswick. He may have hoped to teach in one of the two medical schools there, but the only available position was a new chair of metallurgy in Brunswick's Collegium Carolinum, a school for prospective officials. Crell petitioned Duke Carl for the post on 19

---

9. Crell, letters to J. Black (22 January 1779, 24 October 1782), Edinburgh University Library. Crell later reported that he studied with Black for "almost a year." See J. Black, *Vorlesungen über die Grundlehren der Chemie*, I:ciii.

10. Crell, letter to Black (24 October 1782), Edinburgh University Library. Though, as will be seen, Crell was hostile to alchemical obscurantism, he followed his teacher Beireis in believing that transmutation was possible. Indeed, in the mid-1780s he was apparently contemplating the publication of a treatise on the possibility of transmutation. See Carbonarius, *Beytrag zur Geschichte der höhern Chemie oder Goldmacherkunde* (Leipzig: Hilscher, 1785), p. 442.

11. For Crell's reactions to Cullen's doctrines, see his "Briefe an . . . Beireis," *Gelehrte Beyträge zu den Braunschweigischen Anzeigen*, 11 (1771), 325–332, 369–380, 429–440; and his "Vorrede," in L. Chalmers, *Ein Versuch über die Fieber*, trans. [L. Crell] (Riga: Hartknoch, 1773). Prior to his translation of Black's lectures in 1804, Crell discussed Black's theories on but one occasion. See L. Crell, *Versuche über das Vermögen der Pflanzen und Thiere, Wärme zu erzeugen, und zu vernichten* (Helmstedt: Kühnlin, 1778), pp. 68–70. Black thought that Crell might have disseminated his ideas concerning latent heat on the Continent prior to their publication. See Black, letter to J. Watt (15 March 1780), in J. P. Muirhead, *The Origin and Progress of the Mechanical Inventions of James Watt*, vol. II (London: Murray, 1854), p. 120.

12. F. L. F. Crell (M.D. and Professor of Chemistry at Brunswick), "Some Experiments on Putrefaction," Royal Society, London: *Philosophical Transactions*, 61 (1771), 332–344. He published a German version of the article in Brunswick. See "Einige Erfahrungen, die Faulung betreffend," *Gelehrte Beyträge zu den Braunschweigischen Anzeigen*, 11 (1771), 281–296. According to I. Kaye, librarian at the Royal Society, Crell attended two meetings and submitted his manuscript in November 1770.

January 1771, claiming that he had "devoted special attention to . . . chemistry and particularly metallurgy" during his studies and travels. A few days later, Council President H. B. Schrader von Schliestedt, an old friend of the chemist J. A. Cramer, approved the request. Crell was appointed professor of metallurgy at the Collegium with the modest salary of 200 Rtl.[13]

The next three years were important ones for Crell. His main job was teaching metallurgy and related subjects at the Collegium. In doing so, he acquired both a knowledge of and appreciation for chemistry's relevance to the mining industry.[14] Nonetheless, finding it necessary to augment his salary by practicing medicine,[15] he focused on medical topics in his publications.[16] Crell also became an active member of Brunswick's Masonic Lodge of the Crowned Column. Among his many fraternal friends were several men who assisted him in his subsequent endeavors.[17]

In early 1774 Crell had an opportunity to better his position—his father's successor at Helmstedt, Fabricius, was dying. Hopeful of securing the chair, Crell moved to Helmstedt and began lecturing.[18] Soon after Fabricius's fu-

13. The documents bearing on Crell's appointment to Brunswick's Collegium Carolinum are in the Niedersächsisches Staatsarchiv, Wolfenbüttel (2 Alt 14703: 169–187). Using other records in this archive, Dr. J. König has found that Crell received his salary of 200 Rtl from the "Klosterkasse." For Schrader von Schliestedt's relation to Cramer, see the latter's profile in Appendix I.

14. While in Brunswick, Crell offered the following courses: Summer Semester (SS) 1771: mineralogy, the history of minerals, and their chemical preparation for use in manufactories and factories; Winter Semester (WS) 1771/72: metallurgical chemistry using Gellert's text and including a practical introduction to solvents and mineral analysis; SS 1772: practical metallurgical chemistry, continuing the theoretical part of the preceding term and including assaying; WS 1772/73: mineralogy using Cronstedt's text; SS 1773: metallurgical chemistry using Gellert's text and including demonstration experiments; WS 1773/74: announcement withheld pending his recovery from a stubborn illness. See *Gelehrte Beyträge zu den Braunschweigischen Anzeigen*, 11 (1771), 112, 527; 12 (1772), 119, 434; 13 (1773), 118, 551.

15. Crell's need to practice medicine was all the more pressing because he could not, as he complained at least twice, count on receiving his salary on time. See Crell, "Pro Memoria" (19 September 1771, 6 February 1772) in Niedersächsisches Staatsarchiv, Wolfenbüttel (2 Alt 14703: 177–178, 14728). For medicine in Brunswick during Crell's years there, see W. Artelt, "Das medizinische Braunschweig um 1770," *Medizinhistorisches Journal*, 1 (1966), 240–260.

16. Besides his letters to Beireis (1771) and his translation of Chalmer's treatise on fevers (1773), Crell published "Beantwortung einiger Vorurtheile gegen die Einpfropfung der Blatter," *Gelehrte Beyträge zu den Braunschweigischen Anzeigen*, 11 (1771), 689–704. In this article he told of vaccinating a young woman and a child.

17. Crell joined the Lodge of the Crowned Column upon returning from England and was immediately appointed treasurer. He held this post until 1785 and continued to participate in the Lodge long thereafter. The following members subscribed to his chemical journal between 1784 and 1791: Abich, Beireis, Gravenhorst, von Hoym, Römer, du Roi, Sommer, and Zimmermann (see Appendix III). For information on the membership of Crell and others, I am indebted to F. Schröder, Grandmeister, and H. Schierer, Archivist, of the Johannisloge Carl zur gekrönten Säule, Brunswick. For Crell's summary of a speech that he gave in the Lodge in 1806, see P. Zimmermann, "Briefe aus den letzten Jahren der Universität Helmstedt," Braunschweigische Geschichtsverein: *Jahrbuch*, 9 (1910), 201–202.

18. Crell's transitional status in early 1774 is evident from his omission from the lecture catalogues for SS 1774 of both the Collegium Carolinum and Helmstedt University. See *Gelehrte Beyträge zu den Braunschweigischen Anzeigen*, 14 (1774), 109–116, 149–156. However, he arrived in Helmstedt early enough in the summer term to begin a private course on "physical chemistry" (p. 531). The documents bearing on his appointment are not in the Niedersächsisches Staatsarchiv, Wolfenbüttel, presumably because they were transferred during the Napoleonic regime.

neral that summer, he was appointed ordinary professor and assigned responsibility for medical theory and materia medica.[19] These subjects probably reflected Crell's main interests, for in the immediately ensuing years he published several articles on materia medica.[20]

Though Crell's focus was on medicine, he also had a strong interest in natural theology.[21] This interest led him to initiate a short yet revealing correspondence with A. von Haller. On 28 October 1776 he anonymously sent Haller an essay, or "proof," in which he sought to demonstrate the infinitude of God. Early the next year, Haller put a brief notice in the Göttingen *Anzeigen* inquiring about some minor points in the essay and encouraging its author to publish for "the common good."[22] Gratified by this response, Crell wrote Haller on 7 February 1777 to disclose his identity, clarify his views, and solicit advice about publication. His autobiographical remarks are of great interest, for they reveal his aspirations, and apprehensions, shortly before he began publishing his chemical journal:

> My father was a teacher at this academy [Helmstedt University]. Thanks to your collection of outstanding dissertations, he is still remembered among scholars. My grandfather was the late Heister to whom—as both your letters and many [gifts of] alpine plants . . . testify—you had the goodness to show very much friendship.
> 
> Both of my forefathers died before I could enjoy their instruction in medicine. Nevertheless, I finished my studies here, earning my doctor's hat with the dissertation that I have the honor to enclose. For almost three years I visited Strasbourg, Paris, Edinburgh, and London. Then I was called back to Brunswick as teacher of chemistry in the Collegium Carolinum. Three years ago I was transferred here. The many situations in which I have found myself have hindered me from making myself known. So far the only publications of mine that have appeared are an essay in the *Philos. Transactions* for 1771, a

---

19. Crell's appointment was announced in the *Commentarii de Rebus in Scientia Naturali et Medicina Gestis*, 22 (1774), 533.

20. Between 1774 and 1777 Crell published the following articles: "Ueber die Milchversetzungen," *Gelehrte Beyträge zu den Braunschweigischen Anzeigen*, 15 (1775), 697–712; "Ueber einige Mittel wider ein heftiges Erbrechen," *Magazin vor Aerzte*, 5 (1776), 385–389; "Beobachtung über die Wirkung der wässerichten Auflösung des ätzenden Quecksilbersublimats," ibid., 7 (1777), 649–652; "Ein hartnäckiges viertägliches Fieber durch Alaun und Camillen geheilt," ibid., 7 (1777), 652–654; and "Beobachtung eines sehr heftigen durch den Biesam geheilten Tetenus und anderer Krampfzufälle," ibid., 7 (1777), 654–656. The *Magazin vor Aerzte* was edited by Baldinger, who had earlier published a favorable summary of Crell's dissertation. For Baldinger and his magazine, which was the model for Crell's chemical journal, see G. Mann, "Ernst Gottfried Baldinger und sein Magazin für Aerzte," *Sudhoffs Archiv*, 42 (1958), 312–318.

21. Crell's interest in natural theology soon found expression in courses on physiology for law and theology students. See *Gelehrte Beyträge zu den Braunschweigischen Anzeigen*, 15 (1775), 163; 16 (1776), 190.

22. *Göttingische Anzeigen von gelehrten Sachen* (27 January 1777), 96.

translation of Chalmer's "Essay on Fever," a dissertation prepared for a foreigner—de Zinco—that I also take the liberty of enclosing, and some essays in Professor Baldinger's *Magazin* and other periodicals. But I do have a *Chemisches Magazin* in the works.

This has been my fate, and these have been my works as a doctor! However, besides these activities, I have spent many hours exploring those truths which ought to be the most important for every man: the foundations of our wonderful religion. One fruit of these investigations was the proof that I had the honor of sending you. The reason I did not publish it but rather anonymously sought a frank judgment from you, one of the most acute judges, was a lack of self-confidence. Careful study of the work of the greatest minds has caused me to lose all sense of judgment, for I have found that many have missed the truth where they believe it to have been ascertained. Couldn't I, consequently, with my lesser gifts and in a somewhat foreign subject, fall into similar difficulties all the more easily? Couldn't I put forward something known as something new, something insignificant as something important, something incorrect as something true? These are mistakes that I as a teacher could not make publicly lest I weaken the trust of my students and therefore become less useful in my post. Therefore, Your Honor's highly esteemed approval has set me completely at ease. . . .[23]

This letter reveals Lorenz Crell on the eve of his editorial career as a man with an intense desire for renown as a scholar who, at the same time, doubted whether he would ever deserve such recognition. The letter also indicates that, though he dreamed of winning fame by writing on significant metaphysical subjects, he planned to attain it through the publication of more mundane works "as a doctor." Lastly, the letter reports that his next step "as a doctor" would be to establish—surely in imitation of Baldinger's successful *Magazin vor Aerzte*—a "chemical magazine."

## THE FIRST VOLUME (1778)

Preparing the first volume of his "chemical magazine" took Crell longer than he had anticipated. It took so long, in fact, that his former student C. W. Nose, who had promised to submit some articles, impatiently published them on his own. Dedicating the slender volume to Crell on Easter Day 1778,

---

23. Crell, letter to A. von Haller (7 February 1777), Haller Briefsammlung, Bern Stadtbibliothek. Crell was sufficiently encouraged by Haller's response to his views that he published (anonymously) *Die Unendlichkeit des Weltschöpfers aus der Einrichtung der Natur und ontologischen Gründen erwiesen* (Helmstedt: Kühnlin, 1778) and, years later, *Pyrrho und Philalethes*, ed. F. V. Reinhard (Sulzbach: Seidel, 1812).

Nose expressed the hope that his mentor's "chemical journal" would soon appear.[24] The first volume finally did appear late that summer.

Crell's journal bore an imposing title—*Chemisches Journal für die Freunde der Naturlehre, Arzneygelahrtheit, Haushaltungskunst und Manufacturen*, or "Chemical Journal for the Friends of Natural Science, Medicine, Domestic Economy and Manufacturing." It was dedicated to Marggraf, who had contributed so much to "all friends of chemistry in all nations." The dedicatory epistle was followed by a foreword in which Crell gave his reasons for founding a chemical journal and outlined his plans for it. Then came a section (97 pages) of original "chemical treatises." Almost all of these were written by Crell and his "chemical friend" J. C. C. Dehne, a physician in the small town of Schöningen near Helmstedt.[25] In his own articles, Crell described Gahn's little-known method of extracting phosphorus from bones, showed that this method could be used to produce phosphorus from human bones, explained how to make a glassy substance from this phosphorus, presented the results of his analysis of tallow and human fat, described the chemical properties of various fat derivatives, and announced his success in using chalk and Glauber's salt to produce a pure soda that did not have a sulfuric odor. The original treatises were followed by a final section (115 pages) containing articles extracted from the *Philosophical Transactions* of the Royal Society and the *Mémoires* of the Berlin Academy.

From our standpoint, the most interesting part of this first volume is Crell's foreword. "In this century," he began,

> chemistry's extended influence on learning and its great utility
> for the commonweal are so generally recognized that they need
> no proof. The clarification, expansion, and application of this
> science have even attracted the attention of princes who seek to
> promote its growth with honors and rewards.

Thanks to the patient and laborious collection of observations and experiments, chemistry, "just like the other sciences dealing with nature," had made great progress. But, he warned, "our [the German] nation must, if she wishes

---

24. Nose also acknowledged Crell's role in awakening his interest in chemistry, "our goddess (*Heldgöttin*)," in the letter of dedication. See Nose, *Versuch einiger Beyträge zur Chemie* (Vienna: Trattner, 1778). He probably studied chemistry with Crell both at the Collegium Carolinum, where he was a student from 1770 to 1773, and at Helmstedt University, where he was a student from 1773 to 1775. See the respective matriculation records in J. J. Eschenburg, *Entwurf einer Geschichte des Collegii Carolini in Braunschweig* (Berlin and Stettin: Nicolai, 1812); and at the Ehemalige Universitäts-Bibliothek, Helmstedt.

25. Dehne went through pharmaceutical training in Brunswick and served as a dispenser in J. P. Becker's shop in Magdeburg before enrolling in Helmstedt University in early 1772. While in Helmstedt he became acquainted with Crell and took chemistry during SS 1776 from Beireis, whom he later described as his "patron and teacher." He won his M.D. later that year, then proceeded to Schöningen, where he died in 1791. For Dehne's life, see, in addition to autobiographical remarks in his publications, the matriculation records of Helmstedt University in the Ehemalige Universitäts-Bibliothek, Helmstedt; Beireis's report on students taking his public chemistry course (30 September 1776), Niedersächsisches Staatsarchiv, Wolfenbüttel (4 Alt 19 vorl. nr. 50); and materials in the Stadtarchiv, Schöningen. I am grateful to K. Rose, Town Archivist, for a summary of these last materials.

to remain the recognized leader of the other peoples [in chemistry], continue along the toilsome path that led her to fame and hegemony." Unfortunately, his countrymen often failed to report their discoveries because they had no place to publish a single experiment and were unable to carry out enough experiments to fill an entire book. Anxious to end the "irreparable harm inflicted on our science" by the stillbirth of so many discoveries, Crell had decided to set himself up as "a collector of every new experiment of each chemist who has no better means of making it known." His journal would be completely open to "every chemist, whether he concerns himself with the processing of metals, the preparation of drugs, or experimentation for pleasure." He hoped "all experts in and friends of chemistry who have their science and the spread of useful knowledge at heart" would assist him in this "disinterested" endeavor.

After these general remarks, Crell turned to his specific plans for the journal. Each volume would have three parts. The first would be devoted to contributions that either described new phenomena or presented more accurate accounts of familiar phenomena. All articles, except "incomprehensible" descriptions of alchemical processes, would be received with sincere thanks and rewarded with an appropriate honorarium. Crell promised to repeat any experiment that constituted a major discovery or contradicted accepted principles so as to increase its credibility. The second part of each volume would contain translations of the articles dealing with chemistry that appeared in the memoirs of the leading scientific societies. By reading these articles, the chemist could keep abreast of the important chemical discoveries reported in these rare and expensive journals. Such knowledge would reduce the likelihood of his duplicating the research of others and increase his chances of devising "good, important, and new experiments." The third part would contain "impartial reviews" of new works on chemistry. These would help "the enthusiast for chemistry" decide which books to purchase.

Crell went on to disclose his aspirations for the journal:

> I hope that it will not only help enlarge chemistry as a science but also exert some influence on various aspects of everyday life. If the readers should think that the contents do not correspond to the title, I wish they would not judge the journal by this single volume but rather consider what this enterprise, with help, could become.

He concluded by assuring the reader that, since he was not motivated by personal ambition, he welcomed constructive criticism.[26]

Crell's foreword reveals that, in the year and a half since writing Haller, his relationship to chemistry had undergone a marked change. No longer did he regard himself as a doctor with an interest in chemistry. While preparing the first issue, he had come to see himself as a chemist. Like other German

26. Crell, "Vorrede," *Chemisches Journal*, 1 (1778), 9–20.

chemists of the time, he viewed the discipline as a profound and useful branch of natural science and assumed that his countrymen shared this view. Unlike most who had never been abroad, he realized that foreign chemists posed a serious challenge to Germany's age-old leadership in the science. He believed that this challenge could be thwarted and chemistry's growth accelerated if German chemists only had a means of reporting all their discoveries. Thus, in addition to the thirst for personal recognition that he revealed to Haller, Crell was motivated to found a chemical journal by a nationalistic desire to preserve German leadership in chemistry.

## JOURNAL EDITOR (1779–1789)

The reviewers gave the first volume of Crell's *Chemisches Journal* a favorable reception. For instance, Göttingen's professor of chemistry, J. F. Gmelin, declared in the Göttingen *Anzeigen* that "Professor Crell has made a useful and advantageous attempt to found a magazine for chemical experiments, so many of which would otherwise be lost to the great detriment of the whole science...."[27] In Erfurt's *gelehrte Zeitungen*, an anonymous reviewer observed that "we as well as all thinking chemists dedicate our warmest thanks to Crell for this useful undertaking."[28] Baldinger told the readers of his medical journal that Crell's venture should prove to be "particularly advantageous" for chemistry and applauded the "exactitude" of its original treatises.[29] W. H. S. Bucholtz was equally enthusiastic. In the prestigious *Allgemeine deutsche Bibliothek*, he commended Crell for his "theoretical and practical insights" and wished him "health and leisure for continuing this generally useful undertaking." And, echoing Crell's nationalism, he implied that "German chemists" were far superior to the "windy French chemists."[30]

Just as important as the favorable reviews was the response to Crell's call for contributions. In addition to his chemical friend Dehne, five men sent him articles. Although G. Thorey, a pharmacist in Hamburg, had not published on chemistry before, the other four—Gmelin; J. C. Wiegleb, the noted pharmacist-chemist in Langensalza; C. Mönch, an apothecary in Kassel; and J. F. A. Göttling, administrator of Bucholtz's apothecary shop in Weimar—had all done so.[31] Crell had attracted the collaboration of men who were already active in chemistry. By mid-1779 he was ready to publish the second

---

27. *Göttingische Anzeigen von gelehrten Sachen* (21 November 1778), 1134. That the reviewer was Gmelin is revealed in O. Fambach, *Die Mitarbeiter der Göttingischen Gelehrten Anzeigen 1769–1836* (Tübingen: Universitätsbibliothek, 1976), p. 69. For a biographical profile of Gmelin, see Appendix I.
28. *Erfurtische gelehrte Zeitungen* (28 January 1779), 74–76. The reviewer was probably Erfurt's professor of chemistry and botany, W. B. Trommsdorff.
29. *Neues Magazin für Aerzte*, 1, no. 1 (1779), 87.
30. *Allgemeine deutsche Bibliothek*, 39 (1779), 188–194. For a profile of Bucholtz, see Appendix I.
31. To my knowledge, none of these five men was personally acquainted with Crell. For profiles of Wiegleb and Göttling, see Appendix I.

volume of his *Chemisches Journal*. Dedicating it to J. R. Spielmann in Strasbourg, he exclaimed that

> German chemistry, which for the last few decades has been nearing perfection with great strides, owes a considerable part of its growth to you. . . . Everyone, especially the patriotic chemist, who has his science at heart regards you with the truest respect and warmest gratitude. . . .

In the foreword he observed that chemical research was so time-consuming that it would take many more contributors to sustain regular publication. Therefore he reiterated his

> pressing request to all those chemists who allow their natural enthusiasm for the science to be fired by love of the fatherland that they, so far as is within their capacity, strive to preserve and increase the fame of German chemistry with their painstaking contributions.

After describing the contributions to the second volume, Crell announced that either he or "men who had already acquired fame and names in the world" had written the book reviews. Finally, he informed the reader that henceforth he would end every volume with a list of research proposals.[32]

After the first two volumes, Crell's venture flourished. Contributions came in more quickly, and Gmelin helped out by reviewing books and translating articles.[33] Encouraged, Crell adopted the goal of semiannual publication. Though he did not manage to publish the third volume until early 1780, he issued the next three on schedule. Throughout this period, he made but one change in the journal's format. Starting with the sixth volume, he included extracts from his growing correspondence in the section devoted to original treatises. Some chemists, though they were unwilling to compose formal articles for Crell, were happy to send him publishable news in their letters.

Meanwhile, F. Nicolai, editor of the *Allgemeine deutsche Bibliothek*, had commissioned Crell to review most chemical publications, even the *Chemisches Journal*. Since Nicolai's reviewers were identified only by coded initials, Crell had a remarkable opportunity to boost his enterprise.[34] He did not

---

32. Crell, dedication letter and "Vorrede," *Chemisches Journal*, 2 (1779), [i–vii]. My treatment of Spielmann as a German chemist (see Appendix I) and of Strasbourg University as a German university (see Appendix II) parallels Crell's usage.

33. Crell singled Gmelin out for special thanks in 1780. See Crell, "Vorrede," *Chemisches Journal*, 5 (1780), [v].

34. Nicolai evidently had complete confidence in Crell by fall 1779, even though they had only been corresponding for two years. In 1778, however, Nicolai had not followed Crell's suggestion that the first issue of the *Chemisches Journal* be assigned to C. A. Gerhard or J. C. Wiegleb for review. See Crell, letter to Nicolai (30 September [?] 1778), Nicolai Nachlass, Staatsbibliothek Preussischer Kulturbesitz: Handschriftenabteilung, West Berlin. For the various initials used by Crell during his reviewing career, see G. F. C. Parthey, *Die Mitarbeiter an Friedrich Nicholai's Allgemeiner deutscher Bibliothek nach ihren Namen und Zeichen* (Berlin: Nicolai, 1842), p. 38.

let it pass. In reviewing the second volume, he spoke of "this splendid and generally useful journal" and pointed out that "men of decided service to chemistry" had contributed to the project. And in his review of the third volume, he wrote of "this extraordinarily useful journal." He also used his reviews to draw attention to faults in the articles, a clever technique for correcting his contributors without affronting them.[35]

Besides promoting his venture, Crell used his position as reviewer to discourage competition. In his anonymous review of J. A. Weber's new *Physikalisch-chemisches Magazin für Aerzte, Chemisten und Künstler*, he predicted that his rival's periodical would not have "much luck in comparison with Baldinger's magazine for doctors and Crell's journal for chemists and artisans." His review of Weber's second issue was harsher. Signing with different initials, he commented that

> many isolated observations in this magazine will interest the chemist. However, it is to be wished that the author had been somewhat briefer, had not always mixed in his pet theories, and had not so frequently succumbed to his inclination to be witty. Also the reviewer agrees with the opinion expressed by the different [!] reviewer of the first issue that it would be better if the medical articles were submitted to Baldinger's magazine and the chemical ones to Crell's journal.

Weber, possibly disheartened by this reception, did not publish a third issue.[36]

Another potential rival fared better in Crell's reviews, both because he had an influential patron and because he did not try to compete with the *Chemisches Journal*. In 1779 Bucholtz's assistant, Göttling, began publishing the *Almanach oder Taschenbuch für Scheidekünstler und Apotheker*, an annual whose primary purpose was to disseminate chemical information among artisans and pharmacists. Crell reviewed the first volume twice, once in his *Chemisches Journal* and once in the *Allgemeine deutsche Bibliothek*. Both reviews were favorable. However, in the anonymous review, Crell suggested that the *Almanach* should not seek to go beyond its educational mission.[37]

Crell's initial success as a chemical journalist entitled him, he believed, to recognition. In 1779 he took advantage of a student's trip to Edinburgh to open a correspondence with his former teacher Black. After describing his

---

35. *Allgemeine deutsche Bibliothek*, 43 (1780), 473–479; 45 (1781), 117–120.

36. *Allgemeine deutsche Bibliothek*, 44 (1780), 167–169; 46 (1781), 243–252. For a profile of Weber, see Appendix I.

37. *Chemisches Journal*, 4 (1780), 231–232; and *Allgemeine deutsche Bibliothek*, 52 (1782), 423–428. In the earlier review, Crell revealed that the *Almanach*, which appeared anonymously, was edited by Göttling "under the supervision of Bucholtz, whom every German chemist knows and treasures." Crell may also have been inclined to look favorably on the *Almanach* because its first issue gave his *Chemisches Journal* an extremely positive review, concluding "German *Chymiker!* your customary enthusiasm . . . supports the fame of German chemistry with industrious contributions and thereby confirms the opinion of other nations that you are by nature . . . actually destined to be chemists and *Naturforschern*." See the *Almanach oder Taschenbuch für Scheidekünstler und Apotheker*, 1 (1780), 202.

journal and offering to send some mineral specimens, Crell closed by asking to be "honoured with the title of a foreign membre of your medical college." In response Black accepted Crell's offer of minerals and inquired whether he would like to be made a member of Edinburgh's Philosophical Society. Delighted, Crell requested Black

> to intercede for me by the celebrated Philosophical Society, that she might confer upon me the honour of being a member of her. For, as I have had the good fortune, of being admitted into the Academ. Ceasar. Natur. Curiosor., the Academia Scientias Moguntin, in the Society of Friends inquiring into Nature, at Berlin, and as correspondent to the Royal Society of Gotting; I wish nothing more, as to obtain a like favour from a very learned body, which is so deservedly famous in Germany, because of its Medical Essays, and the Essays Phys. & Litterary; a Society, to which I am, in so many members, so highly obliged.

He also repeated his hope of "being adscribed to the foreign Membres of your College of Physicians."[38] Though some years passed before Crell obtained any memberships from Black, he was more successful, as his second letter indicated, in obtaining membership in other learned societies.

He regarded such recognition, as he told a friend for whom he was arranging membership in the Imperial Academy of Natural Curiosities, as important for one's "well-being" and "fame."[39] There may have been some truth in what he said. On 31 August 1780 the new Duke of Brunswick, Carl Wilhelm Ferdinand, made Crell a councillor of mines. This position apparently gave him a good salary with few extra duties.[40]

Meanwhile Crell was contemplating quarterly publication. In early 1780 his collaborator Gmelin informed the Swedish chemist T. Bergman that such a move was imminent.[41] However, it took over a year for the rate of contributions to build up sufficiently and for Crell to find a publisher in Leipzig who could promise him adequate printing and marketing services. In the spring of 1781 all was ready. Giving the quarterly a new title—*Die neuesten*

---

38. Crell, letters to J. Black (22 January 1779, 24 October 1779), Edinburgh University Library.

39. For Crell's opinion about the value of society memberships, see G. Forster, letter to J. K. P. Spener (29 December 1779), in *Georg Forsters Werke*, vol. XIII, *Briefe bis 1783*, ed. S. Scheibe (Berlin: Akademie Verlag, 1978), pp. 266–267. Crell soon had Forster writing to J. Banks to request his (Crell's) election to the Royal Society (pp. 278–279, 328, 330, 368–369). Eventually sponsored by Duke Carl of Brunswick, J. R. and G. Forster, R. Kirwan, J. de Rotencrantz, S. C. Hollmann, and C. Blagden, Crell was elected to the Royal Society on 3 April 1788, the same day as Guyton de Morveau and Lavoisier. See the membership records, Royal Society, London.

40. For Crell's appointment as councillor of mines, see W. Schrader, "Professor Johann Friedrich Crell und Lorenz v. Crell: Zwei Helmstedter Mediziner," *Alt-Helmstedt: Blätter der Heimatkunde für die Stadt und die Landkreis Helmstedt*, 19, no. 2 (March 1955), 2. Crell's gratitude for the Duke's patronage was made manifest in subsequent dedications to his journal.

41. Gmelin, letter to Bergman (20 February 1780), in "Torbern Bergman's Foreign Correspondence," ed. G. Carlid and J. Nordström, *Lychnos-Bibliotek*, 23, no. 1 (1965), 71.

*Entdeckungen in der Chemie*, or "The Latest Discoveries in Chemistry"—,[42] Crell dedicated it to Duke Carl of Brunswick, thanks to whose "most gracious concern" he could devote "a carefree leisure to scientific endeavors." In the foreword, as in earlier forewords, he spoke both as a German and as a chemist. He expected that most of the discoveries appearing in the journal would come from

> my countrymen, since a cold-blooded spirit of research, slow but accurate reflection, and unremitting patience (characteristics that distinguish us Germans from other peoples) are the best characteristics of a chemist. Thanks to these traits our forefathers became the recognized teachers of all nations; thanks to them the fathers of today's chemistry—a Marggraf, a Spielmann, a Gerhard, a Cartheuser—achieved great respect among foreigners; and thanks to them we young chemists can perhaps always keep our fatherland's chemistry ahead of that of other nations.

But, he cautioned, the task would not be easy, "for several nations are competing with us, and almost all have chemists of the first rank."[43]

In the *Allgemeine deutsche Bibliothek*, Crell made a more direct appeal for contributions. After asserting that the journal had and would probably continue to have the advantage over all competitors,[44] he admitted that the editor, for lack of good contributions, often padded issues with "trivial articles." Then, from behind his cloak of anonymity, he argued that

> our German chemists should be more active than they have been so far. If one removed the contributions of four to six industrious collaborators, only a few pages would remain for all the other German contributions. We know many worthy men who could contribute a great deal to this generally useful work but who have not yet done so. Can they be so indifferent to the honor of their fatherland?[45]

Crell had no scruples about playing on his countrymen's cultural nationalism as a means of inducing them to be more active contributors.[46]

---

42. About the same time as he decided on this title, Crell wrote that "almost daily new discoveries are made in this science, [discoveries] which sometimes give it a different appearance, sometimes contradict accepted opinions, and sometimes correct and enlarge formerly accepted principles." See [Crell], review of J. F. Gmelin's *Einleitung in die Chemie* (1780), *Allgemeine deutsche Bibliothek*, 46 (1781), 239.

43. Crell, dedication letter (9 May 1781) and "Vorbericht," *Die neuesten Entdeckungen in der Chemie*, 1 (1781), [i–ix].

44. At the time Crell was nervous about competition from Hanover's publisher Helwing, who, taking advantage of Crell's new title, had announced a *Neues chymisches Journal*. See Crell, letter to Nicolai (14 November 1781), Nicolai Nachlass, Staatsbibliothek Preussischer Kulturbesitz: Handschriftenabteilung, West Berlin.

45. *Allgemeine deutsche Bibliothek*, 49 (1782), 428–429.

46. Even when bowing in the direction of cosmopolitanism, Crell quickly reverted to his patriotic stance. For instance, he declared in 1783 that "the friend of the sciences takes delight in every truth no matter where it springs up, for he is a cosmopolitan. But the cosmopolitan is

Besides soliciting contributions with his prefaces and reviews, Crell assiduously cultivated his correspondents for articles and chemical news. In 1782, for example, he asked Black to send him any dissertations bearing on chemistry published recently in Edinburgh. "I venture," he went on, "to intreat Your Kindness in communicating to me now & then, one of Your discoveries, to[o] small, to be published directly by itself." It would be enough to jot down a few details in a letter. These could easily be extracted for the journal.[47] Black responded with some quotable chemical news, but he left Crell's subsequent letters unanswered. This distressed Crell, for he knew that Black's communications added luster to his journal.[48]

Indeed, he pointedly referred to his one newsworthy letter from Black, as well as to many others from German and foreign correspondents, when he first wrote Bergman in 1783. Toward the end of this letter, he begged the famous Swede for contributions:

> May I be so bold as humbly to request that you be so obliging as to send me some news of your own most highly treasured discoveries as well as those of your countrymen who pursue chemistry with the most fortunate success and with such ardent effort. Yes, I venture to be so audacious as to ask whether you would have the kindness to send a small contribution for my journal. It would be the crown of the whole work. I know, of course, that your investigations are destined for particular dissertations and the memoirs of the Royal Swedish Academy (which has contributed so much to chemistry's new epoch). But surely a small article would not be missed. Anyway I would not mind if you published it elsewhere soon afterwards. And surely you have a great quantity of individual, detached observations (like those appearing as extracts from letters in my journal) which you lack the time and leisure to develop fully and which I would take great joy in making public.

Bergman, unlike Black, was willing to correspond with Crell. However, his premature death soon brought an end to their interchange.[49]

Although Crell's efforts to cultivate Black and Bergman were thwarted by the first's negligence and the second's death, he had considerable success with others. *Die neuesten Entdeckungen* published the observations of over

---

still a man and loves the land in which he lives, the nation to which he belongs. He is jealous of its honor and its fame which, out of love for the fatherland, he wants to be more brilliant than that of all other nations." See Crell, "Vorbericht," *Die neuesten Entdeckungen in der Chemie*, 11 (1783), [v].

47. Crell, letter to J. Black (5 August 1782), Edinburgh University Library.
48. Black, letter to Crell, *Die neuesten Entdeckungen in der Chemie*, 11 (1783), 97–99. Before giving up on Black, Crell wrote him three more letters (24 October 1782, 25 September 1784, and 20 March 1785).
49. Crell, letters to Bergman (5 May 1783, 10 September 1783, October 1783, 7 December 1783, 20 January 1784, 25 February 1784, 26 April 1784), in "Torbern Bergman's Foreign Correspondence," *Lychnos-Bibliotek*, pp. 21–40.

fifty men, most of them Germans, during its three years of existence. Crell's success was also reflected in the number of learned societies that granted him membership. In the last issue of his quarterly, which went to press in early 1784, he listed ten societies after his name.[50] Had news traveled faster, he could have added two more. On 3 January 1784 the Berlin Academy of Sciences named him a nonresident member. And on 21 January 1784 the Royal Swedish Academy at Stockholm granted him membership.[51]

On the same day that he was named to the Swedish Academy, Crell completed a circular announcing the transformation of his journal into a monthly. He explained that he was taking this step so that he could get his contributors' important discoveries into print sooner, thus safeguarding their priority. He planned to publish in Helmstedt rather than Leipzig so that he could give the monthly careful supervision. And he was going to call it the *Chemische Annalen für die Freunde der Naturlehre, Arzneygelahrtheit, Haushaltungskunst und Manufacturen* in order to dispel the notion, created by the title of his quarterly, that he only wanted successful experiments. Besides discussing such practical matters, he assessed the journal's record:

> The many investigations from obliging, patriotic chemists that have enabled me to put out this journal without interruption, the numerous friends and acquaintances that it has procured for me among German and foreign chemists, and the open use that it has received from scholars of the first rank—all these things prove that people regard my undertaking as useful and its continuation as appropriate.

His journal was valuable, he went on, because

> it has generally encouraged a more lively and industrious pursuit of the fatherland's chemistry; it has brought forward hopeful young chemists, whose talents might otherwise not be so well applied or who would have remained unknown and unused by the public; and it has given capable and knowledgeable chemists a place to compete with their ideas.

In addition, his periodical had laid the basis for future prosperity by "making it almost a fashion among the apprentices of the most considerable branches of chemistry—smelting, pharmacy, and chemical manufacturing—to strive for enlightenment and to read chemical publications." Thus Crell believed his journal had received and should continue receiving support and attention

---

50. By 1 January 1784 Crell belonged to the Imperial Academy of Natural Curiosities, the Electoral Mainz Academy of Useful Sciences in Erfurt, the Berlin Society of Scientific Friends, the Brunswick German Society, the Royal Prussian Society of Sciences at Frankfurt an der Oder, the Göttingen Society of Sciences, the Electoral Palatinate Academy of Sciences, the Edinburgh Literary and Philosophical Society, the Royal Danish Society of Sciences in Copenhagen, and the Burghausen Society for the Moral and Agricultural Sciences.

51. For Crell's election to these two academies, see A. Harnack, *Geschichte der Königlich Preussischen Akademie der Wissenschaften zu Berlin*, vol. I (Berlin: Reichsdruckerei, 1900), p. 478; and "Torbern Bergman's Foreign Correspondence," *Lychnos-Bibliotek*, pp. 10, 36, 38.

because it was promoting both German chemistry and the common good.[52]

Within a year of founding the *Chemische Annalen*, Crell was swamped with more contributions than he could handle, even on a monthly basis. To get them published without undue delay, he established a supplemental journal, *Die Beyträge zu den chemischen Annalen*, or "Contributions to the Chemical Annals." He did so fully convinced that

> the active passion of German chemists for our science will not decrease, nor the love for it cool off, since fleeting fashion has never ruled the Germans in the sciences and the love for chemistry has become ingrained in us over the centuries. . . .[53]

Three issues of the *Beyträge* appeared in 1785 and three more in 1786. That year, indeed, the flood of contributions was so great that Crell inserted many into a cheap second edition of *Die neuesten Entdeckungen*. And the following year he published four issues of the *Beyträge*. Then the flood slackened, and he reduced the number of supplemental issues to two in 1788 and one in 1789. Still, his main journal was not wanting for good contributions. In late 1789 Crell rightly claimed that the *Chemische Annalen* had either published or would soon publish reports of the year's most important chemical discoveries, all of which, he noted proudly, were made by Germans.[54]

As Crell's *Chemische Annalen* flourished between 1784 and 1789, his fame continued to grow. For instance, a Prussian official who was visiting German universities in search of promising professors reported that "Mining Councillor Crell is famous as a chemist, thanks to his writings."[55] To some beginning chemists, he seemed a towering figure. J. J. Bindheim, for example, put Crell in a league with Voltaire, asserting that he was a man of such great talent that he could not be emulated.[56] And A. von Humboldt, then a student of cameralistics, described Crell as "the best chemist in Germany."[57] Crell also had the pleasure of being named to one learned society

---

52. Crell's announcement was published in Baldinger's *Medicinisches Journal*, 1, no. 1 (1784), 72–80. Summaries appeared in the *Allgemeine deutsche Bibliothek*, 56 (1784), 618; and the *Erlangische gelehrte Anmerckungen und Nachrichten*, 39 (30 March 1784), 126.

53. Crell, "Vorbericht," *Chemische Annalen*, no. 2 (1785), [ii]. Notwithstanding Crell's confidence, the Germans were susceptible to fashions. In fact, in the early 1780s, the enthusiasm for useful subjects that was so helpful to Crell's journal forced C. M. Wieland to give the *Teutsche Merkur* a "mercantilistic" instead of a literary orientation. See H. Wahl, "Geschichte des Teutschen Merkur: Ein Beitrag zur Geschichte des Journalismus im achtzehnten Jahrhundert," *Palaestra*, 127 (1914), 160–198.

54. The findings that Crell was so proud to publish were Klaproth's discovery of two new earths and Westrumb's discovery of the combustion of metals in chlorine gas. See Crell, "Vorbericht," *Chemische Annalen*, no. 2 (1789), [i–iii].

55. F. Gedike, who was in Helmstedt in June 1789, also remarked that Beireis's classes on chemistry were far more popular than those of Crell, whose delivery was reputed to be "somewhat unpleasant." See "'Der Universitäts-Bereiser' Friedrich Gedike und sein Bericht an Friedrich Wilhelm II," ed. R. Fester, *Archiv für Kulturgeschichte*, 1st Supp. (1905), 6.

56. J. J. Bindheim, *Rapsodien der philosophischen Pharmakologie, nebst einer Anleitung zur theoretisch-praktischen Chemie* (Berlin: Mylius, 1785), pp. 8, 65.

57. Humboldt, letter to W. G. Wegener (ca. 1 May 1789), in *Die Jugendbriefe Alexander von Humboldts 1787–1799*, ed. I. Jahn and F. G. Lange (Berlin: Akademie Verlag, 1973), p. 21.

after another. By 1789 he could list himself as a member of well over twenty societies on his journal's title page (see Figure 4).[58]

## CRELL'S ACHIEVEMENT

As the twelfth monthly issue of the *Chemische Annalen* was going to press in late 1784, Crell mused that

> patriotic enthusiasm for the fatherland's chemistry filled my heart even before I began to write on the subject. With pleasure I observed its wide influence; with inner satisfaction I studied every work that helped preserve its superiority. . . . Subsequently I have striven to stimulate the fatherland's pursuit of chemistry. My efforts seem not to have been without success. . . .[59]

And a year later he boasted that

> my journals have served to intensify the inherent, I might almost say, inborn propensity of the Germans for chemistry and, besides this general encouragement, to awaken in many a warm love for this science. Not only do scientists, physicians, metallurgists, and apothecaries now concern themselves with chemical publications and choose them as their favorite subjects for conversation, but there has also arisen a close and trusting friendship for chemistry within all orders, even the fair sex. If considerations of space or modesty did not forbid, I would expound on the many new discoveries that first appeared in these journals, the names of meritorious men who made their reputations here, the respect that my friends in neighboring nations have accorded these writings, and finally the happy prospects of the glorious harvest that will be reaped if the many

---

58. Crell's ostentatious parade of his honors evoked criticism from some contemporaries. For instance, on 6 January 1786, Royal Hanoverian Botanist F. Ehrhart wrote C. W. Scheele that "you have long been worthy of membership in all possible learned societies, at least ten times sooner than Crell, whose greatest service has been that he profiteers on the hard work of others, fills his purse, and praises to the heavens all that sell him their wares. . . . *Sed mundus vult d.* [But the world wants to be deceived]." See J. Nordström, "Nagra bortglömda brev och tidskriftsbidrag av Carl Wilhelm Scheele," *Lychnos* (1942), 232–233. Likewise, G. Forster thought that Crell did not live up to his reputation. In a letter to S. T. Sömmering (17 October 1787), he observed that "I am far from regarding him as a truly great *Chemiker*, at least as a great *Praktiker* in this science." See *Georg Forster: Werke*, ed. G. Steiner, vol. IV (Frankfurt am Main: Insel, 1970), p. 481. Similar misgivings about the value of Crell's experimental contributions probably motivated opposition to his election to the St. Petersburg Academy in 1786. However, Crell's opponents were outvoted by the narrow margin of seven to six. See *Procès-verbaux des séances de l'Académie impériale des sciences depuis sa fondation jusqu'a 1803*, vol. IV (St. Petersburg: Akademiia nauk, 1911), p. 52.

59. Crell, "Vorbericht," *Chemische Annalen*, no. 2 (1784), [i].

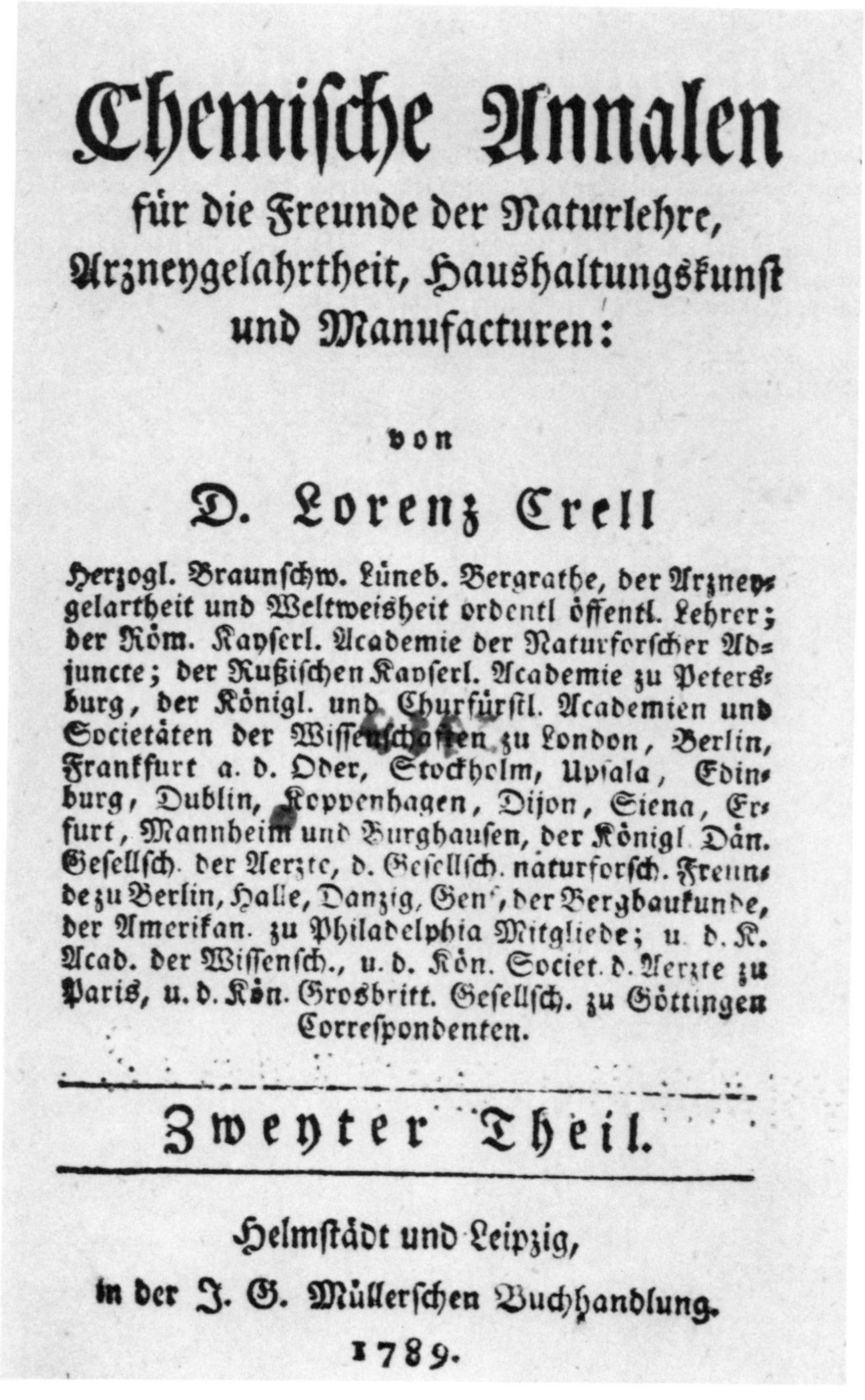

# Chemische Annalen

für die Freunde der Naturlehre, Arzneygelahrtheit, Haushaltungskunst und Manufacturen:

von

## D. Lorenz Crell

Herzogl. Braunschw. Lüneb. Bergrathe, der Arzneygelartheit und Weltweisheit ordentl öffentl. Lehrer; der Röm. Kayserl. Academie der Naturforscher Adjuncte; der Rußischen Kayserl. Academie zu Petersburg, der Königl. und Churfürstl. Academien und Societäten der Wissenschaften zu London, Berlin, Frankfurt a. d. Oder, Stockholm, Upsala, Edinburg, Dublin, Koppenhagen, Dijon, Siena, Erfurt, Mannheim und Burghausen, der Königl. Dän. Gesellsch. der Aerzte, d. Gesellsch. naturforsch. Freunde zu Berlin, Halle, Danzig, Genf, der Bergbaukunde, der Amerikan. zu Philadelphia Mitgliede; u. d. K. Acad. der Wissensch., u. d. Kön. Societ. d. Aerzte zu Paris, u. d. Kön. Großbritt. Gesellsch. zu Göttingen Correspondenten.

### Zweyter Theil.

Helmstädt und Leipzig,
in der J. G. Müllerschen Buchhandlung.
1789.

Fig. 4. Title page of Crell's *Chemische Annalen*, with a characteristic list of his titles and memberships.

seeds which have been sown on the wholesome German soil bear ripe fruit.[60]

Crell was justifiably proud. To be sure, he himself had not really enlarged understanding of any significant chemical phenomenon. But, motivated by desire for fame, eagerness to serve the commonweal, love for all things German, and enthusiasm for chemistry, he had developed the first successful chemical journal. He had, that is, created a discipline-oriented periodical which could serve as a forum for German chemists.

60. Crell, *Auswahl aller eigenthümlichen Abhandlungen und Beobachtungen aus den neuesten Entdeckungen in der Chemie*, vol. I (Leipzig: Weygand, 1786), p. [vi].

# 6 THE FORMATION OF THE GERMAN CHEMICAL COMMUNITY

Decades before Crell began appealing to his colleagues' patriotism, German chemists, like other German men of learning, had begun to link their national and intellectual identities.[1] Among the first to voice this new cultural nationalism was G. H. Burghart, who claimed in 1749 that many "German chemists" enjoyed "the respect, the renown, and even the envy of the arrogant Italians, French, and English." In his opinion "German chemists" deserved this respect because they had "a fundamental knowledge of chemistry and its basic doctrines unlike foreigners, especially the flighty French."[2] Four years later, R. A. Vogel, Göttingen University's first professor of chemistry, gave an inaugural lecture on "the experiments and discoveries with which the Germans have perfected chemistry during this century." Declaring that "German chemists" had enriched the science with "several hundred new and advantageous discoveries," he boasted that "Germany" had produced "more skilled chemists than England, France, Sweden, and Italy combined."[3] In 1761 J. G. Lehmann justified his description of A. S. Marggraf as "the greatest German chemist" by pointing out that even the overbearing French were interested in his colleague's work.[4] And in 1770 V. Rose, Sr. wrote of "the many important new truths, which have recently been discovered and accepted by self-reliant German chemists."[5]

1. For evidence of cultural nationalism among natural philosophers and popularizers of science in mid-eighteenth-century Germany, see W. Schatzberg, "Scientific Themes in the Popular Literature and the Poetry of the German Enlightenment, 1720–1760," *German Studies in America*, 12 (1973), 92, 94, 96–98, 100, 103, 112, 172; and A. von Haller, "Patriotische Ermunterungen an die Deutschen, ihre eigene Nation mehr zu schätzen," *Hannoverische Anzeigen: Hannoverische Beyträge zum Nutzen und Vergnügen*, 1 (1759), 1105–1116.
2. H. F. Teichmeyer, *Medicinische und Chemische Abhandlung von Seignettischen Saltze*, trans. G. H. Burghart (Breslau and Leipzig: Pietsch, 1749), pp. 25–26. For Burghart, whose father lived and studied with Stahl, see F. Börner, *Nachrichten von den vornehmsten Lebensumständen und Schriften jeztlebender berühmter Aerzte und Naturforscher in und um Deutschland*, vol. II (Wolfenbüttel: Meissner, 1752), pp. 495–535.
3. For a summary of Vogel's inaugural address, see the *Göttingische Anzeigen von gelehrten Sachen* (26 November 1753), 1283–1284. For a biographical profile of Vogel, see Appendix I.
4. J. G. Lehmann, "Vorrede," in A. S. Marggraf, *Chymische Schriften* (Berlin: Weber, 1761), I:[ii]. For a biographical profile of Lehmann, see Appendix I. Lending credence to Lehmann's estimate of Marggraf's reputation in France, H. G. McCann informs me that Marggraf ranked fourth between 1760 and 1771 and again between 1772 and 1784 among the chemists referenced by authors of French articles dealing with chemistry.
5. [V. Rose, Sr.], review of J. H. Hagen's *Chymische Betrachtungen über die Herkunft der*

While chemists in Germany were increasingly likely to speak as "German chemists" between the 1740s and 1770s, they were not yet interacting as a community of peers whose collective judgment established the merits of all contributions and contributors. Rather, they oriented themselves to local intellectuals or to chemists throughout Europe. Both their parochialism and cosmopolitanism were manifest in an acrimonious dispute between J. T. Eller and J. H. Pott during the 1750s. Initially the two addressed their diatribes to Berlin's intelligentsia, admixing their arguments with titillating gossip. Then Pott, realizing that he had lost in Berlin, appealed to the chemists in the Paris Academy for arbitration.[6] Just as little sense of community was displayed by the German chemists who commented during the 1760s and 1770s on the relative merits of J. F. Meyer's theory of pinguic acid and J. Black's theory of fixed air. The controversy was desultory, lacking the focus and intensity that characterizes disputes among scientists seeking the support of a common audience.[7]

To create a forum for German chemists was, we have seen, one of Crell's goals in founding his journal. This chapter first shows that the *Chemische Annalen* did indeed come to serve as the forum for the many Germans who, thanks to the prior emergence of social support for chemistry, were predisposed to follow, apply, or advance the science. Then it analyzes the referencing of Crell's contributors, finding that they tended to draw information from and relate their results to the work of a few of their fellows. Finally, it adduces evidence that German chemists, in regular communication with one another before a common audience, were beginning to feel a sense of community by 1789, just before the antiphlogistic revolution in Germany.

## A FORUM FOR GERMAN CHEMISTRY

Unlike editors of modern scientific journals, Crell could not rely on grants, advertisements, and library subscriptions for funds. He had to meet his ex-

---

*feuerbestandigen vegetabilischen Laugensalze* (1768), *Allgemeine deutsche Bibliothek*, 12, no. 2 (1770), 319. For further evidence of Rose's cultural nationalism, see ibid., 11, no. 2 (1770), 26, 28, and supp. 1–12 (1771), 294, 301. For a brief biography of Rose, see the profile of his son, V. Rose, Jr., in Appendix I.

6. The three central publications in the dispute between Eller and Pott were J. H. Pott, *Animadversiones physico-chymicae circa varias hypotheses et experimenta D. Dr. et Consiliar. Elleri: Physicalisch Chymische Anmerckungen über verschiedene Sätze und Erfahrungen des Herrn Hofr. D. Ellers* (Berlin, 1756); [J. G. Lehmann, et al.], *Kurtze Untersuchung der wahren Ursachen, welche den Professor der Chymie bey dem königl. Collegio Medico-Chirurgico zu Berlin, Herrn Joh. Heinrich Pott verleitet, seine so genannte Animadversiones, Theodor Eller einverleibet worden* . . . ([Berlin], 1756); and J. H. Pott, *Fortsetzung seiner physicalisch-chymischen Anmerckungen über des Hrn. Geheimen Rath D. Ellers verschiedene Sätze und Erfahrungen* ([Berlin], 1756). For a description of the controversy, see B. von Freyberg, "Johann Gottlob Lehmann (1719–1767): Ein Arzt, Chemiker, Metallurg, Bergmann, Mineraloge und grundlegender Geologe," *Erlanger Forschungen: Reihe B: Naturwissenschaften*, 1 (1955), 75–81. For profiles of Eller and Pott, see Appendix I.

7. My characterization of the German controversy over pinguic acid and fixed air is based largely on reviews appearing in the *Allgemeine deutsche Bibliothek*. The scientific as well as social dimensions of this controversy deserve close analysis.

TABLE 12: Number of Subscribers to the *Chemische Annalen*, 1784–1789

|                | 1784 | 1785 | 1786 | 1787 | 1788 | 1789 |
|----------------|------|------|------|------|------|------|
| Number new     | 424  | 123  | 72   | 40   | 24   | 27   |
| Number renewing| 0    | 324  | 341  | 351  | 359  | 361  |
| Total number   | 424  | 447  | 413  | 391  | 383  | 388  |

Source: Based on the cumulative list of subscribers presented in Appendix III.
Note: Of the 710 subscribers between 1784 and 1789, 294 (41 percent) subscribed for one to two years, 148 (21 percent) for three to four years and 268 (38 percent) for five to six years.

penses entirely from sales. At first he depended on his publisher and the book dealers to reach the market. In late 1783 when he was preparing to shift to a monthly schedule, he reconsidered this arrangement. He decided to publish the *Chemische Annalen* at his own cost and to market it on a subscription basis. By doing so, he could garner a larger share of the revenues from sales.[8] Moreover, subscription service would enable him to maintain regular contact with his readers. Crell's call for subscriptions elicited a favorable response. From 1784 on, he issued a new list of his subscribers every year, perhaps as a means of giving them the pleasure of seeing their interest in chemistry receive public acknowledgment.[9]

These annual lists reveal that Crell mustered more than enough subscribers to finance his journal. In its first year of publication the *Chemische Annalen* had over 420 subscribers. The number registered a slight gain in 1785, then gradually declined during the rest of the decade (see Table 12). Still, at 3 Rtl a year per subscriber, Crell's annual receipts from subscriptions remained above 1,100 Rtl throughout the 1780s. Since he could also count on some revenues from trade sales, he must have had a surplus.[10]

8. In making his plans, Crell sought "merchandising advice" from his friend F. Nicolai, for whom he had earlier enlisted subscribers. See Crell, letter to Nicolai (3 December 1783), Nachlass Nicolai, Staatsbibliothek Preussischer Kulturbesitz: Handschriftenabteilung, West Berlin. He decided to offer subscription for monthly service at 3 Rtl per year, about 30 percent more than it cost to buy the journal semi-annually at bookstores. See *Erlangische gelehrte Anmerckungen und Nachrichten*, 39 (1784), 126.

9. For the development of subscriptions as a means of selling and buying books in Germany, see R. Wittmann, "Subscribenten- und Pränumerantenverzeichnisse als lesersoziologische Quellen," Wolfenbütteler Arbeitskreis für Geschichte des Buchwesens: *Schriften*, 1 (1977), 125–159. As Wittmann points out, scholars have only begun to exploit subscription lists as a source of information about German readers during the Enlightenment. Under the direction of P. J. Wallis, the exploitation of British lists has already made considerable progress. See his "The Social Index: A New Technique for Measuring Social Trends," University of Newcastle upon Tyne: School of Education: Project for Historical Biobibliography: *PHIBB*, no. 184 (1978).

10. The median production-run of 54 German periodicals published between 1750 and 1800 was 1,000 copies, with a range from 500 to 3,000 copies. The cost of producing 815 copies of one journal was about 175 Rtl and of producing and distributing 1000 copies of another was about 800 Rtl. See J. Kirchner, *Die Grundlagen des Deutschen Zeitschriftenwesens mit einer Gesamtbibliographie der Deutschen Zeitschriften bis zum Jahre 1790*, vol. I (Leipzig: Hiersemann, 1928), pp. 42–47, 54, 69, 86–87. So, presuming that Crell produced at least 500 copies, his net income (after honoraria, production, and postage) was probably 500 Rtl or more.

Besides revealing how Crell financed the *Chemische Annalen*, the subscription lists indicate the character of the journal's audience. Altogether there were 710 subscribers between 1784 and 1789. Well over a hundred subscribed from outside Germany. While many of these men were German by ancestry or birth, some were Russians, Scandinavians, Dutch, French, Italians, and Britons. Whatever their nationality, the subscribers from abroad did more for Crell's journal than help fill its coffers. It was a strong endorsement that chemists and friends of chemistry ordered the *Chemische Annalen* from as far away as Catharinenburg and Philadelphia.

For every foreign subscriber, there were four subscribers who lived in Germany. Large majorities of these men belonged to the bourgeoisie, resided in territories where Protestantism was the prevailing religion, dwelt in medium-sized and large towns, and worked, or planned to work, as pharmacists and physicians. But nonnegligible minorities were nobles, residents of Catholic regions, inhabitants of lesser burgs, and mining administrators or science teachers (see Table 13). Crell's journal evidently reached a broad cross section of chemistry's latent audience in Germany. By doing so, we may suppose, it united Germans interested in following chemistry into what contemporaries called the "chemical public."[11]

While the subscribers provided Crell's journal with revenues and an interested audience, the contributors were even more crucial to its success. Crell, as we have seen, appreciated their importance. Time and again he called upon his readers and correspondents for contributions, appealing to their love of self, country, and chemistry. His exhortations elicited sufficient response, as we have seen, for him to switch from semiannual to quarterly publication in 1781 and from quarterly to monthly publication in 1784. By this year over fifty men were contributing to his journal.[12] The number rose above sixty in 1785 and remained there for the rest of the decade. All in all, the work of 179 named contributors appeared in the *Chemische Annalen* and its companion journal the *Beyträge zu den chemischen Annalen* between 1784 and 1789. While many were ephemeral or dilatory, a score sent in materials often enough to be regarded as collaborators in Crell's venture (see Table 14).

One in every five contributors, like the subscribers, resided outside Germany. Some—e.g., J. J. Bindheim in Moscow, J. T. Lowitz in St. Petersburg, C. W. Scheele in Köping, J. H. Hassenfratz in Paris, and J. C. Doll-

---

11. That Crell's subscribers were representative of chemistry's audience in late eighteenth-century Germany has been shown in my "Chemistry's Enlightened Audience," *Studies on Voltaire and the Eighteenth Century*, 53 (1976), 1069–1086. To my knowledge, the earliest use of the phrase "chemical public" was in a review of C. W. Nose's *Versuch einiger Beyträge zur Chemie* (1778), which, as we saw in the last chapter, was dedicated to Crell and announced his journal. See *Nürnbergische gelehrte Zeitung* (1778), 761. For occurrences of the phrase in the 1780s, see the *Allgemeine Literatur-Zeitung*, no. 3 (September 1786), 566; no. 1 (February 1787), 302.

12. My discussion of Crell's contributors is based on the *Chemische Annalen*, its supplement, the *Beyträge zu den chemischen Annalen*, and the standard biographical dictionaries. D. von Engelhardt has promised to index the *Chemische Annalen* (1784–1804) in a continuation of his "Die chemischen Zeitschriften des Lorenz von Crell," in *Indices naturwissenschaftlich-medizinischer Periodica bis 1850*, ed. A. Geus.

TABLE 13: Germany's Subscribers to the *Chemische Annalen*, 1784–1789

|  | Number | Percentage |
|---|---|---|
| Social class |  |  |
| Bourgeoisie | 515 | 91 |
| Nobility | 49 | 9 |
| Religion of residence |  |  |
| Protestantism | 498 | 90 |
| Catholicism | 57 | 10 |
| Size of residence |  |  |
| Over 20,000 | 142 | 28 |
| 5,000–20,000 | 216 | 38 |
| Under 5,000 | 187 | 34 |

| Occupational area | Bourgeois subscribers | | Noble subscribers | |
|---|---|---|---|---|
|  | Number | Percentage | Number | Percentage |
| Pharmacy | 258 | 52 | 2 | 4 |
| Therapy | 129 | 26 | 4 | 8 |
| Mineral products | 46 | 9 | 11 | 22 |
| Science | 22 | 4 | 1 | 2 |
| Other | 43 | 9 | 31 | 63 |

Source: Based on Appendix III.
Note: Of Crell's 710 subscribers between 1784 and 1789, 564 began subscribing as residents of Germany. As the years passed, a few external subscribers came to Germany and a few internal subscribers went abroad. The data represent subscriber characteristics at the time of first subscribing. The percentages in each main category have been based on the number for whom characteristics could be ascertained. Information could not be obtained regarding the prevailing religion in the residence of 9 subscribers, the population of the residence of 19 subscribers, and the occupation of 17 bourgeois subscribers.

TABLE 14: Number of Contributors to the *Chemische Annalen*, 1784–1789

|  | 1784 | 1785 | 1786 | 1787 | 1788 | 1789 |
|---|---|---|---|---|---|---|
| Number new | 32 | 32 | 27 | 23 | 17 | 15 |
| Number returning | 24 | 31 | 44 | 48 | 55 | 48 |
| Total number | 56 | 63 | 71 | 71 | 72 | 63 |

Source: Based on the contents of the *Chemische Annalen* (1784–1789) and the *Beyträge zu den chemischen Annalen* (1785–1789). The *Beyträge* has been included because it was comprised of materials intended for the *Chemische Annalen*. Crell has been counted as a contributor.
Note: Of the 179 contributors between 1784 and 1789, 121 (67 percent) contributed for one or two years, 37 (21 percent) for three or four years, and 21 (12 percent) for five or six years.

fuss in London—were of German origin. Others—e.g., B. R. Geijer in Stockholm, J. C. de la Métherie in Paris, L. B. Guyton de Morveau in Dijon, M. Landriani in Milan, L. V. Brugnatelli in Pavia, C. Blagden in London, and R. Kirwan in London then Dublin—were of other nationalities. The contributors living abroad rarely gave Crell first access to articles, but they did send news about discoveries and debates in their respective countries. In doing so, they enabled the German chemical public to keep up with developments throughout Europe.

TABLE 15: Comparison of Germany's Subscribers and Contributors to the *Chemische Annalen*, 1784–1789

|  | Subscribers (percentage) | Contributors (percentage) |
|---|---|---|
| Social class |  |  |
| Bourgeoisie | 91 | 94 |
| Nobility | 9 | 6 |
| Religion of residence |  |  |
| Protestantism | 90 | 88 |
| Catholicism | 10 | 12 |
| Size of residence |  |  |
| Over 5,000 | 66 | 79 |
| Under 5,000 | 34 | 21 |
| Occupational area of bourgeois subscribers and contributors |  |  |
| Pharmacy | 52 | 41 |
| Therapy | 26 | 26 |
| Mineral products | 9 | 10 |
| Science | 4 | 15 |
| Other | 9 | 9 |

Sources: Based on Table 13, the sources for Table 14, and, for the 141 men who resided in Germany when they began contributing to Crell's journal, the various biographical dictionaries listed in the introduction to Appendix I.
Note: The percentages in each main category have been based on the number for whom characteristics could be ascertained.

TABLE 16: Germany's Most Active Contributors to the *Chemische Annalen*, 1784–1789

| | | | | Contributing characteristics | | |
|---|---|---|---|---|---|---|
| Name | Age in 1789 | Residence in 1789 | Occupational Areas in 1789 | Number of years | Number of contributions | Number of pages |
| Wiegleb, J. C. | 67 | Langensalza | Pharmacy, science | 6 | 21 | 190 |
| Klaproth, M. H. | 46 | Berlin | Pharmacy, science | 6 | 21 | 138 |
| Heyer, J. C. H. | 43 | Brunswick | Pharmacy | 6 | 14 | 172 |
| Gmelin, J. F. | 41 | Göttingen | Science | 6 | 13 | 237 |
| Achard, F. C. | 36 | Berlin | Science | 5 | 20 | 235 |
| Westrumb, J. F. | 36 | Hameln | Pharmacy | 6 | 68 | 369 |
| Gren, F. A. C. | 29 | Halle | Science | 5 | 16 | 140 |
| Hermbstaedt, S. F. | 29 | Berlin | Science, technology | 5 | 22 | 133 |

Source: See Table 14.
Note: Those men have been included whose "activity product," i.e., (number of years contributing) × (number of contributions) × (number of pages contributed), exceeded 10,000. The next cluster of contributors had activity products around 3,000.

As the contributors outside Germany were giving a cosmopolitan air to the *Chemische Annalen*, those living in Germany were making the journal into a genuine forum for German chemistry. Between 1784 and 1789, over 140 residents of Germany published in Crell's journal, thereby attesting to its openness. Like the German subscribers, these men tended to be commoners and Protestants who lived in medium-sized and large towns. They also tended to earn their livings in pharmacy and medicine. And they too had nobles, Catholics, mining administrators, and science teachers in their midst (see Table 15).

While the number and variety of Crell's German contributors attested to the openness of the *Chemische Annalen*, their combined stature manifested the journal's importance. Among Crell's contributors were nearly all the recognized chemists residing in Germany.[13] Moreover, among Crell's more active contributors (see Table 16) were the four most prominent German chemists of the late 1780s—F. C. Achard, M. H. Klaproth, J. F. Westrumb, and J. C. Wiegleb.[14] By favoring Crell with a steady flow of manuscripts, these men alone made the *Chemische Annalen* indispensable reading for all chemists and friends of chemistry in Germany.

In sum, circulating throughout Germany and presenting the work of renowned as well as ordinary German chemists, Crell's journal came to serve as the forum for German chemistry during the 1780s. No longer did it make sense for German chemists to orient themselves to local intellectuals who, though generally interested in their work, could not provide informed criticism and appreciation. No longer did it make sense to orient themselves to chemists throughout Europe who, though able to provide expert evaluation, were likely to be unwilling or slow to respond. By publishing in the *Chemische Annalen*, they could rapidly reach an audience with the expertise and means for providing responses that were both knowledgeable and timely.

## ATTENTION PATTERNS AMONG GERMAN CHEMISTS (1784–1789)

In their articles and letters, Crell's contributors often referred to persons whose publications, achievements, or activities seemed worthy of notice. An analysis of the collective referencing of the German contributors can be used, therefore, to ascertain how attention was distributed among German chemists. To carry out such an analysis, however, it is necessary to modify modern techniques of citation analysis. By twentieth-century standards, the number of articles appearing in Crell's journal was small. Moreover, the references varied greatly in the amount of information they conveyed. Hence, for each of the 134 contributors living in Germany in 1789, I have determined how many of the remaining 133 contributors cited or mentioned him at least once in the *Chemische Annalen* and *Beyträge zu den chemischen Annalen* between 1784 and 1789. To judge from the results, Crell's contributors distributed

---

13. For instance, of the twenty-two chemists living in Germany whom J. F. Gmelin included on a list of leading eighteenth-century chemists in 1789, twenty contributed to Crell's journal between 1784 and 1789. The two noncontributors were N. J. von Jacquin in Vienna, whose primary interest was botany, and C. F. Wenzel in Freiberg, whose administrative duties were evidently so burdensome that he could no longer engage in research. For Gmelin's list, see his *Grundriss der allgemeinen Chemie* . . . , vol. I (Göttingen: Vandenhoeck & Ruprecht, 1789), pp. xxvii–xxviii.

14. For the reputations of these men, see the analysis of referencing by Crell's contributors later in this chapter and Appendix I.

TABLE 17: Referencing by Germany's Contributors to the *Chemische Annalen*, 1784–1789

| | Number of referencers | Number of referencees | Names of referencees with 4 or more referencers |
|---|---|---|---|
| Periphery | 0 | 49 | |
| | 1 | 27 | |
| | 2–3 | 24 | |
| | 4–7 | 19 | 4: Gren[a,b], C. A. Hoffmann, Lasius, Unger. 5: Andreae[a], Gerhard[a], Girtanner[a], Heyer[a,b], Schiller. 6: W. H. S. Bucholtz[a], Danz, Hagen[a], Lichtenberg, Trebra, Voigt, Westendorff. 7: Gmelin[a,b], Hahnemann, Werner. |
| | 8–15 | 13 | 8: Dehne. 9: Achard[a,b], Hermbstaedt[a,b], Storr, Weigel[a]. 10: Born[a], Ilsemann[a], J. C. F. Meyer[a]. 11: Crell[a], Ferber. 12: Klaproth[a,b], Leonhardi[a]. 14: Göttling[a]. |
| Center | 16–31 | 2 | 22: Westrumb[a,b]. 31: Wiegleb[a,b]. |

Source: See Table 14.
Note: Attention has been restricted to the referencing behavior of the 134 contributors to the *Chemische Annalen* between 1784 and 1789 who resided in Germany in 1789.
[a]See Appendix I.
[b]One of Germany's eight most active contributors to the *Chemische Annalen*, 1784–1789. See Table 16.

their attention among one another in a periphery-center pattern (see Table 17).[15]

At the periphery of the contributors' attention were some hundred unreferenced and rarely referenced contributors—popularizers, utilizers, drudges, novices, and able chemists with low publication rates. Somewhat more visible were nineteen contributors between the periphery and core, who received a sufficient number of references to preclude incidental associations as the source of their visibility. This intermediate group included some well-known mineralogists—e.g., F. W. H. von Trebra, J. C. W. Voigt, A. G. Werner—and one respected experimental physicist—G. C. Lichtenberg—as well as several chemists. In comparison with the peripheral and intermediate figures, fifteen Germans received explicit notice from a fair number of Crell's contributors. Crell, J. G. Leonhardi, and C. E. Weigel owed their prominence largely to their editorial and translating activities. The remaining twelve were known mainly for their efforts to apply and advance chemical knowledge. J. C. C. Dehne was visible on account of his investigations of antimonial drugs. I. von Born, J. J. Ferber, and G. C. C. Storr were noted for their efforts to use chemistry in mineralogy and the mineral products industry. J. C. Ilsemann, J. C. F. Meyer, and especially M. H. Klaproth were recognized for their analytical skill. And F. C. Achard, J. F. A. Göttling, S. F. Hermbstaedt, J. F. Westrumb, and J. C. Wiegleb were conspicuous as broad-ranging and productive chemists.

15. H. G. McCann has found the same periphery-center pattern of attention among French and British chemists. See his *Chemistry Transformed: The Paradigmatic Shift from Phlogiston to Oxygen* (Norwood, N.J.: Ablex, 1978), pp. 103–116. Studies of modern science have also found this pattern. See especially D. Crane, *Invisible Colleges: Diffusion of Knowledge in Scientific Communities* (Chicago: University of Chicago Press, 1972).

TABLE 18: Referencing by Germany's Fifteen Most Visible Contributors to the *Chemische Annalen*, 1784–1789

|  | Number of referencers | Number of referencees | Names of referencees with 3 or more visible referencers |
|---|---|---|---|
| Periphery | 0 | 83 | |
|  | 1 | 26 | |
|  | 2 | 10 | |
|  | 3 | 7 | Gmelin[a,b], Hagen[a], Hahnemann, Hermbstaedt[a,b], Ilsemann[a], J. C. F. Meyer[a], Storr. |
|  | 4 | 3 | Gren[a,b], Crell[a], Leonhardi[a]. |
|  | 5 | 3 | Göttling[a], Klaproth[a,b], Westrumb[a,b]. |
|  | 6 | 0 | |
|  | 7 | 1 | Wiegleb[a,b]. |
| Center | 8 | 1 | Achard[a,b]. |

Source: See Table 14.
Note: Attention has been limited to those men who were referenced by eight or more contributors between 1784 and 1789 (see Table 17).
[a]See Appendix I.
[b]One of Germany's eight most active contributors to the *Chemische Annalen*, 1784–1789. See Table 16.

Like all 134 contributors, these fifteen relatively prominent contributors also distributed their attention in a periphery-center pattern (see Table 18). However, they relocated several men within this pattern, evidently because they put a higher value on originality in chemistry. In their eyes, Born, Dehne, Ferber, and Weigel were among the discipline's peripheral cultivators. Meanwhile, J. F. Gmelin, F. A. C. Gren, K. G. Hagen, and S. Hahnemann were placed alongside Crell, Hermbstaedt, Ilsemann, Leonhardi, Meyer, and Storr in the intermediate region, which was now occupied entirely by chemists. And Achard, Göttling, Klaproth, Westrumb, and Wiegleb emerged as the central figures in German chemistry.

But, it may well be asked, did the fifteen chemists whom the most visible contributors placed in the intermediate and central regions comprise more than an aggregate of men with similar interests? The fact that they attracted this attention says little about their social solidarity. In fact, to judge from referencing among these chemists in Crell's journal, Hahnemann, Hagen, Ilsemann, Leonhardi, Göttling, Achard, and probably Storr were not closely integrated with the others. By the same token, the density of references, especially reciprocal references, among Meyer, Klaproth, Gren, Wiegleb, Gmelin, Hermbstaedt, Crell, and Westrumb strongly suggests that these eight men constituted an interactive collectivity (see Table 19).

Looking beyond Crell's journal, we find that these eight chemists were linked not only by published references but also by prior personal contacts. Living in Protestant towns within two hundred miles of Berlin, engaged in medical-technical-scientific occupations, and imbued with a common enthusiasm for chemistry, they had gradually made one another's acquaintance (see Appendix I). At the very least, personal familiarity must have increased their curiosity about one another's work. Occasionally, their contacts exerted a

TABLE 19: Referencing among Germany's Most Visible Chemists (ca. 1789)

Name of Referencee

| Name of Referencer | Hahnemann | Hagen | Ilsemann | Leonhardi | Göttling | Achard | Meyer | Klaproth | Storr | Gren | Wiegleb | Gmelin | Hermbstaedt | Crell | Westrumb |
|---|---|---|---|---|---|---|---|---|---|---|---|---|---|---|---|
| Hahnemann | | | | | | | | | | | | | | | |
| Hagen | | | | | | | | | | | | | | r | |
| Ilsemann | | | | | | | | | | | | | | | |
| Leonhardi | r | | | | | | | | | | | | | | |
| Göttling | | | r | | | r | | r | | | | | | | |
| Achard | | | | | | | | | | | | | | | |
| Meyer | | | | | | r | | R | r | R | | | | | |
| Klaproth | | | | | r | R | | | | r | | | | | R |
| Storr | | r | | r | r | r | | | | | r | r | | r | R |
| Gren | | r | | r | r | r | r | | | | r | | R | R | R |
| Wiegleb | | | | | r | R | | | | | | R | | | |
| Gmelin | | | r | r | r | r | r | r | | | r | | | R | R |
| Hermbstaedt | | r | | | r | | | r | | R | R | | | R | R |
| Crell | r | r | | r | | | | r | | R | r | R | R | | R |
| Westrumb | r | r | r | r | r | r | r | R | R | R | r | R | R | R | |

more profound influence. In 1767/68, as we have seen, Westrumb, who was then a young apprentice in Hanover's Hof-Apotheke, was inspired by his close association with the studious journeyman Klaproth to go beyond a superficial knowledge of chemistry. Around 1780 Hermbstaedt attended Wiegleb's new chemical-pharmaceutical boarding school in Langensalza, learning enough about chemistry to begin independent research soon afterwards. And in 1782/83 Gren found Crell's assistance invaluable in making the transition from a pharmaceutical to an academic career. Though no more than three of these eight chemists ever assembled in the same room, their various personal ties surely facilitated their consolidation as the leadership group for German chemistry.

## A SENSE OF COMMUNITY

As German chemists became accustomed to using Crell's journal and as a few of their prominent fellows coalesced into an interactive leadership group, the realization grew that chemists in Germany constituted a national community. In 1785, almost voicing such an awareness, C. W. Nose observed that his friend's journal had inaugurated

> an epoch in German chemistry by bringing about a more rapid and advantageous spread of information, by awakening or busying so many good heads, and by serving as a rallying point where further light could be shed on the subject.[16]

The following year, one contributor expanded on the theme, writing that Crell had

> already created a kind of association among Germany's chemists and natural scientists. Couldn't one go beyond this . . . to set up a meeting during the summer in a pleasant place where every natural scientist and chemist in attendance could meet a number of specialists in and friends of his favorite subject from all ends of Germany? Everybody would come with proposals, projects, questions, experiments, and any other scientific concern they might have. People would agree upon plans, establish personal relations, overcome prejudices against one another, etc. Perhaps more ambitious activities would emerge from such beginnings. It would be . . . a unique national undertaking.

Though this anonymous "chemist" suggested a general scientific meeting, probably to ensure adequate attendance, Crell thought a meeting for chemistry alone might be feasible. He labeled the suggestion a "proposal for a meeting of friendly chemists."[17]

Awareness of increasing cohesion among German chemists was also expressed by an anonymous reviewer of the *Chemische Annalen* in the *Allgemeine Literatur-Zeitung*. In 1787 he remarked that,

> notwithstanding the recognized worth and utility of this journal, we must express the wish, which we have heard from several experts in chemistry, that Crell would exercise greater care in deciding which treatises to publish. Alongside excellent articles there often stand others that would hardly be missed by the *chemischen Publikum*, for they have been written by men who

---

16. [Nose], review of Westrumb's *Kleine physikalisch-chemische Abhandlungen* (1785), *Allgemeine deutsche Bibliothek*, 64 (1785), 488.

17. "Vorschlag: Ueber eine zu verabredende Zusammenkunft freundschaftlicher Chemisten," *Chemische Annalen*, no. 1 (1786), 179–180. Nothing came of this proposal. See L. Eck, "Ein erster Versuch zur Gründung eines 'Vereins deutscher Chemiker' und zur Veranstaltung von Hauptversammlungen," *Zeitschrift für angewandte Chemie*, 44 (1931), 631–632.

are obviously still unacquainted with the first principles of the science.

A year later the same reviewer reiterated his hope that Crell would be more selective. So long as Crell included the work of authors who "had not yet mastered the rudiments of chemistry and natural philosophy," it would appear that he was more concerned with stuffing his pages than with expanding knowledge.[18] This reproach must have stung Crell. Yet in repeatedly calling for higher standards, the reviewer was proclaiming that the *Chemische Annalen* belonged not just to Crell but to all German chemists.

The growing sense of community among German chemists even began to influence their rhetoric in scientific controversies. In 1787, hoping to attract more attention to a theory he had developed, Gren accused his German colleagues of being too ready to accept the views of foreigners.[19] The following year Westrumb responded to Gren with the claim that

> I am not ruled by Anglomania, nor by Gallomania. . . . There are few who have such a warm love for German chemistry and German chemists as I do, few with so warm a prejudice for what originates in the fatherland and so cold [a feeling] for what originates abroad as I. But this inner love for my fatherland does not blind me to the good from abroad.[20]

Westrumb, whose later response to Lavoisier's theory belied his virtuous stance, certainly felt a strong sense of community with his fellow German chemists. Indeed, soon afterwards when he became embroiled in a dispute with Girtanner, he sought arbitration, not from French chemists as Pott had done three decades before, but rather from "one of our meritorious German chemists, such as Göttling, Heyer, Hermbstaedt, Klaproth, or Wiegleb."[21]

Through using Crell's journal as a forum, German chemists came to regard the collectivity of German chemists as their primary audience. They coalesced into a national discipline-oriented community—the German chemical community. Then in the mid-1780s, they began to sense their new social solidarity. One or another suggested an annual chemical meeting in Germany, requested tighter control of the community's forum, urged greater indepen-

---

18. Review of the *Chemische Annalen* (1786), *Allgemeine Literatur-Zeitung*, no. 1 (February 1787), 302; review of the *Chemische Annalen* (1787), ibid., no. 3 (July 1788), 160. The same reviewer was probably responsible for a later expression of the same sentiment. See review of the *Chemische Annalen* (1789), ibid., no. 1 (March 1790), 593.

19. Gren, "Versuche und Beobachtungen über die Entstehung der fixen und phlogistischen Luft," *Beyträge zu den chemischen Annalen*, 2 (1787), 319. The year before Gren had chided his fellow German chemists for neglecting their countryman Lehmann's pioneering investigations of wolfram, thereby allowing foreigners to reap the glory of discovering tungsten. See J. J. and F. d'Elhuyar, *Chemische Zergliederung des Wolframs*, trans. F. A. C. Gren (Halle: Waisenhaus, 1786), p. 8.

20. Westrumb, *Kleine physikalisch-chemische Abhandlungen*, vol. II, pt. 2 (Leipzig: Müller, 1788), p. [v].

21. Westrumb, "Einige Versuche über die Auflöslichkeit des Eisens im blossen Wasser," *Chemische Annalen*, no. 2 (1788), 307.

dence from foreign influences, and appealed to German chemists for arbitration. Though the members of this nascent community were peers, they were not equals. The German chemical community had a periphery-core structure. Melding with the larger chemical public, the periphery consisted of scores of unrecognized or barely recognized chemical authors. Intermediate between this periphery and the core was a score of more visible users and devotees of chemistry. And at the community's core were some eight German chemists with relatively great visibility and mutual awareness. Led by these core chemists—especially the editor Crell, the activist Westrumb, and the veteran Wiegleb—German chemists were a thriving and increasingly self-aware national chemical community in 1789.

Though primarily oriented toward their peers within the German chemical community, German chemists also paid attention to chemists outside Germany. Indeed, Crell's German contributors were as likely to refer, in ascending order, to F. Fontana, J. Priestley, B. G. Sage, L. B. Guyton de Morveau, A. L. Lavoisier, A. Baumé, and R. Kirwan as they were to refer to many German chemists.[22] They were united, but not isolated. Yet during the 1780s, few Germans recognized that Lavoisier was building an irresistible case for his revolutionary chemical system. Only in 1789, with the publication of Lavoisier's *Traité* and the emergence of the first German antiphlogistonists, did the possibility of revolution become clear. In the next two chapters, we see how the reception of Lavoisier's theory in Germany was mediated by the German chemical community and how the community's first crisis completed its formation by making German chemists fully aware of the existence and importance of their new social solidarity.

22. Eight of Crell's German contributors referenced Fontana; 10, Priestley; 11, Sage; 12, Guyton de Morveau and Lavoisier; 15, Baumé; and 18, Kirwan. Only foreign chemists living in 1789 have been included on this list.

# 7 THE "FRENCH CHEMISTRY"

In June 1789 Crell shared a disturbing letter from Paris with the readers of his *Chemische Annalen*. Herr E—— reported that

> I have witnessed a most remarkable drama here, one which to me as a German was very unexpected, and quite shocking. I saw the famous M. Lavoisier hold a ceremonial auto-da-fé of phlogiston in the Arsenal. His wife . . . served as the sacrificial priestess, and Stahl appeared as the *advocatus diaboli* to defend phlogiston. In the end, poor phlogiston was burned on the accusation of oxygen. Do you not think I have made a droll discovery? Everything is literally true. I will not say whether the cause of phlogiston is now irretrievably lost, or what I think about the issue. But I am glad that this spectacle was not presented in my fatherland.[1]

No such event could have occurred in Germany, for Lavoisier's antiphlogistic system did not yet have any adherents there. Despite warnings from abroad, German chemists still dismissed his system as a passing French fad. In the three years following Herr E——'s letter, however, they came to realize that the "French chemistry," as some called the theory, could not be ignored. Though most members of the German chemical community remained loyal to phlogiston, several Germans at the community's periphery and a few closer to its center embraced the antiphlogistic system. Their conversions plunged the newly formed community into a state of crisis. The present chapter, after briefly recounting the antiphlogistic revolution in France, analyzes the reception of the French chemistry in Germany from mid-1785, when Lavoisier first denounced phlogiston, to mid-1792, when the German chemical community was deep in turmoil over its theoretical direction.

Until the late nineteenth century, the German antiphlogistic revolution was generally portrayed as a conflict between nationalistic partisans of Stahl and objective advocates of Lavoisier, that is, as a contest between prejudice and science.[2] Then G. W. A. Kahlbaum and A. Hoffmann advanced an alterna-

---

1. E——, letter to Crell, *Chemische Annalen*, no. 1 (1789), 519. "Can we Germans," Wiegleb remarked about this incident two years later, "do anything but laugh from the depths of our hearts? Perhaps the scenario can soon be changed." See J. C. Wiegleb, *Geschichte des Wachstums und der Erfindungen in der Chemie in der neueren Zeit*, vol. II (Berlin and Stettin: Nicolai, 1791), pp. 478–479.

2. A. N. Scherer, "Geschichte der Ausbreitung der neuern Theorie: Teutschland," *Archiv*

tive interpretation. They argued that most German phlogistonists were sober scientists who felt duty-bound to oppose Lavoisier's theory so long as it was untested in some respects and flawed in others. By contrast, the German antiphlogistonists were enthusiasts whose advocacy of the French system was motivated by its logical coherence, explanatory power, and heuristic value. Hence, in Kahlbaum and Hoffmann's view, the antiphlogistic revolution was a struggle in which imaginative chemists were forced by their cautious colleagues to provide convincing evidence for their radical theory.[3] Though better researched and more sophisticated than its positivistic predecessor, this interpretation did not settle the matter.[4]

The present chapter, while conceding the importance of scientific considerations in motivating the German chemists' theoretical allegiances, advances evidence that this is but one side of the story. Nationalistic sentiments definitely promoted resistance to Lavoisier's theory. At the same time, cosmopolitan or Francophile proclivities evidently encouraged receptivity to this revolutionary doctrine. Attitudes toward the authority of the German chemical community also seem to have influenced theoretical allegiances. The stronger a scientist's identification with the community, the more likely he was to defend or conform to its prevailing phlogistic faith. Conversely, so long as a scientist was sufficiently interested to take any position at all, the weaker his affiliation with the community, the more likely he was to be attracted to the Lavoisierian bandwagon. These findings are in accord with T. S. Kuhn's scheme of scientific revolutions. As Kuhn maintains, scientists seem especially susceptible to extrascientific influences when confronted with a choice between incommensurable paradigms.[5] Indeed, as Kuhn's stu-

*für die theoretische Chemie*, 1 (1800), 3–34; J. B. Trommsdorff, *Versuch einer allgemeinen Geschichte der Chemie*, facsimile of the 1806 ed., pt. 3 (Leipzig: Zentral-Antiquariat, 1965), pp. 109–111; H. Kopp, *Geschichte der Chemie*, vol. I (Brunswick: Vieweg, 1843), pp. 341–342; and A. Ladenburg, *Vorträge über die Entwicklungsgeschichte der Chemie in den letzten hundert Jahren* (Brunswick: Vieweg, 1869), p. 14.

3. G. W. A. Kahlbaum and A. Hoffmann, "Die Einführung der Lavoisier'schen Theorie im besonderen in Deutschland," *Monographieen aus der Geschichte der Chemie*, 1 (1897).

4. While some historians have accepted Kahlbaum's interpretation—see, for instance, J. R. Partington, *A History of Chemistry*, vol. III (London: Macmillan, 1962), pp. 492–493; and H. Schimank, "Theorien der Chemie des 18. Jahrhunderts im Urteil der Zeitgenossen," *Sudhoffs Archiv*, supp. 7 (1966), 152—others have continued to attribute a role to nationalism in the German resistance to Lavoisier—see, for instance, G. Urdang, *Goethe and Pharmacy* (Madison, Wis.: American Institute of the History of Pharmacy, 1949), p. 65; and M. P. Crosland, *Historical Studies in the Language of Chemistry* (London: Heinemann, 1962), p. 208. In reviewing Crosland's book, R. Schmitz called for "modern German studies" of the matter—*Sudhoffs Archiv*, 47 (1963), 500. Recently H. Vopel has argued that the contestants' support and training played a decisive role in determining their positions. Lavoisier's German opponents, he contends, lacked the facilities and physical-mathematical outlook needed for replication and appreciation of his experiments. By contrast, Lavoisier's German allies, who in Vopel's opinion were in relatively favorable positions and had better educations, could more easily test and evaluate the revolutionary theory. See H. Vopel, "Die Auseinandersetzung mit dem chemischen System Lavoisiers in Deutschland am Ende des 18. Jahrhunderts" (Diss., Leipzig University, 1972). Although Vopel's study of the German antiphlogistic revolution is far more thorough than that of Kahlbaum and Hoffmann, he does not make a compelling case for his economic-educational differentiation of the protagonists.

5. T. S. Kuhn, *The Structure of Scientific Revolutions*, 2d ed. (Chicago: University of Chicago Press, 1970), pp. 144–159, 198–204.

dent E. Frankel has suggested, a scientific community may remain complacent in the face of anomalies unless and until one or more members, motivated by extrascientific as well as scientific considerations, engender a crisis mentality by vociferously advocating a new paradigm.[6]

## LAVOISIER AND THE ANTIPHLOGISTIC REVOLUTION IN FRANCE

Between the early 1760s and mid-1780s, as various gases were distinguished from one another and their chemical properties investigated, chemists throughout Europe recognized that existing theory did not account for the new findings. Most believed that it would be possible to explain the results of pneumatic chemistry by making this or that refinement in the concept of phlogiston. Lavoisier, by contrast, soon concluded that mere tinkering would never make sense of the new data. A revolutionary theory was needed. Ultimately he convinced chemists that he was correct.[7]

It was in 1766 that Lavoisier began thinking about the chemical properties of air. Six years later he carried out several remarkable experiments involving the combustion of phosphorus and sulfur and the reduction of lead calx. By November 1772 he was claiming that his findings were among the most noteworthy obtained "since Stahl;" and in February 1773 he was predicting that his work would bring about "a revolution in physics and chemistry."[8] However, during the next decade his ideas were constantly evolving in response to a flood of new discoveries. Not until 1783, after he had repeated Cavendish's production of water from pure air (oxygen) and inflammable air (hydrogen), did Lavoisier settle on the essentials of his system.

Lavoisier's system differed from competing theories in three major ways. First, it paid close attention to the weights of the reactants and products in chemical reactions. Second, it portrayed oxygen gas, not the burning substance, as the source of the heat accompanying combustion. And third, it regarded metals, carbon, sulfur, and phosphorus as simpler substances than the products resulting from their calcination or combustion.

6. E. Frankel, "Corpuscular Optics and the Wave Theory of Light: The Science and Politics of a Revolution in Physics," *Social Studies of Science*, 6 (1976), 141–184.

7. Presented as background, this short section is essentially derivative. For a comprehensive but somewhat chaotic description of the original publications, see J. R. Partington, *A History of Chemistry*, vol. III. For a thoughtful portrayal of Lavoisier's scientific development and a valuable review of the secondary literature, see H. Guerlac, "Lavoisier, Antoine-Laurent," *Dictionary of Scientific Biography*, ed. C. C. Gillispie et al., vol. VIII (New York: Scribner's, 1973), pp. 66–91. For a recent quantitative analysis of the responses to Lavoisier's theory in the French and British periodical literature down to 1795, see H. G. McCann, *Chemistry Transformed: The Paradigmatic Shift from Phlogiston to Oxygen* (Norwood, N.J.: Ablex, 1978). McCann argues that young chemists enjoying personal contacts with Lavoisier were most likely to be receptive to his theory and older chemists lacking such contacts were most likely to be resistant. His case for this prosaic thesis is no more compelling than that based on familiar literary evidence, primarily because he took articles rather than individuals as his ultimate unit of analysis.

8. H. Guerlac, *Lavoisier—The Crucial Year: The Background and Origin of His First Experiments on Combustion in 1772* (Ithaca, N.Y.: Cornell University Press, 1961), pp. 223, 230.

Confident of the truth of his new system, Lavoisier set out to persuade others of its merits. He began with a striking experiment. Having synthesized water from hydrogen and oxygen, he sought to complete the proof of its compound nature. To do so, he passed steam over red-hot iron. As he expected, he obtained inflammable air. Moreover, the iron, which was chemically altered on its surface, increased in weight. His interpretation was that the hot iron decomposed the water by combining with its oxygen and releasing its hydrogen in the form of inflammable air. During the years 1784 and 1785 Lavoisier carried out several public demonstrations of the synthesis and decomposition of water. He also presented several memoirs to the Paris Academy of Sciences, showing that his theory solved several of chemistry's outstanding problems. In early 1785 his efforts to win adherents began to succeed. P. S. Laplace and other Parisian mathematicians rallied to his side. In April the prominent chemist C. L. Berthollet announced his support for Lavoisier's views.[9] Heartened by these successes, Lavoisier made a frontal attack on the concept of phlogiston. In a memoir read before the Academy in June 1785, he criticized phlogistonists for adhering to contradictory theories and denounced phlogiston as a figment of their imagination.

During the next two years, support for Lavoisier's theory continued to grow within the French cultural area. J. B. van Mons, a young chemist in Brussels, published a book on the "antiphlogistic chemistry."[10] A. F. Fourcroy, an influential Parisian chemist, began proselytizing for Lavoisier's system in his courses and texts.[11] In early 1787 the French antiphlogistic revolution reached its turning point with the full conversion of L. B. Guyton de Morveau, an important chemist of Dijon who was visiting Paris to decide whether to base his proposals for a reform of chemical terminology on the new doctrines.[12] In the ensuing months, Lavoisier, Guyton de Morveau, Berthollet, Fourcroy and others created a vocabulary for chemistry that mirrored their view of chemical reality. These men named all compounds according to their constituents and the simple gases, such as oxygen and hydrogen, according to their properties. In April 1787 Lavoisier read the preliminary discourse of the *Méthode de Nomenclature Chimique* to the Academy. Henceforth, he and his allies used the new nomenclature in all their publications, compelling chemists interested in their work to learn how to think within the antiphlogistic framework.

Lavoisier and his circle did not stop there. In 1788 they cooperated in a

---

9. H. E. LeGrand, "The 'Conversion' of C.-L. Berthollet to Lavoisier's Chemistry," *Ambix*, 22 (1975), 58–70.

10. J. R. Partington, *A History of Chemistry*, III:491.

11. W. A. Smeaton, "Fourcroy, Antoine François de," *Dictionary of Scientific Biography*, vol. V (1972), pp. 89–93. For a hostile report from a German who attended Fourcroy's lectures in 1786, see *Magazin für die Naturkunde Helvetiens*, 1 (1787), 241–242.

12. W. A. Smeaton, "Guyton de Morveau, Louis Bernard," *Dictionary of Scientific Biography*, vol. V (1972), pp. 600–604. Prior to 1787, as Smeaton has recently reiterated, Guyton had already enbraced many tenets of Lavoisier's system. See *Annals of Science*, 36 (1979), 666–668. Soon after giving Lavoisier his unqualified support, Guyton explained his conversion to Crell. See the *Chemische Annalen*, no. 2 (1787), 54–55.

detailed refutation of R. Kirwan's version of the phlogiston theory. Later that year, inspired by Crell's *Chemische Annalen* and desiring a congenial forum for their work, they decided to found a journal for chemistry, the *Annales de Chimie*.[13] And in the spring of 1789, Lavoisier published a systematic account of his theory in the *Traité Elémentaire de Chimie*.

So far as Lavoisier and his friends were concerned, the publication of the *Traité* marked the end of the antiphlogistic revolution in France. Fourcroy, for instance, claimed that Lavoisier's experiments had brought about a revolution in chemistry some years before and asserted that those who still supported phlogistic theories were either behind the times or lacking in skill.[14] Full of pride, Lavoisier informed his aged acquaintance B. Franklin that

> French scientists are divided at present between the old and new doctrines. On my side I have Messrs. de Morveau, Berthollet, de Fourcroy, Laplace, Monge and generally the physicists of the Academy. The scientists of London and England are also gradually dropping Stahl's theory, but German chemists adhere to it strongly. There is then a revolution in an important branch of human knowledge accomplished since you left Europe.[15]

Although he recognized the existence of opposition, Lavoisier regarded the antiphlogistic revolution as an accomplished fact.

## THE INITIAL GERMAN RESPONSE (MID-1785 TO MID-1789)

While the antiphlogistic revolution had run its course in France by mid-1789, it had not yet begun in Germany. Chemists there presumed that one or another version of the phlogiston theory would succeed in resolving the puzzles that had emerged from pneumatic chemistry. Crell and others were promoting Kirwan's system, which identified phlogiston with inflammable air.[16]

---

13. For the founding and subsequent history of the *Annales*, see S. Court, "The *Annales de Chimie*, 1789–1815," *Ambix*, 19 (1972), 113–128. Crell was proud to have been imitated by the "most respected and famous men of such an enlightened nation." See his "Vorbericht," *Chemische Annalen*, no. 2 (1789), [vi].

14. H. Guerlac, "The Chemical Revolution: A Word from Monsieur Fourcroy," *Ambix*, 23 (1976), 1–4. In a similar spirit, G. Monge referred to the French chemists who opposed Lavoisier as "*incrédules*." See Monge, letter to J. Watt (10 December 1789), in J. P. Muirhead, *The Origin and Progress of the Mechanical Inventions of James Watt*, vol. II (London: Murray, 1854), pp. 237–238.

15. Lavoisier, letter to B. Franklin (2 February 1790), in D. I. Duveen and H. S. Klickstein, "Benjamin Franklin (1706–1790) and Antoine Laurent Lavoisier," *Annals of Science*, 11 (1955), 123.

16. In 1787 Kirwan listed Crell, Wiegleb, Westrumb, Hermbstaedt, and W. Karsten as Germans who agreed with him in identifying the base of inflammable air with phlogiston. See his "Versuch über das Phlogiston, und die Bestandtheile der Säure," *Beyträge zu den chemischen Annalen*, 3 (1787), 133. His support in Germany was more extensive than he realized. In Göttingen, for instance, the chemist J. F. Gmelin, the physicist G. C. Lichtenberg, and the naturalist J. F. Blumenbach were all adherents of his views. See [Gmelin], review of Kirwan's *Physisch-chemische Schriften* (vol. III, 1788), *Allgemeine deutsche Bibliothek*, 88 (1789), 184–187; Lichtenberg, letter to Blumenbach (15 June 1786), in *Georg Christoph Lichtenberg:*

F. A. C. Gren was championing a theory based on the hypothesis that phlogiston had negative weight.[17] Very few German chemists regarded Lavoisier's system as a contender in this competition. And not a single one was a proponent of antiphlogistic doctrines.[18]

The slowness of German chemists to appreciate that Lavoisier's system threatened all phlogistic theories cannot be attributed to geography or language.[19] Soon after Lavoisier attacked phlogiston on the floor of the Paris Academy, German chemists were informed of his position. The news first arrived from Britain. In mid-1785 Kirwan wrote Crell, advising phlogistonists to reconcile their differences so that they could respond to Lavoisier's challenge. He hoped German chemists would "undertake a basic investigation of this theory [that metals, sulfur, and charcoal do not contain phlogiston], for if it should be established, all that we have hitherto regarded as well founded will be overturned."[20] In 1786 J. C. de la Métherie urged Crell and

---

*Schriften und Briefe*, ed. W. Promies, vol. IV (Munich: Hanser, 1967), p. 676; and Blumenbach, *Institutiones Physiologicae* (Göttingen: Dieterich, 1787), p. 112. Vopel also identifies C. F. Bucholz, J. F. A. Göttling, M. H. Klaproth, J. G. Leonhardi, F. A. Richter, and G. L. C. Rousseau as Kirwan's adherents. See Vopel, "Die Auseinandersetzung mit dem chemischen System Lavoisiers in Deutschland," pp. 70–76.

17. Gren, who regarded Kirwan's system as the prevailing theory, devoted considerable energy to discrediting it as he advanced his own views. See Gren, "Versuche und Beobachtungen über die Entstehung der fixen und phlogistischen Luft," *Beyträge zu den chemischen Annalen*, 2 (1787), 296–330, 424–444; 3 (1788), 229–259; and his "Ueber Luft, Brennstoff und Metallkalke," in Westrumb, *Kleine physikalisch-chemische Abhandlungen*, vol. III, pt. 1 (Leipzig: Müller, 1789), pp. 415–479. For Gren's theory, also see J. R. Partington and D. McKie, "Historical Studies on the Phlogiston Theory," *Annals of Science*, 3 (1938), 1–58.

18. In addition to Lavoisier, Kirwan also remarked on the steadfastness of German chemists. See Kirwan, letter to Crell, *Chemische Annalen*, no. 2 (1789), 221. In saying that there were no antiphlogistonists in Germany by mid-1789, I am excluding the town of Strasbourg, which in chemistry as in other things was becoming increasingly French in orientation. See F. L. Ford, *Strasbourg in Transition, 1648–1789* (Cambridge, Mass.: Harvard University Press, 1958). At least two residents of Strasbourg embraced Lavoisier's doctrines before mid-1789: Baron P.-F. de Dietrich (1743–1793) and F. L. Schurer (1764–1794). The date of Dietrich's conversion is not known. See D. I. Duveen and H. S. Klickstein, "A Letter from Guyton de Morveau relating to Lavoisier's attack against the *phlogiston* theory (1778); with an account of de Morveau's conversion to Lavoisier's doctrines in 1787," *Osiris*, 12 (1956), 364. For Schurer's conversion, see his *Synthesis oxygenii experimentalis confirmata* (Strasbourg: Heitz, colophon 19 January 1789); and S. F. Hermbstaedt's review of Schurer's book, *Bibliothek der neuesten physisch-chemischen, metallurgischen, technologischen und pharmaceutischen Literatur*, 3, no. 3 (1791), 315.

19. To be sure, communication barriers did prevent German chemists from learning of Lavoisier's private doubts about phlogiston as soon as his countrymen, some of whom recognized his growing skepticism as early as 1778. On 15 January 1778, for instance, P. J. Macquer wrote Guyton de Morveau of his fear that Lavoisier would succeed in overthrowing the "theory of phlogiston or combined fire." See D. I. Duveen and H. S. Klickstein, "A letter from Guyton de Morveau relating to Lavoisier's attack against the *phlogiston* theory (1778)," *Osiris*, p. 347. Thanks to a lively cross-channel communication, some British chemists became aware of Lavoisier's doubts in the early 1780s. Thus, on 24 June 1782, J. Priestley wrote B. Franklin asking him to inform the Duke of la Rochefoucauld "that my experiments are certainly inconsistent with Mr. Lavoisier's supposition of there being no such thing as *phlogiston*." See D. I. Duveen and H. S. Klickstein, "Benjamin Franklin (1706–1790) and Antoine Laurent Lavoisier," *Annals of Science*, p. 119.

20. Kirwan, letter to Crell, *Chemische Annalen*, no. 2 (1785), 337.
21. Métherie, letter to Crell, *Chemische Annalen*, no. 1 (1786), 330–331.

"other German chemists [to] give their judgment on this important issue."[21] Any German chemist whose concern about the direction of chemistry was aroused by these general warnings could have pieced together Lavoisier's system from his articles in the Paris Academy's *Mémoires* or, lacking this expensive journal, from translations published by Crell and C. E. Weigel.[22] But there were few German chemists with this much concern.[23] And there was none who, having learned how Lavoisier explained chemical phenomena without recourse to phlogiston, took the theory seriously enough to subject it to the searching scrutiny proposed by Kirwan.

German chemists took so little interest in Lavoisier's theory because they viewed it as just another French speculation. In 1787 J. G. A. Höpfner and S. F. Hermbstaedt both published a letter from a German visiting Paris who, after listing some of the main points in Lavoisier's system, commented

> you may wonder, good friend, whether he proves all these beautiful things? Sure he proves them, just like everybody proves things in France—that is, he talks perpetually and proves nothing. His oxygen system is his spider's web which, though it is woven with artistry and care, is useless.[24]

The next year saw several such comments. Hermbstaedt derided Lavoisier's numerous followers "among the French" as "haughty" men who "opposed the principles of sound reason."[25] Reviewing the new nomenclature, J. F. Gmelin remarked that "in Germany at least, this system is not yet regarded as proven or firmly grounded, despite the ability of those who have introduced it. Consequently the terminology runs the danger of falling in a heap with the system."[26] J. G. Leonhardi suggested that Lavoisier and his friends denied the existence of phlogiston not "from conviction, but rather from obstinacy and love of novelty."[27] Gren wrote of "the deceptive and appealing

---

22. For the availability of Lavoisier's books and articles in French and German, see D. I. Duveen and H. S. Klickstein, *A Bibliography of the Works of Antoine Laurent Lavoisier* (London: Dawson, 1954). Summaries of Lavoisier's views were also available by 1788 in A. F. Fourcroy, *Handbuch der Naturgeschichte und der Chemie*, trans. P. Loos, with notes by J. C. Wiegleb, vols. I–II (Erfurt: Keyser, 1786); and M. van Marum, *Beschreibung einen ungemein grossen Elektrisier-Maschine*, trans. H. W. C. Eschenbach, 1st continuation (Leipzig: Schwickert, 1788), pp. 47–72.

23. While traveling in England and Holland during 1787/88, Göttling recognized the need for German chemists to become familiar with Lavoisier's theory. See his "Lavoisier's Theorie über Verbrennung, das Athemholen der Thiere, Entstehung der Säuren, und Verkalkung der Metalle, als kurze Uebersicht zusammengetragen," *Almanach oder Taschenbuch für Scheidekünstler und Apotheker*, 10 (1789), 79–120, which appeared in the fall of 1788.

24. Letter to Höpfner, *Magazin für die Naturkunde Helvetiens*, 1 (1787), 240–244; reprinted in the *Bibliothek der neuesten physisch-chemischen . . . Literatur*, 1, no. 2 (1787), 54–55. In reprinting the letter, Hermbstaedt deleted some inaccurate statements about Lavoisier's theory and the assertion that all theorizing was a waste of time.

25. Hermbstaedt, review of R. Kirwan's *An Essay on Phlogiston* (1787), *Bibliothek der neuesten physisch-chemischen . . . Literatur*, 1, no. 3 (1788), 274.

26. [Gmelin], review of L. B. Guyton de Morveau et al.'s *Méthode de Nomenclature Chimique* (1787), *Göttingische Anzeigen von gelehrten Sachen* (5 January 1788), 15–16.

27. P. J. Macquer, *Chymisches Wörterbuch; oder, Allgemeine Begriffe der Chymie nach alphabetischer Ordnung*, ed. and trans. J. G. Leonhardi, 2d German ed. of 2d French ed., vol. II (Leipzig: Weidmann, 1787), p. 287.

sophistries of the French *Chemisten.*"[28] And J. C. Wiegleb made light of "the completely improbable hypotheses of Lavoisier."[29]

Kahlbaum and Hoffmann notwithstanding, nationalistic antipathy to French culture certainly conditioned the initial response of the German chemical community to Lavoisier's system. It made most German chemists averse to learning the theory. And it deterred the few who did so from going on to submit the theory to rigorous testing.

## FROM CONSENSUS TO SECTARIANISM (MID-1789 TO MID-1792)

Starting in the summer of 1789, more and more German chemists recognized that the antiphlogistic system could not be dismissed. Lavoisier's publication of the *Traité* made it easy to learn his views. And Lavoisier's triumph in France and gains in Britain and the Netherlands made it clear that his theory had a possibility of emerging triumphant. As time passed, the new French theory began to make inroads in Germany itself. In the summer of 1791, Crell informed an Italian colleague that, though "the old system does not want for defenders such as Westrumb, Klaproth, Gren, and Wiegleb, . . . the antiphlogistic theory has found partisans in Hermbstaedt, Girtanner, Link, etc."[30] About the same time Wiegleb decided to compose a lengthy defense of phlogiston, because "Lavoisier's new chemical system has recently received the approval of several [men] in Germany."[31] And C. Girtanner observed that the controversy had divided "the chemists into two different sects."[32] By the spring of 1792, H. F. Link felt sufficiently confident to crow that

> the antiphlogistic theory . . . was originally despised and misunderstood by German chemists. As is only just, it is now getting revenge—the number of its adherents grows daily, many of its most important opponents now give it their approval, and it threatens the old chemistry with a complete overthrow, at least a total transformation.[33]

A month later Hermbstaedt proudly announced that M. H. Klaproth favored Lavoisier's system.[34] The conversion of Germany's most respected analyst

---

28. Gren, "Ueber Luft, Brennstoff und Metallkalke," in J. F. Westrumb, *Kleine physikalisch-chemischen Abhandlungen*, III, pt. 1, p. 417.
29. A. F. Fourcroy, *Handbuch der Naturgeschichte und der Chemie*, p. 393.
30. Crell, letter to L. V. Brugnatelli (28 August 1791), *Annali di Chimica*, 3 (1791), 47.
31. Wiegleb, "Beweissgründe des geläuterten Stahlischen Lehrbegrifs vom Phlogiston, und der Grundlosigkeit des neuen chemischen Systems der Franzosen," *Chemische Annalen*, no. 2 (1791), 388.
32. [Girtanner], review of W. Higgins's *A Comparative View of the Phlogistic and Antiphlogistic Systems* (1791), *Göttingische Anzeigen von gelehrten Sachen* (20 October 1791), 1681.
33. A. L. Lavoisier, *Physikalisch-chemische Schriften*, vol. IV, ed. H. F. Link (Greifswald: Röse, preface 1 March 1792, publ. 1792), p. iii.
34. A. L. Lavoisier, *System der antiphlogistische Chemie*, trans. S. F. Hermbstaedt, vol. I (Berlin and Stettin: Nicolai, foreword 18 April 1792, publ. 1792), p. [x].

established that the French theory had real prospects of emerging from the contest with the German chemical community's allegiance.

But why, once they could easily familiarize themselves with Lavoisier's system, did most German chemists remain phlogistonists? And why did a few German chemists, and several friends of chemistry, decide to become antiphlogistonists? Kahlbaum and Hoffmann demonstrated that such decisions were variously influenced by doubt about the underpinnings of the French theory or by respect for its achievements and promise. But they failed to demonstrate that theoretical allegiances were independent of other influences. Considered in the context of Kuhn's interpretation of scientific revolutions, this failure is not surprising. If scientists experience rival theories as incommensurable paradigms, their allegiances are likely to be subject to influences that are extrinsic to the theories. This hypothesis is supported by a close look at phlogistic loyalties and antiphlogistic conversions in Germany between mid-1789 and mid-1792.

**Phlogistic Loyalties**

Cultural nationalism, which, as we have seen, influenced most German chemists to dismiss Lavoisier's theory between mid-1785 and mid-1789, continued to exercise an influence in the ensuing years. In late 1789 Gren claimed that his version of the phlogiston theory had "rescued the German Stahlian theory of phlogiston."[35] In 1790 Crell was hoping that a defense "for our phlogiston, which the French want to steal from us" would soon be forthcoming from "a *vollgültigen deutschen Chemisten*."[36] In 1791 Leonhardi maintained that "the whole new French system still rests on very shaky ground."[37] In early 1792 Wiegleb commented that it was remarkable that "a German chemist" should be "a proselyte of the French."[38] And that summer, J. F. Westrumb, who had earlier demeaned "the often lightweight discoveries of our neighbors," proclaimed that he was a "zealous adherent of the German or Stahlian school."[39]

Besides cultural nationalism, peer pressure within the German chemical community motivated phlogistic loyalties. The chemist who adhered to prevailing opinion could expect that his work would be evaluated according to customary criteria. By contrast, the chemist who failed to conform could anticipate a severe reception to all his endeavors. In 1791 peer pressure compelled the antiphlogistonist Hermbstaedt to concede that it would have been

---

35. F. A. C. Gren, *Systematisches Handbuch der gesammten Chemie*, vol. II, pt. 2 (Halle: Waisenhaus, preface 20 November 1789, publ. 1790), p. 78.
36. Crell, letter to K. G. Hagen, in A. Hagen, "K. G. Hagen's Leben und Wirken," *Neue preussische Provinzial-Blätter*, 9 (1850), 84; and L. C., review of H. F. Link's *Einige Bemerkungen über das Phlogiston* (1790), *Chemische Annalen*, no. 2 (1790), 478.
37. Macquer, *Chymisches Wörterbuch*, vol. VII (1791), p. 33.
38. [Wiegleb], review of Hermbstaedt's *Systematischer Grundriss der allgemeinen Experimentalchemie* (1791), *Allgemeine deutsche Bibliothek*, 106 (1792), 217–220.
39. Westrumb, letter to Crell (1 October 1790), *Chemische Annalen*, no. 2 (1790), 522; and "Einige Bemerkungen, verschiedene Gegenstände der neuen Chemie betreffend," ibid., no. 2 (1792), 3.

"impertinent" to have based his new textbook solely on Lavoisier's system.[40] The following year it induced young J. B. Trommsdorff to declare that "I am no antiphlogistonist" when, for the sake of brevity, he called a new neutral salt discovered by Berthollet "*sel oxygène*."[41] Indeed, the pressure to conform was such that most of the community's ambitious young recruits remained phlogistonists throughout the first phase of the German antiphlogistic revolution.[42]

Not only cultural nationalism and peer pressure but also status anxiety inclined some of the German chemical community's informal leaders to stand by phlogiston. They felt that a threat to doctrines they had been propounding for years was a threat to their legitimacy as leaders. Of course, they could hardly give direct expression to such feelings. Still in 1791, as the jeopardy became more palpable, they revealed their vested interest in phlogiston. Wiegleb, the community's doyen, urged his colleagues to avoid "discovery mania" and "excessive partiality for innovations."[43] Crell, the community's convener, argued against the new terminology on the grounds that young chemists still needed the old terminology to read the indispensable works of their forerunners.[44] And Westrumb, whose recent discovery that powdered metals burn in chlorine gas had reinforced his high stature in the community, assured his fellow chemists that "we need fear no reform of our concepts, no overthrow of our painstakingly built system."[45]

Thus, those German chemists who decided to remain loyal to phlogiston were not motivated by skepticism alone. They were also influenced by pride in the German origin of the concept of phlogiston, by repugnance for the French origin of the antiphlogistic system, by sensitivity to prevailing opinion among their peers, and by fear of losing hard-earned positions of leadership.

40. Hermbstaedt, *Systematischer Grundriss der allgemeinen Experimentalchemie*, vol. I (Berlin: Rottmann, foreword 26 April 1791, publ. 1791), p. vi.

41. Trommsdorff, "Ueber das neue Neutralsalz aus dephlogistisirter Salzsäure und Pflanzenalkali," *Chemische Annalen*, no. 1 (1792), 423.

42. Among the younger recruits to chemistry who remained loyal to phlogistic doctrines during the early 1790s were J. B. Richter (b. 1762), G. F. Hildebrandt (b. 1764), F. Wurzer (b. 1765), J. B. Trommsdorff (b. 1770), W. A. E. Lampadius (b. 1772), and C. H. Pfaff (b. 1773). For biographical profiles of these men, all of whom eventually achieved some renown as chemists, see Appendix I.

43. Wiegleb, *Geschichte des Wachstums und der Erfindungen in der Chemie in der neuern Zeit*, II:424, 502. His disciple Hermbstaedt complained that if this attitude prevailed, if "sophistry continued to distort new discoveries . . . as it had during the last decade, chemistry would not attain its proper goal in 1000 years." See his review of Wiegleb's book, *Bibliothek der neuesten physisch-chemischen . . . Literatur*, 4, no. 1 (1792), 121.

44. Crell, "Ueber die Nothwendigkeit einer chemisch-technischen Sprach-Veränderung, und ihre Gesetze," *Chemische Annalen*, no. 1 (1791), 335.

45. Westrumb, *Geschichte der neu entdeckten Metallisirung der einfachen Erden* (Hanover: Helwing, preface 5 March 1791, publ. 1791), p. 142. For Westrumb's discovery that metals burn in chlorine, see his "Neue Bemerkungen über einige merkwürdige Erscheinungen durch die dephlogistisierte Salzsäure," *Chemische Annalen*, no. 1 (1790), 3–21, 109–129. Like Westrumb, Crell believed this discovery provided powerful confirmation for phlogiston. See Crell, letter to G. V. M. Fabbroni (4 March 1790), Library of the American Philosophical Society, Philadelphia; and Crell, "Vorbericht," *Chemische Annalen*, no. 2 (1790), iii.

## Antiphlogistic Conversions

If cultural nationalism, peer pressure, and status concerns helped motivate phlogistic loyalties, we would expect the early German followers of Lavoisier to have been men who were less susceptible than most German chemists to such influences. This expectation is borne out by the pattern of antiphlogistic conversions in Germany.

During the years 1789 and 1790, Lavoisier's theory received a warmer welcome in Vienna than elsewhere in Germany. In fact, the antiphlogistic revolution passed its turning point there by the summer of 1790 (see Table 20).[46] One reason for Lavoisier's rapid acceptance in Vienna was that cosmopolitanism characterized the cultural life of the polyglot Hapsburg capital.[47] Accordingly, I. von Born, N. J. von Jacquin, and J. A. Scherer, initiators of the Hapsburg antiphlogistic movement, were not taken aback by the system's provenance. If anything, Jacquin, a Dutchman of French ancestry, and Born, a partisan of the French Revolution, may have counted the theory's French origin as a point in its favor.[48] Allied to the initiators' cosmopolitanism, or Francophilia, was an indifference to prevailing opinion in the German chemical community. Though all three had names as chemists, Jacquin was better known as a botanist and Born as a mineralogist. Though all three probably kept up with Crell's journal, Jacquin and Scherer were not even among its occasional contributors. Their marginality within the community gave them considerable license. Jacquin, for instance, had been sufficiently independent to question the presence of phlogiston in metals during the 1780s and so isolated as to receive no comment whatsoever from beyond the walls of Vienna.[49] Thus, not only immersed in a cosmopolitan cultural environment but also unconstrained by loyalty to or pressure from the German

---

46. J. Haubelt has drawn attention to the early acceptance of Lavoisier's theory in the Hapsburg Empire. See his "Les Idées de Lavoisier en Europe centrale au xviii$^e$ siècle," XII$^e$ Congrès international d'histoire des sciences, 1968: *Résumés des communications*, p. 92. In 1793–1794 Viennese chemists were the first in Europe to use the new French nomenclature in an official pharmacopoeia. See K. Ganzinger, "Die Übernahme von Lavoisiers neuer chemischer Nomenklatur in das österreichische Arzneibuch von 1794," *Sudhoffs Archiv*, 58 (1974), 303–311.

47. For lively descriptions of Vienna, see F. Nicolai, *Beschreibung einer Reise durch Deutschland und die Schweiz im Jahre 1781*, vol. III (Berlin and Stettin: Nicolai, 1784); H. Sander, *Beschreibung seiner Reisen*, vol. II (Leipzig: Jacobaer, 1784), pp. 465–519; and R. Townson, *Travels in Hungary with a Short Account of Vienna in the Year 1793* (London: Robinson, 1797), pp. 2–31.

48. One indication of their cosmopolitanism was that Born, Jacquin, and Scherer were still publishing much of their work in Latin and French. For Born's radicalism, see, for instance, R. Townson, *Travels in Hungary*, pp. 410–421. For profiles of Born and Jacquin, see Appendix I.

49. Jacquin based his argument against the presence of phlogiston in metals primarily on the fact that mercury calx could be reduced without a reducing agent. See his *Anfangsgründe der medicinisch-practischen Chymie* (Vienna: Wappler, 1783), pp. 322–329, 335–337. Though he doubted that phlogiston was a component of metals, Jacquin did regard it as a constituent of sulfur and other combustibles (pp. 289–299). The only persons to comment on his views in print were his students N. C. Molitor and Scherer, both of whom adopted his position. See J. Ingen-Housz, *Vermischte Schriften physisch-medicinischen Inhalts*, trans. N. C. Molitor, 2d ed., vol. II (Vienna: Wappler, 1784), pp. 27–32, 101–103; and J. A. Scherer, *Geschichte der Luftgüteprüfungslehre für Aerzte und Naturfreunde*, vol. I (Vienna: Wappler, 1785), pp. 30–32, 74–82, 112–113.

chemical community, Born, Jacquin, and Scherer seem to have found it easy to recognize the favorable outlook for and scientific merits of Lavoisier's system. Once they had become antiphlogistonists, lesser figures within their orbits in Vienna, Schemnitz, and Prague seem to have found it even easier to follow their lead.[50]

Meanwhile, uninfluenced by the Hapsburg antiphlogistonists who made but one indirect effort to spread their views,[51] converts were slowly appearing in Protestant Germany (see Table 21). Like their Hapsburg counterparts, these converts seem not to have been troubled by the French origin of the new theory. Most were living in major intellectual centers which, though far less cosmopolitan than Vienna, were far more receptive to French ideas than the towns inhabited by the majority of German chemists.[52] In fact, F. Wolff in Berlin was so receptive to French influences that he exclaimed, "if the new French discoveries can be completely confirmed, our present system will undergo a revolution just as significant for the chemist and physicist as the French political revolution is for the friend of mankind."[53] It seems likely, therefore, that attachment to cosmopolitan ideals or enthusiasm for the French Revolution counterpoised whatever sympathies the converts in Protestant Germany may have entertained for the cultural nationalism that was so strong within the German chemical community.

More important than the converts' feelings about the new theory's French provenance were their feelings about prevailing opinion among German chemists. Like the Hapsburg antiphlogistonists, several converts in Protestant Germany had no more than peripheral affiliations with the German chemical community. J. T. Mayer, J. F. Blumenbach, C. F. Hindenburg, and G. S. Klügel were established scientists whose work centered outside chemistry. Link, Wolff, E. W. Martius, A. von Humboldt, and F. Baader were still in the process of defining their interests and making their names. On account of their marginality, these nine men seem not to have felt beholden to or

---

50. There is considerable evidence of personal ties among the Hapsburg antiphlogistonists. Jacquin was a former teacher of Born, Mikan, Prochaska, Meidinger, and Scherer. Born was a patron and masonic brother of Ruprecht and Haidinger. Scherer was a former student of Mikan. For further information, see the biographical sources listed in Table 20.

51. Thinking that his friend Ruprecht had reduced the simple earths to metals, Born evidently planned to use Ruprecht's experiments in support of Lavoisier's theory once he had gained acceptance for them among German chemists. The stratagem was never implemented, however, because Klaproth and Westrumb demonstrated that Ruprecht's "new" metals were really compounds of iron (from the crucible) and phosphoric acid (from the bone ash used as a reducing agent). For a fairly complete history of the episode, see Westrumb, *Geschichte der neu entdeckten Metallisirung der einfachen Erden*.

52. Berlin, Göttingen, Leipzig, and Halle were much more cosmopolitan than such places as Helmstedt (Crell's residence), Langensalza (Wiegleb's residence), and Hameln (Westrumb's residence). Even Freiberg, with its famous mining school, was fairly cosmopolitan. In fact, one of Humboldt's close friends there was the Spanish antiphlogistonist A. del Rio. See *Die Jugendbriefe Alexander von Humboldts 1787–1799*, ed. I. Jahn and F. G. Lange (Berlin: Akademie Verlag, 1973), pp. 161–162, 165.

53. F. L. Schurer, *Abhandlung vom Säurestoff und seiner Verbindung mit andern Körpern*, trans. [F. Wolff] (Berlin: Petit, 1790), pp. ix–x. This anonymous translation was inspired by and dedicated to Hermbstaedt, who probably concurred with Wolff's pro-French sentiments.

TABLE 20: Acceptance of the Antiphlogistic Theory in Hapsburg Germany, 1789–1792

| Convert's name | Conversion date | Place | Age | Primary occupation | Presence in Crell's Journal, 1784–1789[a] | | |
|---|---|---|---|---|---|---|---|
| | | | | | Contributions | Chemical publications reviewed | References received |
| I. von Born[b] | 1789/90 | Vienna | 48 | Mining councillor | 3 | 1 | 10 |
| N. J. von Jacquin[c] | 1789/90 | Vienna | 63 | Professor of botany and chemistry | 0 | 2 | 0 |
| J. A. Scherer[c] | 1789/90 | Vienna | 35 | Physician | 0 | 2 | 0 |
| F. von Tihavsky[d] | 1790 | Vienna | 29 | Lieutenant in the artillery | 0 | 0 | 0 |
| K. von Meidinger[e] | 1790/91 | Vienna | 40 | Minor official | 0 | 0 | 0 |
| A. von Ruprecht[f] | 1790 | Schemnitz | ca. 35 | Professor of chemistry | 0 | 2 | 0 |
| C. Haidinger[f] | 1790 | Schemnitz | 34 | Professor of mining | 0 | 1 | 0 |
| J. G. Mikan[g] | 1790 | Prague | 48 | Professor of botany and chemistry | 0 | 0 | 0 |
| G. Procháska[g] | 1790 | Prague | 41 | Professor of anatomy and physiology | 0 | 0 | 0 |

[a]Based on the contents of the *Chemische Annalen* (1784–1789) and the *Beyträge zu den chemischen Annalen* (1785–1789). "Chemical publications reviewed" refers to articles as well as books. "References received" indicates the number of Crell's German contributors living in 1789 who referred, at least once, to the person in a contribution made between 1784 and 1789.

[b]Born: He announced his conversion to Lavoisier's system in his *Catalogue méthodique et raisonné de la Collection des Fossiles de Mlle Éléonore de Raab*, vol. I (Vienna: Degen, 1790), p. [vii]. However, as J. Haubelt has suggested, he probably embraced the antiphlogistic system in 1789. See J. Haubelt, "Bornovy běličské pokusy," *Dějiny věd a techniky*, 3 (1968), 158–168. For a biographical profile of Born, see Appendix I.

[c]Jacquin and Scherer: Their adherence to Lavoisier's theory was announced by Scherer in A. P. Nahuy, *Chymische Abhandlung von der Entstehung des Wassers aus der Verbindung der Grundstoff der reinen und brennbaren Luft*, trans J. A. Scherer (Vienna: Wappler, foreword 1 June 1790, publ. 1790), dedication letter and pp. iii–viii. For a biographical profile of Jacquin, see Appendix I, and for Scherer, see *Neuer Nekrolog der Deutschen*, 22, no. 1 (1844), 355–359.

[d]Tihavsky: He used the new nomenclature in his "De metallis e Terris obtinendis," in *Collectaneis ad botanicam, chemicam et historiam naturalem spectantibus*, ed. N. J. von Jacquin, vol. IV (Vienna: Wappler, 1790), pp. 3–36. I am indebted to Winter, Director of the Kriegsarchiv in Vienna, for biographical information concerning Tihavsky (1761, Leopoldstadt–1822, Vienna).

[e]Meidinger: The earliest evidence of his commitment to the antiphlogistic system was his "Pränumerationsnachricht, an die Liebhaber der Chemie" (24 December 1791), *Allgemeine Literatur-Zeitung: Intelligenzblatt* (February 1792), 150–152. For Meidinger, See C. Wurzbach, *Biographisches Lexikon des Kaiserthumes Oesterreich*, vol. XVII (Vienna: Hof- und Staatsdruckerei, 1867), pp. 277–278.

⁷Ruprecht and Haidinger: Both revealed their adherence to the antiphlogistic system in published letters. See Ruprecht, letter to Born, *Chemische Annalen*, no. 2 (1790), 11; and Haidinger, letter to Born (20 May 1790), in *Bergbaukunde*, ed. I. von Born and F. von Trebra, vol. II (Leipzig: Goeschen, 1790), p. 461. There is no adequate biography of Ruprecht. Probably born in Eggenberg, he enrolled in the Schemnitz Mining School in 1772, studied in Freiberg and Scandinavia at state expense from 1777 to 1779, served as professor of chemistry in Schemnitz from 1779 to 1792, and spent his remaining years as a high mining official in Vienna. See *Gedenkbuch zur hundertjährigen Gründung der königl. ungarischen Berg- und Forst-Akademie in Schemnitz 1770–1870*, ed. G. Faller (Schemnitz: Joerges, 1871), pp. 27–28, 32, 110; *Festschrift zum hundertjährigen Jubiläum der königl. Sächs. Bergakademie zu Freiberg*, vol. I (Dresden: Meinhold, 1867), p. 228; and V. A. Eyles, "The Evolution of a Chemist," *Annals of Science*, 19 (1963), 163. According to H. L. Mikoletzky, Director of the Finanz- und Hofkammarchiv in Vienna, Ruprecht died in 1814. For Haidinger, see C. Wurzbach, *Biographisches Lexikon*, vol. VII (1861), pp. 206–208.

⁸Mikan and Procháska: J. Haubelt has found evidence that these two Prague professors accepted Lavoisier's theory in 1790. See his "Les Idées de Lavoisier en Europe centrale au xviiᵉ siècle," XIIᵉ Congrès international d'histoire des sciences, 1968: *Résumés des communications*, p. 92. For Mikan and Procháska, see D. Kirndorfer, *Die Personalbibliographien der Professoren . . . der medizinischen Fakultät der Karl- Ferdinands-Universität in Prag . . . 1749–1800* (Diss., Erlangen University; Erlangen: Hogl, 1971), pp. 39–43, 56–61. For Procháska, also see V. Krutz, "Procháska, Georgius," *Dictionary of Scientific Biography*, ed. C. C. Gillispie et al., vol. XI (New York: Scribner's, 1975), pp. 158–160.

TABLE 21: Acceptance of the Antiphlogistic Theory in Protestant Germany, 1789–1792

| Convert's name | Conversion date | Place | Age | Primary occupation | Presence in Crell's Journal, 1784–1789[a] | | |
|---|---|---|---|---|---|---|---|
| | | | | | Contributions | Chemical publications reviewed | References received |
| H. F. Link[b] | 1789 | Göttingen | 22 | Medical student | 2 | 0 | 0 |
| S. F. Hermbstaedt[c] | 1789 | Berlin | 29 | Private science lecturer | 22 | 2 | 9 |
| F. Wolff[d] | 1790 | Berlin | 24 | Science teacher | 0 | 0 | 0 |
| C. Girtanner[e] | 1790 | Göttingen | 29 | Freelance author | 2 | 0 | 5 |
| J. T. Mayer[f] | 1791 | Erlangen | 38 | Professor of physics | 0 | 0 | 0 |
| E. W. Martius[g] | 1791 | Erlangen | 35 | Journeyman pharmacist | 1 | 0 | 1 |
| J. F. Blumenbach[h] | 1791 | Göttingen | 39 | Professor of natural history | 0 | 0 | 1 |
| A. von Humboldt[i] | 1791 | Freiberg | 22 | Mining student | 0 | 0 | 0 |
| F. Baader[j] | 1792 | Freiberg | 27 | Mining student | 0 | 1 | 0 |
| M. H. Klaproth[k] | 1792 | Berlin | 49 | Professor of chemistry | 21 | 0 | 12 |
| C. F. Hindenburg[l] | 1792 | Leipzig | 51 | Professor of physics | 0 | 0 | 0 |
| G. S. Klügel[l] | 1792 | Halle | 58 | Professor of physics | 1 | 0 | 0 |

[a]See note a, Table 20.
[b]Link: The earliest indication of his conversion was an announcement of his plans to translate the *Traité*. See the *Allgemeine Literatur-Zeitung: Intelligenzblatt* (5 August 1789), 793–794. Soon afterwards (26 August 1789), he defended three antiphlogistic propositions during his M.D. examination. See his *Flora Goettingensis* (Göttingen: Grape, 1789/90). His friend F. A. A. Meyer dedicated a poem entitled "Lob des Phlogiston" to him. See the *Teutsche Merkur* (October 1789), 100–104. For a biographical profile of Link see Appendix I.
[c]Hermbstaedt: As early as 20 August 1789, he was trying to interest F. Nicolai in publishing a translation of the *Traité* made under his direction by a student who had just returned from a year's study in Paris. See Hermbstaedt, letter to Nicolai, in H. Vopel, "Die Auseinandersetzung mit dem chemischen System Lavoisiers in Deutschland am Ende des 18. Jahrhunderts" (Diss., Leipzig University, 1972), p. 55. Soon afterwards he proclaimed his conversion to Lavoisier's theory in a review of Kirwan's *Essai sur la Phlogistique* . . . (1788), *Bibliothek der neuesten physisch-chemischen . . . Literatur*, 2, no. 3 (1789), 278–279. For a biographical profile of Hermbstaedt, see Appendix I.
[d]Wolff: He first expressed his antiphlogistic sympathies in F. L. Schurer, *Abhandlung vom Sauerstoff und seiner Verbindung mit andern Körper*, trans. [F. Wolff] (Berlin: Petit, 1790), which was dedicated to his "friend" and "teacher" Hermbstaedt. For Wolff, see J. E. Hitzig, *Verzeichniss im Jahre 1825 in Berlin Lebender Schriftsteller* (Berlin: Dümmler, 1826), pp. 304–305.

ᵉGirtanner: He revealed his commitment to the antiphlogistic system in his "Sur l'irritabilité, considérée, comme principe de vie dans la nature organisée," *Observations sur la physique,* 36 (June 1790), 422–440. This article was probably written in Göttingen, shortly before he left on a trip to Britain and France. In London (ca. 1 June 1790) Girtanner discussed his notion that oxygen was the basis of the life-force with A. von Humboldt. See *Die Jungendbriefe Alexander von Humboldts 1787–1799,* ed. I. Jahn and F. G. Lange (Berlin: Akademie Verlag, 1973), 236–237. For a biographical profile of Girtanner, see Appendix I.

ᶠMayer: He first indicated his adherence to Lavoisier's theory in a letter to Gren, *Journal der Physik,* 3 (1791), 19–20, and more fully in the foreword (May 1791) to his *Ueber die Gesetze und Modificationen des Wärmestoffs* (Erlangen: Palm, 1791), pp. [iii–iv]. For Mayer, See T. Gastauer, *Die Personalbibliographien des Lehrkörpers der philosophischen Fakultät zur Erlangen von 1743 bis 1806* (diss., Erlangen University; Erlangen: Hogl, 1968), pp. 93–103.

ᵍMartius: In his autobiography he claimed that he defended Lavoisier's theory in a pharmaceutical examination that he took before H. F. Delius, who died in October 1791. See his *Erinnerungen aus meinem neunzigjährigen Leben* (Leipzig: Voss, 1847), pp. 135–137.

ʰBlumenbach: He revealed his partiality for the antiphlogistic theory in a favorable review of B. Higgins's *A Comparative View . . .* (1789), *Medicinische Bibliothek,* 3, no. 3 (1791), 442–450. For Blumenbach, see W. Baron, "Blumenbach, Johann Friedrich," *Dictionary of Scientific Biography,* ed. C. C. Gillispie et al., vol. II (New York: Scribner's, 1970), pp. 203–205.

ⁱHumboldt: He first disclosed his conversion to Lavoisier's system in a letter to D. L. G. Karsten (26 November 1791). See *Die Jungendbriefe Alexander von Humboldts 1787–1799,* pp. 161–162. For Humboldt, see K.-R. Biermann, "Humboldt, Friedrich Wilhelm Heinrich Alexander von," *Dictionary of Scientific Biography,* vol. VI (1972), pp. 549–555.

ʲBaader: The first evidence of his antiphlogistic leanings appears in Humboldt's letter to P. Usteri (7/10 January 1792), in *Die Jugendbriefe Alexander von Humboldts 1787–1799,* p. 165. Soon afterwards Baader declared his appreciation for "the great work of revolution" started by Lavoisier. See his "Ideen über Festigkeit und Flüssigkeit zur Prüfung der physikalischen Grundsätze des Herrn Lavoisier," *Journal der Physik,* 5 (1792), 247. For Baader, see H. Grassl, "Baader, Benedikt Franz Xavier von," *Neue Deutsche Biographie,* vol. I (Berlin: Duncker & Humblot, 1953), pp. 474–476.

ᵏKlaproth: His preference for the antiphlogistic theory was announced by Hermbstaedt in the foreword (18 April 1792) to A. L. Lavoisier, *System der antiphlogistische Chemie,* trans. S. F. Hermbstaedt, vol. I (Berlin & Stettin: Nicolai, 1792), p. [x]. For a biographical profile of Klaproth, see Appendix I.

ˡHindenburg and Klügel: Their support for Lavoisier was announced by Hermbstaedt in an article written during the spring of 1792. See his "Ueber Oxygen und Phlogiston," *Chemische Annalen,* no. 2 (1792), 211. For Hindenburg, see K. Haas, "Hindenburg, Carl Friedrich," *Dictionary of Scientific Biography,* vol. VI (1972), pp. 403–404; and for Klügel, see J. Folta, "Klügel, Georg Simon," *ibid.,* vol. VII (1973), pp. 404–405.

bound by the chemical community's prevailing allegiance to phlogiston. Humboldt, for instance, predicted that even though he had become an antiphlogistonist, Gren "will not be very angry—I unfortunately know too little chemistry to make him so."[54] Little was holding them back, therefore, from basing their decisions on their personal impressions of the worth and prospects of Lavoisier's theory.

Much more closely affiliated with the German chemical community than these nine men was the convert Girtanner, a free-lance author of independent means. Prior to enlisting under Lavoisier's banner, he had made various contributions to Crell's journal and even squabbled with Westrumb.[55] Nonetheless, he took such delight in advancing new ideas and challenging popular beliefs that he would have had no qualms about defying prevailing opinion among German chemists, including his former teacher Gmelin. As the physicist G. C. Lichtenberg quipped, "first he wrote about venereal diseases, then [against the French] Revolution and [for] the French chemistry."[56]

In contrast to all these men, the two remaining converts in Protestant Germany—Hermbstaedt and Klaproth—were leading members of the German chemical community. Both had dedicated themselves to chemistry. Both were among Crell's most active contributors. Both enjoyed the respect of their fellow German chemists. Yet, both became antiphlogistonists in advance of their community.

It was in the summer of 1789 that Hermbstaedt (Figure 5) parted company with his peers, including his former teacher Wiegleb and his friend Klaproth. To judge from his testimony, he did so because he was overwhelmed by the truth of Lavoisier's theory:

> Why is everybody so opposed to a theory that rests on plain facts? Why does everybody try so forcefully to obscure the great truth that lies hidden beneath these facts? Why instead does everybody attempt to preserve a doctrine that is so rarely capable of explaining the physical phenomena as is the theory of phlogiston? I freely confess that until now, when I opposed the theory of the nonexistence of an inflammable principle, yes even when I sought to prove its inadequacy in my lectures, I did so for no other reason than that I did not yet have an exact knowledge of the entire theory. Oh, since the same thing could happen to other Germans as happened to me, I implore my countrymen, I pressingly implore them—study this excellent theory, test it, judge it. And when everyone has done so, I am

---

54. Humboldt, letter to Gren's brother-in-law, D. L. G. Karsten, (26 November 1791), in *Die Jugendbriefe Alexander von Humboldts 1787–1799*, p. 162.

55. Crell once characterized his contributor Girtanner as "an insightful and self-reliant chemist." See his review of Girtanner's *Dissertatio inauguralis chemica* (1782), *Die neuesten Entdeckungen in der Chemie*, 10 (1783), 148. Girtanner's quarrel with Westrumb was over the solubility of iron in pure water. See their articles in the *Chemische Annalen*, no. 1 (1788), 195–200; no. 2 (1788), 206–218, 300–307.

56. Lichtenberg, letter to A. G. Kastner (25 January 1792), in *Lichtenbergs Briefe*, ed. A. Leitzmann and G. Schüddekopf, vol. III (Leipzig: Dieterich, 1904), p. 42.

Fig. 5. S. F. Hermbstaedt, the most active German proponent of Lavoisier's theory.
Source: Bild Nr. 35369, Deutsches Museum, Munich

completely convinced that there will be nothing more that can properly hold us back from accepting it. Why should one not recognize the truth as truth? I am convinced of this theory's truth and only have a few experiments to perform in order to imbed it in my innermost being, in order to be able to speak only in its terms. Should these experiments confirm the theory, I will defend it until everybody sets aside all prejudices and accepts it as I do.[57]

No sooner had Hermbstaedt embraced Lavoisier's theory than he set himself the task of winning over the German chemical community. So eager was he to champion the theory that in the fall of 1789 he pressured Link, who had already announced plans to translate the *Traité*, to relinquish the translation in his favor. Then in December, he proclaimed his publication plans in the German intelligentsia's leading newspaper.[58] The rapidity with which Hermbstaedt became an antiphlogistic advocate, and the subsequent energy with which he pursued this self-appointed task,[59] suggest that there may have been something about his situation that made him particularly receptive to Lavoisier's theory. In fact, he was an ambitious man who, though he had publicly renounced his pharmaceutical career in 1786,[60] had not yet managed to obtain an academic position. His lack of success was not for want of effort. In 1788, for instance, he had sought the succession to the Berlin Medical-Surgical College's chair of chemistry and pharmacy, but was turned down by a commission that included not only the chair's occupant, C. H. Pein, but also Klaproth.[61] Frustrated in his attempts to get ahead, Hermbstaedt was probably ready to seize any opportunity that presented itself. In such a frame of mind, he could easily have sensed that adoption and espousal of a theory

57. Hermbstaedt, review of R. Kirwan's *Essai sur la Phlogistique* (1788), *Bibliothek der neuesten physisch-chemischen . . . Literatur*, 2, no. 3 (1789), 278–279.
58. *Allgemeine Literatur-Zeitung: Intelligenzblatt* (23 December 1789), 1239. Vopel, drawing upon the Hermbstaedt-Nicolai correspondence and other sources, has traced Hermbstaedt's subsequent endeavors to ensure that his translation, 1,016 copies of which were published by Nicolai in early 1792, had a favorable reception. See Vopel, "Die Auseinandersetzung mit dem chemischen System Lavoisiers in Deutschland," pp. 56–58.
59. Besides publishing the German translation of the *Traité* and encouraging F. Wolff to translate Schurer's exposition of the new system, Chaptal's antiphlogistic textbook, and the French critique of Kirwan, Hermbstaedt wrote a host of signed and unsigned reviews supporting Lavoisier's theory and a textbook that presented the phlogistic and antiphlogistic doctrines.
60. "I have completely abandoned pharmacy and am now devoting myself solely to chemistry," Hermbstaedt declared in his "Anmerckungen über einige zusammengesetzte Arzneimittel und die ungewisse Zubereitung derselben," *Neues Magazin für die gerichtliche Arzneikunde und medizinische Polizei*, 2 (1786), 71.
61. Although Hermbstaedt's petition was formally denied on the grounds that his knowledge of materia medica was deficient, M. Stürzbecher believes that other considerations, perhaps Hermbstaedt's infringement on the Medical-Surgical College's monopoly on medical instruction in Berlin, must have been decisive. See M. Stürzbecher, "Die Vorlesungsankündigungen von Sigismund Friedrich Hermbstädt am Collegium medico-chirurgicum und der Universität in Berlin," in *Pharmazie und Geschichte: Festschrift für Günter Kallinich zum 65. Geburtstag*, ed. W. Dressendörfer et al. (Straubing and Munich: Donau, 1978), pp. 189–191. Hermbstaedt was also denied membership in the Berlin Academy in 1788. See G. E. Dann, *Martin Heinrich Klaproth (1743–1817)* (Berlin: Akademie Verlag, 1958), pp. 39–40.

with so many strengths and such good prospects might redound to his advantage. In short, Hermbstaedt's ambition may well have increased his receptivity to the antiphlogistic system.

More than two and a half years after Hermbstaedt's conversion, his senior colleague Klaproth (Figure 6) sided with Lavoisier. In switching to the antiphlogistic system, he was in no way jeopardizing his reputation which, rather than being linked to the fate of phlogistic doctrines, rested on his precise analytical work. Still, it must have been a difficult step. In becoming an antiphlogistonist, he was not only turning away from his peers, including his longtime friend Westrumb, but he was also straying from the empirical standard to which he ordinarily held himself. He was committing himself before he had repeated all Lavoisier's experiments.[62] He was persuaded to take this plunge by a low-key yet relentless campaign from Hermbstaedt. He was invited, for instance, to provide a detailed critique of his junior colleague's translation of the *Traité*. He acceded to the request, thereby giving Hermbstaedt ample opening to plead Lavoisier's cause.[63] Then in early 1792, as his skepticism was being overcome, Klaproth was probably beset with further antiphlogistic advocacy from young Humboldt, who had recently become a convert in Freiberg.[64] Finally in April, he conceded to Hermbstaedt that he believed Lavoisier's system must be true if, as he expected, Lavoisier's experiments proved to be reliable. And in June he asked the Berlin Academy to assist him in acquiring the equipment needed to perform the necessary tests.[65]

Like the phlogistonists, therefore, those Germans who allied themselves with Lavoisier before mid-1792 took their stance for more than strictly scientific reasons. Almost all, by virtue of their residence in relatively cosmo-

---

62. In deference to Klaproth's feelings on this score, Hermbstaedt asserted that, because the apparatus needed to test Lavoisier's system was beyond the means of "a *deutschen Privatmannes*," German decisions for and against this system had hithertofore been based on "nothing more than empty *Raisonnement*." See A. L. Lavoisier, *System der antiphlogistische Chemie*, I: [iii].

63. For indications of Hermbstaedt's interaction with Klaproth over the *Traité* translation, see the *Bibliothek der neuesten physisch-chemischen . . . Literatur*, 3, no. 2 (March 1791), x; and Lavoisier, *System der antiphlogistische Chemie*, I:[x].

64. There is no direct evidence that Humboldt had close contacts with Klaproth before the latter's conversion. However, in June he was considering dedicating a book on halurgy to Klaproth as "a silent means of forging him more closely to the new chemistry." See his letter to H. C. Freiesleben (5 June 1792), in *Die Jugendbriefe Alexander von Humboldts 1787–1799*, p. 195. For the relations of Humboldt and Klaproth, see W.-H. Hein, "Alexander von Humboldt und Martin Heinrich Klaproth," *Beiträge zur Geschichte der Pharmazie*, 29, no. 2 (1977), 9–15.

65. Although G. E. Dann stresses Klaproth's independence in deciding to adopt to Lavoisier's theory—see his "Klaproths Wandlung zum Antiphlogistiker," Karl-Marx-Universität Leipzig: *Wissenschaftliche Zeitschrift: Mathematisch-naturwissenschaftliche Reihe*, 5 (1955/56), 49–53—, I think it likely that Hermbstaedt's cultivation of Klaproth helped propel him into the German chemical community's antiphlogistic avant-garde. For Klaproth's subsequent investigation of the empirical underpinning of Lavoisier's theory, see Dann, ibid.; the next chapter; and Klaproth, letters to M. van Marum, (12 November 1792, 29 December 1792, 14 April 1793, 12 November 1793, and 25 January 1794), in the Van Marum Collection, Part I: The Letters (Hollandische Maatschappij der Wettenschappen), microfiche ed. by Micro-Methods Limited (East Ardsley, Wakefield, England), Library of the American Philosophical Society, Philadelphia.

Fig. 6. M. H. Klaproth, the most prominent German convert to Lavoisier's theory.
Source: *Neue allgemeine deutsche Bibliothek*, 7 (1793)

politan centers or their political sympathies, seem to have been less susceptible to German cultural nationalism and more receptive to ideas originating in France. All but three, by virtue of being outside or peripheral to the German chemical community, seem to have been less sensitive to the peer pressure and status concerns that bolstered phlogistic loyalties and, conversely, more responsive to portents of antiphlogistic victory. Once these men perceived merit in Lavoisier's system, they had little, if any, reason to hesitate about enlisting under his banner. Likewise, despite closer affiliations with the German chemical community, Girtanner was evidently so eager to make a splash and Hermbstaedt so eager to get ahead that neither was about to be deterred by prevailing opinion among German chemists. Hence, as soon as these two sensed that Lavoisier must be correct and victorious, they set themselves up as his German champions. In comparison with all the other early converts, Klaproth was deliberate in his approach to the issue. Even so, he acted with uncharacteristic haste. Responding to pressure from Hermbstaedt, he committed himself before he had ascertained that the theory's empirical foundations were absolutely solid.

Klaproth's conversion in 1792 ended the first phase of the German antiphlogistic revolution. No longer, as they had done from mid-1785 to mid-1789, could German chemists laugh off Lavoisier. No longer, as they were wont to do after mid-1789, could they presume that phlogistic doctrines would ultimately prevail. The German chemical community had been thrown into a state of crisis by the conversions of men on its margins, of Girtanner, of Hermbstaedt, and especially of Klaproth. The converts' independence had undermined the authority of Crell, Westrumb, and Wiegleb. The converts' arguments had given rise to profound doubts concerning the direction of chemistry in Germany. And the converts' partisanship, while increasing the German chemists' awareness of their interdependence, had strained their social solidarity. The stage was set, as we now see, for the decisive conflict in the German antiphlogistic revolution—the debate over the reliability of Lavoisier's account of the reduction of mercury calx per se.

# 8 THE NOTORIOUS REDUCTION EXPERIMENT

In the year following Klaproth's conversion, German chemists came to regard the reduction of mercuric oxide as the crucial experiment for determining the truth or falsity of the entire antiphlogistic system. "The dispute here," Crell reported to an Italian colleague in October 1793, "is above all over a single experiment; that is to say, whether or not the oxide of mercury prepared by fire naturally yields pure air."[1] Crell and other German phlogistonists questioned Lavoisier's account of the experiment, while German antiphlogistonists staunchly defended its reliability. The present chapter recounts this central episode in the German antiphlogistic revolution.[2]

As the story unfolds, it lends strength to the thesis that German chemists had come to constitute a national discipline-oriented community by the late eighteenth century. In tempo and intensity the debate over the reduction experiment was quite unlike the desultory controversy between the proponents of pinguic acid and fixed air two decades earlier. The phlogistic and antiphlogistic spokesmen oriented themselves neither to local intellectuals, whose confidence was easy to win, nor to foreign chemists, whose shift to Lavoisier was nearing completion, but rather to their fellow German chemists. In sympathy with their peers' empiricism, they tacitly agreed to focus on the accuracy of Lavoisier's description of the reduction experiment. This is not to say that the rival spokesmen limited themselves to empirical matters. Too much was at stake. They could not refrain from affirming their own competence, judgment, and impartiality. Nor could they resist raising insidious questions about their rivals' ability, discernment, and integrity. Before long, the contestants were deeply suspicious of one another. At loggerheads, they enlisted witnesses for their experiments, even requested physicists to carry out independent tests. Though the debate was a struggle for the German chemical

---

1. Crell, letter to G. V. M. Fabbroni (4 October 1793), Library of the American Philosophical Society, Philadelphia.
2. In contrast to A. N. Scherer, who assigned a decisive role to the dispute over mercuric oxide, G. W. A. Kahlbaum and A. Hoffmann virtually ignored this debate in their history of the German antiphlogistic revolution. See Scherer, "Friedrich Albrecht Carl Gren," *Allgemeines Journal der Chemie*, 2 (1799), 394–395; and Kahlbaum and Hoffmann, "Die Einführung der Lavoisier'schen Theorie im besonderen in Deutschland," *Monographieen aus der Geschichte der Chemie*, 1 (1897). H. Vopel describes the dispute over the reduction experiment but, ignoring its social-psychological dimension, fails to recognize its decisive role in the antiphlogistic revolution. See H. Vopel, "Die Auseinandersetzung mit dem chemischen System Lavoisiers in Deutschland am Ende des 18. Jahrhunderts" (Diss., Leipzig University, 1972), pp. 108–124.

## LAVOISIER AND MERCURIC OXIDE

Prior to the 1770s, the compound that we know as mercuric oxide was regarded as two different substances because there were two methods for preparing it. The direct but tedious method was to simmer mercury in an open vessel. Gradually the oxide formed on the surface as a red powder. So prepared, mercuric oxide was called *mercurius praecipitatus per se*, mercury calx per se, and other similar names, because the mercury seemed to be making the transformation by itself. The second method, though indirect, required much less effort. One dissolved mercury in nitric acid, evaporated the liquid to obtain mercuric nitrate, and then decomposed this nitrate by heating to obtain mercuric oxide. When prepared in this way, it was called *mercurius praecipitatus ruber*, red precipitate of mercury, and other such names.[3]

During the 1770s and 1780s, chemists began thinking of mercury calx per se and the red precipitate of mercury as one and the same substance. French chemists led the way. P. Bayen and Lavoisier observed that, when reduced by heat alone, both the calx and the precipitate evolved gas as they turned into mercury. In suggesting that the substances were identical, Bayen pointed out that they yielded the same quantity of gas; Lavoisier, that they yielded the same kind of gas.[4]

The identity of mercury calx per se and the red precipitate of mercury was not the only conclusion drawn from the reduction experiment. After all, here was a calx, a metal supposedly bereft of its phlogiston, that could be returned to its metallic form simply by heating. Success did not depend on the presence of charcoal or any other source of phlogiston. In the years 1775 and 1776, Bayen and Lavoisier concluded that phlogiston did not participate in the reduction of mercury calx.[5] In the ensuing years Lavoisier went further.

---

3. For the history of mercuric oxide before the 1770s, see *Gmelins Handbuch der anorganische Chemie*, 8th ed., system no. 34, pt. A, issue 1 (Weinheim/Bergstrasse: Verlag Chemie, 1960), pp. 41–46; and C. E. Perrin, "Prelude to Lavoisier's Theory of Calcination: Some Observations on *Mercurius calcinatus per se*," *Ambix*, 16 (1969), 140–141.

4. For P. Bayen's argument that the calx per se and red precipitate are essentially the same, see his "Essais Chymiques ou expériences faites sur quelques précipités de Mercure, dans la vue de découvrir leur nature," *Observations sur la physique*, 5 (1775), 156. Lavoisier did not say that the two substances were identical in 1776, but he implied as much. See his "Mémoire sur l'existance de l'air dans l'acide nitreux, & sur les moyens de décomposer & de recomposer cet acide," Académie des Sciences, Paris: *Mémoires* (1776, publ. 1779), 671–680. By 1790 the identity of calx per se and red precipitate was widely accepted. In Germany, for instance, "Wiegleb, Bucholtz, Westendorff, and others" regarded them as identical substances. See P. J. Macquer, *Chymisches Wörterbuch; oder, Allgemeine Begriffe der Chymie nach alphabetischer Ordnung*, ed. and trans. J. G. Leonhardi, 2d German ed. of 2d French ed., vol. IV (Leipzig: Weidmann, 1789), p. 365. I stress the point because the identity of these calces subsequently became an issue in the German debate over mercuric oxide.

5. Bayen, "Essais chymiques . . . ," *Observations sur la physique*, 5 (1775), 154, 157; and Lavoisier, "Mémoire sur l'existance de l'air dans l'acide nitreux . . . ," Académie des Sci-

He used experiments with mercury calx to argue that all metallic calces, rather than being simple substances, contained what he later called "oxygen." He used them to show that oxygen was a component of nitric acid. And he used them to establish that atmospheric air consisted of two gases with different chemical properties.[6] In short, Lavoisier gave experiments with the red oxide of mercury, as mercuric oxide was called in the new nomenclature, a prominent place in his arguments.

## THE PHLOGISTIC CHALLENGE (1790–JULY 1792)

The first German chemist to question Lavoisier's account of the reduction of mercury calx was young Gren (Figure 7). In early 1790 he informed Crell that

> the most important and novel thing I have to tell you is that I have discovered that the red calx of mercury, if it has been calcinated in open vessels, does not give off any dephlogisticated air [oxygen] when reduced in closed vessels; that consequently this mainstay of Lavoisier's system is abolished; that no fresh metal calx contains air; that embodied air is actually a chimera. . . . The antiphlogistonists in France will probably not be able to save their system no matter how heatedly they defend it.

Besides publishing in the *Chemische Annalen*, Gren issued his challenge in two other places. In the first issue of his new *Journal der Physik*, he promised that he would soon

> show that [the reduction] experiment, upon which the antiphlogistonists' theory is chiefly built, is false if the mercury calx is completely fresh or has previously been glowing in an open vessel, but [that it] is probably true if the precipitate has absorbed moisture from the air. Hence the artificial theory that de la Métherie proposes [to explain away the reduction experiment] is not needed to rescue the Stahlian phlogiston.

And in his chemistry text, he maintained that dephlogisticated air could not be obtained from mercury calx that was still hot from its preparation. Those chemists who had reported obtaining the gas must, Gren thought, have used calx that had absorbed water, either during prolonged storage or during preparation by the nitric-acid method. As evidence of their negligence, he

---

ences, Paris: *Mémoires* (1776, publ. 1779), 680. Also see C. E. Perrin, "Prelude to Lavoisier's Theory of Calcination . . . ," *Ambix*, 16 (1969), 146–151.

6. Lavoisier, "Mémoire sur la nature du Principe qui se combine avec les Métaux pendant leur calcination, & qui en augmente le poids," Académie des Sciences, Paris: *Mémoires* (1775, publ. 1778), 520–526; "Mémoire sur l'existance de l'air dans l'acide nitreux . . . ," ibid. (1776, publ. 1779), 671–680; and "Expériences sur la Respiration des Animaux, & sur les changements qui arrive à l'air en passant par leur poumon," ibid. (1777, publ. 1780), 185–194.

Fig. 7. F. A. C. Gren, the first German phlogistonist to challenge Lavoisier's experimental reliability.
Source: *Neue allgemeine deutsche Bibliothek*, 22 (1796)

pointed out that they had failed to specify that they had used *fresh* mercury calx per se.[7]

Gren's challenge was slow to elicit comment, probably because his assertions were not substantiated by experimental details and because the veracity of Lavoisier's system was not yet a major issue in Germany. Late in 1790 Göttling in Jena noted in his *Almanach* that Gren's "important discovery" undermined a "complete section of the Lavoisierian system."[8] Half a year later Westrumb in Hameln, with his visitor Crell as a witness, took up Gren's claim. Like Gren, he failed to obtain any vital air when he reduced mercury calx per se. But unlike Gren, he did obtain water vapor.[9] Meanwhile, unaware of Westrumb's investigations, Gren reiterated his challenge. He denied that "red mercury calx yields dephlogisticated air if first heated in an open vessel until glowing," describing a reduction experiment in which he obtained a little water but no gas whatsoever. He hoped that the experiment would be repeated by "chemists who have the enlargement of our knowledge at heart and who have been provided by fate with more leisure and a better position than mine." And once again he surmised that those who had produced dephlogisticated air from mercury calx per se were using calx that had absorbed water from the atmosphere.[10]

In late 1791 Wiegleb in Langensalza commented on the reduction experiment in an extended critique of the French chemical system. He believed that Gren's account of the experiment might well be the correct one, that Lavoisier probably had not used fresh mercury calx per se. But further experiments were needed to settle the matter.[11] About the same time, Westrumb informed Gren that he still regarded water as a component of calces. "Otherwise," he

---

7. For Gren's initial challenge to Lavoisier's account of the experiment, see Gren, letter to Crell, *Chemische Annalen*, no. 1 (1790), 432–433; *Journal der Physik*, 1 (1790), 133fn; and his *Systematisches Handbuch der gesammten Chemie*, vol. II, pt. 2 (Halle: Waisenhaus, preface 20 November 1789, publ. 1790), pp. 189–195, 204. In failing to obtain any gas or vapor from mercury calx, Gren was confirming his view that calces were simple substances, a notion that he had cherished since 1786. See the translation of his 1786 dissertation, "Versuche und Beobachtungen über die Entstehung der fixen und phlogistischen Luft," *Beyträge zu den chemischen Annalen*, 3 (1788), 240–246.

8. *Almanach oder Taschenbuch für Scheidekünstler und Apotheker*, 11 (1791), 291.

9. Toward the end of 1791, Westrumb provided Gren with a brief description of the experiments carried out in Crell's presence. See Westrumb, letter to Gren (30 November 1791), *Journal der Physik*, 5 (1792), 46. But, on account of the press of business, he did not find time to present a detailed account of these experiments until the following spring. See Westrumb, "Einige Bemerkungen, verschiedene Gegenstände der neuen Chemie betreffend," *Chemische Annalen*, no. 2 (1792), 31–36. In obtaining water vapor from mercury calx, Westrumb was confirming his view that water was a component of calces, a notion that he had cherished since 1786. See his "Auch ein Beytrag zur Chemie von Luft- und Wassererzeugung," *Beyträge zu den chemischen Annalen*, 1, no. 4 (1786), 47–48, 59–62.

10. Gren, "Prüfung der neuern Theorien über Feuer, Wärme, Brennstoff, und Luft," *Journal der Physik*, 3 (1791), 479–482.

11. Wiegleb, "Beweissgründe des geläuterten Stahlischen Lehrbegrifs . . . ," *Chemische Annalen*, no. 2 (1791), 442. Writing anonymously, Wiegleb was somewhat more assertive. He claimed that, if Gren were correct in saying that mercury calx did not yield dephlogisticated air, the whole French system must soon collapse. See [Wiegleb], review of F. L. Schurer's *Abhandlung vom Säurestoff und seiner Verbindung mit andern Körpern* (trans. [F. Wolff], 1790), *Allgemeine deutsche Bibliothek*, 106 (1792), 447.

went on, "I maintain with you that one cannot force a *single bubble* of air out of fresh metallic calces prepared in fire, and I insist that anyone who maintains the opposite has never obtained air from mercury calx per se. . . ." Westrumb also agreed with Gren that experiments using mercury calx prepared by the nitric-acid method were irrelevant.[12]

Finally in June 1792, over two years after Lavoisier's reliability had first been challenged by Gren, Westrumb (Figure 8) brought the issue to a head. In a letter to Gren he told of using some fresh mercury calx per se that he had been preparing for the last nine months to perform the "notorious" reduction experiment. The calx, which he insisted was not so easy to prepare as that used by "the friends of oxygen," produced water vapor once it began to glow, but not "a single bubble of air." Assuring Gren that both local witnesses and outsiders could vouch for his reliability, Westrumb gave him permission to use the letter as he saw fit. Gren wasted no time in taking advantage of this offer. He incorporated Westrumb's letter into a "report for natural scientists and chemists" and sent it to the *Intelligenzblatt* of the *Allgemeine Literatur-Zeitung*.[13]

Westrumb's report caused a sensation among German phlogistonists. Wiegleb, who had been cautious in his acceptance of Gren's original challenge, happened to read it while he was reviewing some recent antiphlogistic works. He felt that the controversy would soon end. Westrumb had once again demonstrated "the tireless striving of our German chemists to find truth." His experiment demolished Lavoisier's whole system:

> . . . if vital air cannot be produced from mercury calx, then the participation of this [gas] in the calcination of metals is not proven, then the calcination of glowing metals when steam flows over them is not to be ascribed to a decomposition of water, then the compound nature of water collapses, then there is no reason for maintaining that carbonic acid is produced during the reduction of metallic calces with charcoal powder, then oxygen is a chimera, then the simple nature of carbon is a whim, then the whole new French chemical system is a monster of the power of imagination.[14]

Similar visions dazzled other German phlogistonists. Gren exulted that the

---

12. Westrumb, letter to Gren (30 November 1791), *Journal der Physik*, 5 (1792), 46. Gren must have been delighted to get this letter for he had recently been subjected to severe criticism from abroad. See C. L. Berthollet, "Observations sur quelques faits que l'on a opposés à la doctrine anti-phlogistique," *Annales de Chimie*, 11 (October 1791), 16–17; and J. B. van Mons, letter to Gren, *Journal der Physik*, 5 (1792), 48–49. In responding to his foreign critics, Gren insisted on the necessity of using freshly prepared calx per se. See his letter to van Mons (25 February 1792), *Annales de Chimie*, 13 (April 1792), 67–69; and "Etwas zur Vertheidigung gegen Herrn Berthollet," *Journal der Physik*, 5 (1792), 274–276.
13. Gren, "Nachricht für Naturforscher und Chemisten," *Allgemeine Literatur-Zeitung: Intelligenzblatt* (4 July 1792), 677–679. Gren later published Westrumb's letter in his *Journal der Physik*, 6 (1792), 32–34.
14. [Wiegleb], review of A. L. Lavoisier's *System der antiphlogistische Chemie* (trans. S. F. Hermbstaedt, 1792), *Neue allgemeine deutsche Bibliothek*, 2 (1793), 600; and review of C. Girtanner's *Anfangsgründe der antiphlogistischen Chemie* (1792), ibid., 3 (1793), 563.

Fig. 8. J. F. Westrumb, the foremost German critic of Lavoisier's reliability.
Source: Bild-Archiv, Österreichische Nationalbibliothek, Vienna

failure of fresh mercury calx per se to give off vital air was "naturally the death blow to the new French system of oxygen."[15] In a letter to Gren, Westrumb exclaimed that "just as 1 August 1774 was the birthday of the antiphlogistic chemistry, so I hope [7 June 1792] will be its death day." He also wrote with evident satisfaction of the "crisis" provoked by his experiments and expressed the hope that the outcome would be favorable to the "friends of German chemistry."[16]

Thanks, therefore, to widespread regard for Westrumb's experimental

15. Gren, "Neue Bestätigung durch Versuche, dass der in Feuer bereitete Quecksilberkalk keine Lebensluft bey seiner Wiederherstellung für sich in Glühen liefert," *Journal der Physik*, 6 (1792), 29; and letter to Girtanner, excerpted in Girtanner, "Nachricht für Naturforscher und Chemisten," *Allgemeine Literatur-Zeitung: Intelligenzblatt* (29 August 1792), 871–872.

16. Westrumb, letter to Gren, *Journal der Physik*, 6 (1792), 212; and *Kleine physikalisch-*

skills as well as to a growing sense that the antiphlogistic theory merited serious consideration, the issue of Lavoisier's reliability assumed great importance among German chemists during the summer of 1792. The challenge to his reliability was clear. Whereas Lavoisier had maintained that mercuric oxide produced oxygen when it was reduced by heating, Westrumb and Gren claimed that it failed to yield a single bubble of air. Whereas Lavoisier and his allies had assumed that the method of preparing and the length of storing mercuric oxide were of no consequence, Westrumb and Gren insisted on the importance of carrying out the test with fresh mercury calx per se.

## THE ANTIPHLOGISTIC RESPONSE (AUGUST–OCTOBER 1792)

Before the appearance of Westrumb's report in the *Intelligenzblatt*, German antiphlogistonists were so busy disseminating information about their adopted theory that they gave little attention to the phlogistic challenge. In late 1790 J. T. Mayer, professor of physics in Erlangen, had written Gren suggesting that fresh calx be weighed before and after cooling to determine whether it actually absorbed anything from the atmosphere as it cooled.[17] And during the first half of 1792, Hermbstaedt, now professor of chemistry and pharmacy in Berlin, had touched on Gren's claims in his *Bibliothek der neuesten physisch-chemischen . . . Literatur* and in an article for Crell's *Chemische Annalen*. He reported that he had obtained "the purest vital air from freshly prepared mercury calx" in experiments witnessed not only by his students but also by M. Herz (professor of philosophy in Berlin). He acknowledged that he had used calx prepared by the nitric-acid method, asserting that he could not see why mercury calx per se, which was so difficult to prepare, should give different results.[18]

Westrumb's claims stirred the antiphlogistic spokesmen into action, focusing their attention on the reduction experiment. Girtanner in Göttingen was the first to comment. On 12 August 1792 he responded to personal letters from Gren and Westrumb as well as to their published report by sending his own "report for natural scientists and chemists" to the *Intelligenzblatt*. He suggested that hydrogen, originating from dust deposited on Westrumb's calx during its long preparation, had combined with oxygen from the calx to produce Westrumb's water vapor. In any case, Girtanner argued, "such a firmly

---

*chemische Abhandlungen*, vol. IV, pt. 1 (Hanover: Hahn, preface October 1792, publ. 1793), pp. xvii–xviii. As will be seen below, W. A. E. Lampadius and G. F. Hildebrandt were also impressed by Westrumb's experiments.

17. Mayer, letter to Gren, *Journal der Physik*, 3 (1791), 19–20.

18. Hermbstaedt, review of J. A. Scherer's "Scrutinum Hypotheseos Principii Inflammabilis" (1790), *Bibliothek der neuesten physisch-chemischen . . . Literatur*, 4, no. 1 (1792), 66; and "Ueber Oxygen und Phlogiston," *Chemische Annalen*, no. 2 (1792), 211–215. In an addendum to the article in Crell's journal, Hermbstaedt claimed that his recent experiments with [purchased] calx per se vindicated his assumption that the method of preparing the calx was of no importance. Though this article was published well after Westrumb's experiments, it must have been written before news of them reached Berlin.

grounded system as the antiphlogistic system, which rests on hundreds of experiments, cannot be overthrown by a single experiment."[19]

The antiphlogistonists in Berlin, possibly because they recognized a golden opportunity to further their cause, were readier to accept the claim that the reduction experiment was crucial. They reduced mercury calx to mercury several times during September 1792. The first report of their work was sent to the *Intelligenzblatt* by J. Peschier, a Genevan who was studying pharmacy in Berlin. On a recent visit to Halle, Peschier had heard Gren inveigh against those who believed oxygen could be obtained from mercury calx per se. Back in Berlin, he reported Gren's remarks to Klaproth. Thinking to settle the matter, Klaproth had Peschier reduce mercury calx in his laboratory. They conducted two trials in the presence of Hermbstaedt and Gren's brother-in-law, the mining administrator D. L. G. Karsten. In the first test they used calx per se from London, the purity of which was evident from both its crystallization and its failure to yield white vapors upon heating. In the second test they used calx per se that Klaproth himself had prepared and sealed in a bottle a few years earlier. Both times they obtained oxygen. And in the first experiment, the only one carried out with enough calx to permit use of the balance the weight of the gas was approximately equal to the difference in weight between the calx and resultant mercury.[20]

Soon after Peschier sent his report to the *Intelligenzblatt*, Hermbstaedt sent longer reports to the *Journal der Physik* and the *Chemische Annalen*. He seems to have been eager to increase confidence in the results, for he assigned Peschier the role of observer and elevated Klaproth to the role of experimenter. "One can trust," he asserted, "that such an accurate worker as Klaproth did this experiment with the greatest exactitude. Moreover, the completely impartial witnesses who were present [Hermbstaedt (!) and Karsten] can testify that there was not a single error." Westrumb must have made some mistake. To be sure, Girtanner's hypothesis that dust had settled on Westrumb's calx was implausible. It seemed more likely that there had been a leak in Westrumb's pneumatic apparatus, for he should have obtained, at the very least, the air in the retort. Hermbstaedt also suggested that the water obtained by Westrumb and also by Klaproth (!) came from the water over which the gas was collected rather than from the mercury calx. The water must be extraneous; otherwise Klaproth would not have found that the weight of the original mercury calx was equal to the sum of the weights of the resultant oxygen and mercury. Why, Hermbstaedt wondered, had Westrumb failed to report the weights observed in his experiment? Until such issues were resolved, he was unwilling to grant that Westrumb had shaken "the base

---

19. Girtanner, "Nachricht für Naturforscher und Chemisten," *Allgemeine Literatur-Zeitung: Intelligenzblatt* (29 August 1792), 871–872; reprinted in the *Journal der Physik*, 6 (1792), 416–418.

20. Peschier, "Résultat de deux experiences sur le dégagement du gas oxigêne par la reduction de l'oxide de mercure rouge autrement dit mercure précipité per se," *Allgemeine Literatur-Zeitung: Intelligenzblatt* (20 October 1792), 1022–1024; reprinted in the *Journal der Physik*, 6 (1792), 420–422.

pillar of the antiphlogistic system." Besides reporting Klaproth's experiment and criticizing Westrumb's work, Hermbstaedt told of his own endeavors. He had recently experimented with fresh calx per se prepared by the Berlin pharmacist Behrens. The trial was, as many witnesses could testify, successful. He also announced that he was beginning to prepare some calx in his own laboratory.[21]

In mid-October 1792, as later reports revealed, Hermbstaedt impatiently harvested the fresh calx that had formed on the surface of the simmering mercury. He then carried out the reduction experiment several times in the presence of Gren's former student, D. L. Bourguet, and other witnesses. Each time he obtained pure air, but never any water. Exuberant, he wrote Westrumb a long letter, reporting recent work in Berlin and raising questions about experimental techniques and results. And he sent Gren some fresh calx, urging him to do the experiment himself. Soon afterwards he began preparing a much larger quantity of calx for future experiments.[22]

## REFORMULATION OF THE CHALLENGE (SEPTEMBER–DECEMBER 1792)

Meanwhile the phlogistonists were busy. Upon reading Girtanner's critique in the *Intelligenzblatt*, Westrumb sent his reaction to Gren. It would have been difficult for Girtanner, he believed, to "have spoken more dangerously for oxygen and for the fidelity and veracity of the antiphlogistonists." Since dust could just as easily have fallen on their calx, why had they failed to obtain water during their reduction experiments? After a few more thrusts and parries, Westrumb enumerated all the precautions he had taken to prevent contamination.[23] Girtanner's critique also provoked one "Christ. Antiphlogisticus" to send a sarcastic note to the *Intelligenzblatt*. The anonymous writer, perhaps Gren, complimented Girtanner on his "completely brand-new chemical discovery" that dust was the basis of hydrogen. This discovery aptly revealed the "true magnificence" of the "modern and *philosophical* system of the French chemistry."[24]

---

21. Hermbstaedt, "Einige Anmerkungen über die Entbindung der Lebensluft (gas oxygêne), aus für sich verkalkten Quecksilber, durch blosses Glühen; nebst Untersuchung derjenigen Einwendungen, welche Hr. Prof. Gren, und der Hr. Bergcomm. Westrumb diesem Versuche entgegengesetzt haben" (28 September 1792), *Journal der Physik*, 6 (1792), 422–429; and "Bemerkungen über die Entbindung der Lebensluft . . . ," *Chemische Annalen*, no. 2 (1792), 387–398. In a later report of the experiments in Klaproth's laboratory on 16 September 1792, Hermbstaedt left out all reference to Peschier. See C. W. Scheele, *Sämmtliche physische und chemische Werke*, ed. S. F. Hermbstaedt, vol. I (Berlin: Rottmann, foreword February 1793, publ. 1793), pp. 179–180.
22. The fullest account of Hermbstaedt's work with mercuric oxide during the fall of 1792 was presented in his disciple J. F. H. Suersen's "Einige Bemerkungen über die Entbindung der Lebensluft aus dem für sich verkalkten Quecksilber," *Chemische Annalen*, no. 1 (1793), 421–424. Hermbstaedt mentioned sending a letter to Westrumb (6 October 1792) and a package to Gren (12 October 1792) in the *Chemische Annalen*, no. 2 (1792), 434; no. 1 (1793), 339.
23. Westrumb, letter to Gren, *Journal der Physik*, 6 (1792), 418–419.
24. Christ. Antiphlogisticus, "Ein ganz nagelneue chemische Entdeckung," *Allgemeine Literatur-Zeitung: Intelligenzblatt* (22 September 1792), 943–944.

Although the phlogistonists could spurn a gadfly like Girtanner, they could not ignore the reports from Berlin. On 21 and again on 22 October 1792, shortly after receiving Hermbstaedt's letter, Westrumb reduced fresh calx two times in the presence of several witnesses. The results were staggering. With mercury rather than water in his pneumatic apparatus, he collected much pure air and only a drop or two of water. Dejected, he sent the bad news to Gren. A week later, however, a jubilant Westrumb was writing Gren and Crell. He had been misled by an assistant who had neglected to dry the apparatus used for heating the calx. Residual water in the retort must have been the source of the pure air. To test this idea, Westrumb had heated fresh calx until ten percent volatilized and then sprinkled it with water. As he expected, pure air was generated in abundance. Having detected his error, he repeated the experiment with the calces of several other metals. This work convinced him that water was not an essential component of calces. He now agreed with Gren that calces were simple substances. Any water or air yielded by them must originate from absorbed water or air. Hence, mercury calx per se produced varying amounts of air depending upon how much it had absorbed. Calx that had been exposed for six months yielded more air than calx that had been sealed in a bottle for six months, and this in turn yielded more air than completely fresh calx. He still needed to perform the decisive experiment of reducing fresh glowing mercury calx per se, but he fully expected that such calx would not produce any air whatsoever.[25]

Evidently chastened by his discovery that mercury calx per se yielded oxygen under some circumstances, Westrumb asked Gren and Crell to hold back his reports until his findings had been confirmed. In late November 1792, he mailed samples of fresh calx that had been heated until forty percent had volatilized to Hermbstaedt and J. B. Trommsdorff, a young pharmacist in Erfurt who had recently supported Gren's theory of calces. Sending the calx to Hermbstaedt must have been a gesture of defiance, for in a letter to Gren he implied that he only expected "to hear the truth" from Trommsdorff.[26]

In December 1792 Gren reviewed the dispute over the reduction experiment in his *Journal der Physik*. He gave short shrift to Girtanner, saying that the dust hypothesis had been "grabbed out of the air." Next he took up Hermbstaedt's account of Klaproth's experiment. He first called attention to a trivial error in calculation. Then full of gusto, Gren argued that Hermbstaedt had followed his master too closely. Like Lavoisier, Hermbstaedt had

---

25. Westrumb, letters to Gren (23, 30 October 1792), *Journal der Physik*, 7 (1793), 148–149; and letter to Crell, *Chemische Annalen*, no. 1 (1793), 164. News of Westrumb's varying results was also communicated to scientists in Göttingen. See Lichtenberg's diary entries for 26 October, 5 November 1792 in *Georg Christoph Lichtenberg: Schriften und Briefe*, ed. W. Promies, 2nd ed., vol. II (Munich: Hanser, 1975), pp. 764–765.

26. Westrumb reported sending out the samples of calx in a letter to Crell, *Chemische Annalen*, no. 1 (1793), 110; and a letter to Gren (4 December 1792), *Journal der Physik*, 7 (1793), 150. For Trommsdorff's support of Gren's theory, see his "Einige Versuche über die Luft und Wassererzeugung aus Metallkalken," *Journal der Physik*, 6 (1792), 214–222. For a biographical profile of Trommsdorff, see Appendix I.

ignored the "hydrostatic" principle when determining the weight of the oxygen. A gas's true weight equaled the apparent weight of the gas plus the weight of the air displaced by the gas. Thus, if Klaproth obtained as much oxygen as Hermbstaedt reported, the true weight of the gas must have been much greater than the difference in weight between the calx and the resultant mercury. Gren concluded his hydrostatic objections on an optimistic note:

> The enthusiasm and prejudice for [Lavoisier's] system, engendered by the fascination for novelty and by the convenience with which its laws can be used as easy formulas, will pass. Then it will be evident that those who dared to examine the system of the oxygenists and to combat the obnoxious aristocratic despotism that puts it forward, are not to be denounced as stubborn or old fashioned.

In fact, as Gren soon admitted, his hydrostatic objections were groundless, for they were based on a careless reading of Lavoisier.

Besides criticizing Girtanner and Hermbstaedt, Gren presented a case for his own position. He reported that J. M. Schiller, an apothecary in Rothenburg ob der Tauber, had recently confirmed his and Westrumb's results "with self-prepared and fresh mercury calx." He also reported that an unnamed informant [Westrumb] had convinced him to modify his stance:

> I now grant that red mercury calx prepared in fire can yield dephlogisticated air when reduced; however, I do not grant that it yields this air by itself, but rather *I maintain that it yields this air only insofar as it contains water and retains this water until it glows.*

Gren believed that mercury calx per se, like all other metallic calces, had a strong affinity for the water in the atmosphere. It would even absorb this water during the calx's preparation if the fire were allowed to subside. Once absorbed, such water could only be expelled by heating the calx until it glowed. Part of the water would be released before the calx glowed; then, at glowing heat, the remainder would be converted to phlogisticated air. Before this phlogisticated air could escape, however, it lost its phlogiston to the mercury calx, reducing the calx to metallic mercury. Hence it was dephlogisticated air that was actually released during reduction. The production of this gas, therefore, did not prove that "the pure calx of mercury contains the basis of vital air as an essential component."[27]

To recapitulate, Hermbstaedt and Klaproth defended Lavoisier's reliability so effectively during September and October 1792 that Westrumb, followed by Gren, soon restated their challenge. They admitted that mercury calx per se released oxygen when it was reduced. They maintained, nonetheless, that

---

27. Gren, "Gesammlete Nachrichten in Betreff des Streits, ob der reine Kalk des Quecksilbers die Basis der Lebensluft als wesentlichen Bestandtheil enthalte," *Journal der Physik*, 6 (1792), 416–432, 442–447.

this oxygen was not an integral component of the calx. Rather, they regarded it as an extraneous by-product of the interaction between the calx and absorbed water. The crucial experiment was, they now thought, to reduce mercury calx per se after all this water had been expelled by prolonged heating of the calx at its temperature of volatilization.

## CHEMICAL SECTARIANISM (JANUARY–FEBRUARY 1793)

### The Antiphlogistonists

In early January 1793, Hermbstaedt began a new round of experiments, using the mercury calx per se that he had been preparing since mid-October. The occasion for his first experiments seems to have been the arrival in Berlin of W. A. E. Lampadius, a phlogistic-minded youth who had been trained in Göttingen by the chemist Gmelin and the physicist G. C. Lichtenberg. Lampadius believed that Westrumb's experiments of June 1792 threatened a "complete overthrow of the new antiphlogistic chemistry." Still, at Hermbstaedt's urging, he participated in two trials of the reduction experiment. Both times, as Lampadius soon informed Gren, "the purest dephlogisticated air" was obtained from calx that had been heated to the glowing point. Besides helping perform the experiment, he sealed two drams of Hermbstaedt's hot calx in a bottle and posted it to Lichtenberg.[28]

Hermbstaedt's effort to persuade Lampadius was just the prelude to an intensive campaign. In an article sent to Gren on 24 February 1793, Hermbstaedt reported that he had reduced "almost still glowing calx" several times. On every occasion, he had obtained pure oxygen gas but never a trace of water. Besides Lampadius and his ally A. von Humboldt, Hermbstaedt's witnesses included P. C. Abildgaard (director of Copenhagen's veterinary school), C. A. W. Berends (professor of therapy in Frankfurt an der Oder), C. Knape (professor of anatomy in Berlin), A. von Mussin-Puschkin (a count from St. Petersburg), E. F. Rettberg (a mining secretary in Berlin), J. D.

---

28. For Lampadius's views on the reduction experiment in the spring and early summer of 1792, see his *Kurze Darstellung der vorzüglichsten Theorien des Feuers dessen Wirkungen und verschiedenen Verbindungen* (Göttingen: Dieterich, foreword July 1792, publ. 1793), pp. vi–vii, 134–135. His appreciation of Westrumb's experiments must have increased as the result of contact with Westrumb's disciple, J. Bischoff, who arrived in Göttingen during the summer or fall of 1792. During the fall, Bischoff carried out some reduction experiments in the Göttingen University laboratory, obtaining results similar to those of his former mentor. See Westrumb, letter to Gren (4 December 1792), *Journal der Physik*, 7 (1793), 150; and Bischoff, "Ein paar Worte über Metallkalke," *Chemische Annalen*, no. 1 (1793), 411–414. Then in the late fall, Lampadius started for Russia. For Hermbstaedt's experiments carried out in Lampadius's presence, see Lampadius, letter to Gren (10 January 1793), *Journal der Physik*, 7 (1793), 148; Lampadius, *Versuche und Beobachtungen über die Elektrizität und Wärme der Atmosphäre angestellt im Jahre 1792 nebst der Theorie der Luftelektrizität nach den Grundsätzen des Hern. de Lüc und einer Abhandlung über das Wasser* (Berlin and Stettin: Nicolai, foreword January 1793, publ. 1793), pp. 197–200; and Suersen, "Einige Bemerkungen . . . ," *Chemische Annalen*, p. 418. For a profile of Lampadius, see Appendix I.

Ribini (secretary to a Viennese aristocrat), and G. F. Sick (professor of veterinary medicine in Berlin).[29]

In addition to reporting his own experiments, Hermbstaedt included F. Wolff's eyewitness report of a successful trial recently carried out by Klaproth and the pharmacist V. Rose, Jr. "Hitherto," Wolff confessed,

> my adherence to Lavoisier's system was moderated by one reservation: whether the experiments were really as claimed, whether prejudice to a system did not cause more or less to be seen than was really present. Unfortunately, we have all too many examples of how prejudice can influence the chemist's crucible. We must maintain our independence from authorities, for the most valuable and excellent men can err. However, I have now seen the experiment with my own eyes, have observed all the circumstances as exactly as possible, and am fully convinced.[30]

This report from Wolff as well as all his own experiments must have given Hermbstaedt the feeling that Lavoisier's empirical reliability was unassailable.

Hermbstaedt also defended Lavoisier's method for determining the weight of gases. Lacking a background in mathematics, he based the defense on letters from his allies Wolff, Humboldt, and Mayer. The trouble with Gren's argument, Wolff pointed out, was that he had thoroughly confused the concepts of absolute weight and relative weight. Lavoisier had made no such mistake. He had first measured the absolute weight of a given volume of atmospheric air by seeing how much weight a rigid vessel lost when it was evacuated. Then he had calculated the density of the air. Knowing that figure, he had been able to ascertain the absolute weight of any other gas once he knew how much a given volume of that gas weighed in the atmosphere. Humboldt, though originally troubled by Gren's argument, indicated his agreement with Wolff's rebuttal to "the new hydrostatic objections that Prof. Gren makes against our antiphlogistic theory."[31] Working independently in Erlangen, Mayer also concluded that Gren's argument was riddled with misunderstandings.

Besides defending Lavoisier's empirical and computational reliability, Hermbstaedt complained about Gren's experimental inactivity. He sent Gren another sample of mercury calx per se so that his rival could "undertake the reduction experiment himself and observe the results with his own eyes." He

---

29. Hermbstaedt, "Rechtfertigung gegen Hrn. Prof. Gren's hydrostatische Einwürfe, den Gehalt an Sauerstoffgas im Quecksilberkalke betreffend," *Chemische Annalen*, no. 1 (1793), 324–349. This article appeared in Crell's journal because Gren refused to publish it.
30. Ibid., 326. For a profile of V. Rose, Jr., who was Hermbstaedt's brother-in-law, see Appendix I.
31. Ibid., 333–334. For Humboldt's original reaction to Gren's hydrostatic objections, see *Die Jugendbriefe Alexander von Humboldts 1787–1799*, ed. I. Jahn and F. G. Lange (Berlin: Akademie Verlag, 1973), p. 236.

believed that "the mere theoretician, who does not experiment, can have no consideration now, *for only arguments based on experience are worth anything in this dispute*!!" Only by using their balances could Gren and the other phlogistonists prove that absorbed water, rather than oxygen, was responsible for a calx's loss in weight during reduction. Hermbstaedt's own experiments with glowing calx had convinced him that absorbed water was not the source of the oxygen gas released during reduction. He was in no mood for theories about water being transformed into oxygen by glowing heat. He was reporting "facts, . . . *not Raisonnement.*"[32]

## The Phlogistonists

Meanwhile, the phlogistonists had found a willing ally in Erfurt. On 21 January 1793, Trommsdorff performed several experiments with his own calx as well as that sent him by Westrumb. Soon afterwards he informed Westrumb of his results by letter and sent reports to Crell and Gren for publication in their journals. Trommsdorff declared that "dogmatic pronouncements, which are especially typical of the French chemists, influence me as little as do authorities; neither can keep me from testing the matter myself and reporting my findings." Like Westrumb, he found that freshly prepared and glowing calx yielded neither vital air nor water. He was sure that Hermbstaedt, Girtanner, Peschier, and others "had made mistakes [in their observations] or failed to experiment with proper care." By contrast, "the two chemists Gren and Westrumb have worked and observed exactly and correctly." So that nobody would think that he had carried out his experiments at his writing desk, Trommsdorff sent Gren a testimonial signed by his witnesses, A. F. Hecker (professor of medicine in Erfurt) and C. E. Meier (a recent M.D. there). He believed that the controversy was "too lively to last for long" and hoped "something would be decided soon now."[33]

In the meantime, Westrumb's morale was again flagging. It was not easy, he wrote Crell, to be treated "as a dirty, dishonest, and hopeless tyro in chemistry." To be sure, his critics "dressed up [their] harsh accusations in civil clothes," but this did not make their charges any less offensive. Westrumb's fighting spirit was restored by Trommsdorff's letter. He soon wrote Crell and Gren, telling them the good news and describing his latest result—calx produced neither air nor water during reduction when it was put into the pneumatic apparatus after twenty percent had volatilized at the glowing point. "Thus," Westrumb concluded his note to Gren, "we have told the truth. Nonetheless, the opponents will [continue to] hide behind new subterfuges,

32. Hermbstaedt, "Rechtfertigung . . . ," *Chemische Annalen*, no. 1 (1793), 340, 343.
33. Trommsdorff's letter to Westrumb was mentioned by Westrumb in a letter to Gren (27 January 1793), *Journal der Physik*, 7 (1793), 150. Also see Trommsdorff, letter to Crell, *Chemische Annalen*, no. 1 (1793), 256–257; and his "Auch einige Versuche mit dem für sich verkalkten Quecksilber," *Journal der Physik*, 7 (1793), 37–42. Trommsdorff's witness, Hecker, was hardly impartial. At least, a few months earlier, he had opined that "the serious German" had no reason to embrace "the chimeras of the French chemists." See his "Plan dieses Journal," *Journal der Erfindungen, Theorien und Widersprüche in der Natur- und Arzneiwissenschaft*, 1, no. 1 (1792), 10, 15.

other modifications of their hypotheses, and who knows where else." Impatient to inform "the whole chemical public" of his position, he sent a brief report to the *Intelligenzblatt*. Calx freed of its hygroscopic water by partial volatilization did not yield vital air. On the other hand, calx embodying absorbed water or sprinkled with water did yield vital air. He assured the readers that "several worthy German men" supported his claims and that "consequently one of the mainstays of the system of oxygen falls to the ground."[34]

Soon after posting this confident notice, Westrumb received unsettling reports from the phlogistonist Gmelin and an antiphlogistonist, probably Hermbstaedt. He immediately sent their findings to Trommsdorff for an opinion. In responding, Trommsdorff agreed with Westrumb that Gmelin's success in obtaining gas from mercury calx per se must have been due to inadequate volatilization prior to the experiment. While he could calmly discuss Gmelin's possible mistake, Trommsdorff could only think that the antiphlogistonist had "bungled." It would, he affirmed, "be too much for all of us to have erred." He would not allow himself to "be shouted down by the opponents until they convince me that Prof. Hecker, Dr. Meier, and I no longer have healthy senses." If the antiphlogistonists could produce vital air from calx after forty percent had been volatilized, he would never do another experiment. He would be "ruined as a chemist." He was sure, however, that he did not stand in any jeopardy, for the antiphlogistonists did their experiments at their "desks."[35]

Unlike Westrumb and Trommsdorff, Gren was not experimenting with mercury calx per se. His involvement remained at the level of commentary. In the February 1793 issue of his *Journal der Physik*, he maintained that every scientist should be interested in the dispute because

> nothing less depends on it than the overthrow of the whole
> neological theory of the antiphlogistonists and consequently of
> the nomenclature based on it, which, to be sure, must become
> an incomprehensible jargon as soon as the facts that serve as its
> basis are proved to be only delusions.

He praised Westrumb for demonstrating that absorbed water was the source of all vital air produced by mercury calx per se. The crucial experiment was to heat calx to the glowing point, then put it in a hot evacuated retort, and then obtain vital air by further heating. To show that the antiphlogistonists had little prospect of success, Gren quoted parts of Westrumb's letters written from 23 October 1792 through 11 February 1793. He also reported that Schiller in Rothenburg had continued to obtain results that were consonant

---

34. For the development of Westrumb's views between mid-January and mid-February 1793, see Westrumb, letters to Crell, *Chemische Annalen*, no. 1 (1793), 109–112, 165; letters to Gren (27 January 1793, 11 February 1793), *Journal der Physik*, 7 (1793), 150–151; and "Nachricht" (February 1793), *Allgemeine Literatur-Zeitung: Intelligenzblatt* (13 March 1793), 1796.

35. Trommsdorff, letter to Westrumb (16 March 1793), *Chemische Annalen*, no. 1 (1793), 248–251; also printed in *Journal der Physik*, 7 (1793), 241–244.

with his and Westrumb's findings. Gren concluded by admitting that his hydrostatic objections were groundless. He thought it "droll" that his error had induced Hermbstaedt to obtain a smaller proportion of vital air in the experiment with Lampadius than had been obtained in Klaproth's experiment of September 1792.[36]

Thus, during January and February 1793, both sides sought to strengthen their positions with more experiments and arguments. Hermbstaedt and Klaproth found, as they expected, that hot, dehydrated mercury calx per se yielded oxygen gas. Westrumb and Trommsdorff found, as they expected, that such calx yielded nothing. Until late February, with the exception of Lampadius's letter to Gren of early January, there seems to have been no direct communication between the rival camps. Then the antiphlogistonists re-established contact. An unnamed antiphlogistonist reported his results to Westrumb, who in turn passed the report on to Trommsdorff. The young Erfurt chemist, having earlier implied that the antiphlogistonists experimented at their writing desks, staked his reputation on the outcome of the reduction experiment. About the same time, Gren finished his second commentary on the debate by insinuating that Hermbstaedt had been tampering with his data. Meanwhile on 24 February, Hermbstaedt sent Gren an article that described recent work in Berlin and lambasted Gren's hydrostatic objections and experimental inactivity. The time was ripe for a decisive confrontation.

## HERMBSTAEDT'S VINDICATION OF LAVOISIER (MARCH–APRIL 1793)

On 26 March 1793, after reading Westrumb's confident note in the *Intelligenzblatt*, Hermbstaedt sent "a notice for natural scientists" to the same periodical. He opened with the claim that he knew "full well how to combine scientific disagreement with high personal respect" and called Westrumb a "worthy man." Then he turned to the reduction experiment. He insisted that whenever he had reduced fresh glowing calx, he had obtained a quantity of vital air whose weight equalled the loss in weight of the calx. He had sent calx to Gren, Lichtenberg, Mayer, Crell, and Westrumb and was awaiting an impartial account of their findings. He had found that Gren's objection to Lavoisier's method of weighing gases was based "*on a hydrostatic error that he had too eagerly embraced.*" Since his article rectifying this error had been rejected by Gren, he would submit it to Crell for the *Chemische Annalen*. Hermbstaedt felt that he had already made some progress in the debate, because Gren and Westrumb now admitted that calx released vital air in some instances. He hoped that his steadfastness, which was "based entirely on experiments performed in cold blood and on impartially judged experiences,"

---

36. Gren, "Fortgesetzte Nachrichten in Betreff des Streits, ob der reine Kalk des Quecksilbers die Basis der Lebensluft als Bestandtheil enthalte," *Journal der Physik*, 7 (1793), 146–153.

would soon enable him to make further headway. Regardless of his success, he believed that experiments would lead "more surely to the goal than all *Raisonnement*."[37]

A few days after sending this notice to the *Intelligenzblatt*, Hermbstaedt received the February 1793 issue of the *Journal der Physik*. No doubt piqued by Gren's insinuations, he decided to settle the matter once and for all. On 3 April, he and Klaproth carried out the reduction experiment twice in the presence of many witnesses. Later that day he posted a report of his "new confirmation of a basic chemical principle" to the *Intelligenzblatt*. In one experiment he used the calx that Westrumb had sent him in December. With everyone watching, he broke the container's seal, removed the calx, heated it until a sixth had volatilized, put it in a hot retort, and began the reduction. Then as luck would have it, the retort broke and sand fell on the calx. The calx and sand were extricated from the broken vessel and, after weighing, placed in a second hot retort. This time the reduction proceeded smoothly, and vital air was obtained. In the other experiment Hermbstaedt used fresh calx that he had just prepared. He followed the same procedures and, without mishap, obtained the same results. Hoping that these experiments would answer Westrumb, Trommsdorff, Schiller, and others, he suggested that 3 April 1793 be regarded "as the second death day of phlogiston." As a guarantee of his veracity, he appended a testimonial signed by all his witnesses. In addition to the familiar names of Humboldt, Abildgaard, Wolff, Rose, Rettberg, and Bourguet, he had the vouching of F. W. von Reden (a count and Prussian mining and smelting director for Silesia), P. Erman (a teacher of French and physics in Berlin), P. A. Lampe (a physician from Danzig), J. F. H. Suersen (a medical student from Kiel), G. Eimbke (a medical student from Hamburg who had studied with Gren), and Struensee (another of Gren's former students).[38]

Hermbstaedt soon followed up his attack by submitting an annotated version of the report to the *Chemische Annalen*. In the notes, he anticipated that Eimbke, Struensee, and Bourguet, all students of Gren and good phlogistonists, would inform their mentor of his and Klaproth's results. He also claimed that he had obtained the very same proportion of vital air in the experiment with Westrumb's calx as Klaproth had obtained in his experiment the preceding September. And he stressed the fact that "thirteen impartial eyewitnesses" (including himself!) could testify that he had not made a mistake. Turning from his experiments to his critics, he decried Trommsdorff's rudeness in charging him with "lying" and advised Hecker and Meier, "estimable men who can certainly observe with cold blood," to pay close heed to his rival's experimental techniques. He dealt with Gren's charge that he had tampered with his data by pointing out that the December 1792 issue of the *Journal der Physik* with its hydrostatic objections had not arrived in Berlin until February

---

37. Hermbstaedt, "Nachricht für Naturforscher," *Allgemeine Literatur-Zeitung: Intelligenzblatt* (13 April 1793), 279–280.

38. Hermbstaedt, "Neue Bestätigung einer chemischen Grund-Wahrheit," *Allgemeine Literatur-Zeitung: Intelligenzblatt* (24 April 1793), 319–320.

1793, long after his experiments with Lampadius had been reported. The reason so little pure air had been collected was that the retort had broken during the experiment. Turning to Westrumb, Hermbstaedt urged him to conduct another round of experiments. If "this worthy man" still obtained water rather than oxygen gas, "then I must believe that some unlucky star has led Prof. Klaproth and me, as well as all the eyewitnesses mentioned above, to err until now." Just in case Westrumb should continue getting the same results, Hermbstaedt called on the physicists Lichtenberg and Mayer to perform the experiment and render their verdicts. In concluding, he promised that neither "simple sophistries" nor "offensive expressions" would ever lead him to abandon his position. Apparently regarding the issue as settled, Hermbstaedt never again bothered to debate the reduction experiment with Trommsdorff, Gren, or Westrumb.[39]

## END OF THE DEBATE (MAY–NOVEMBER 1793)

Hermbstaedt's assault overwhelmed the phlogistic spokesmen. Each responded in his own way. Gren avoided direct comment. A conciliatory letter from J. B. van Mons suggested an easier line of retreat. Replying to van Mons in the July 1793 issue of the *Journal der Physik*, Gren expressed delight with the Belgian's admission that fresh calx yielded water vapor as well as oxygen gas. "No defender of oxygen," he exclaimed, "has ever been so open-hearted with me." He still thought, however, that the only way to establish that oxygen was a component of mercury calx was to obtain the gas from "glowing calx without first letting it cool and become bright red." Turning to more general issues, Gren addressed van Mons's concern that dissension was retarding the progress of chemistry:

> the truth can only be found and ensured through doubt and controversy. In a science based on facts, not faith, progress is only hindered by *jurare in verba magistri*, by belief in authorities, by blind acceptance of dogma.... The new chemistry and its nomenclature actually consists more of matters of faith, than of fact. Accordingly, the more the self-styled heralds and proselytes of this confession seek to acquire new proselytes with lists of famous converts, the more vigorously

---

39. Hermbstaedt, "Neue Bestätigung einer chemischen Grundwahrheit, den Gehalt des Sauerstoffs, im wasserfreyen Quecksilberkalke betreffend," *Chemische Annalen*, no. 1 (1793), 303–314. One indication that Hermbstaedt regarded the matter as settled was that he had his disciple Suersen write a history of his endeavors. See Suersen, "Einige Bemerkungen über die Entbindung der Lebensluft aus dem für sich verkalktem Quecksilber," *Chemische Annalen*, p. 415–426. Despite Hermbstaedt's appeals, neither Mayer nor Lichtenberg ever announced the results of their experiments with the calx sent them by Hermbstaedt in January 1793. Lichtenberg's first experiments—carried out by his students A. G. L. Lentin and K. G. Dengel with Gmelin as a witness—were inconclusive; so Hermbstaedt sent more calx. See Lichtenberg's diary entries (2 March, 27 April 1793), in *Georg Christoph Lichtenberg: Schriften und Briefe*, II: 773, 776; and his letter to Hermbstaedt (8 April 1793), in *Lichtenbergs Briefe*, ed. A. Leitzmann and C. Schüddekopf, vol. III (Leipzig: Dieterich, 1904), pp. 260–261.

must one resist the introduction of this symbolic doctrine into chemistry. . . . Truth does not depend on the number of its devotees. . . . Hence I find it laughable when *Hermbstaedt* in Berlin preaches in lofty tones to his countrymen that "nothing is more important for the new system than the conversion of a *Kirwan* and a *Klaproth.*" Is not that the talk of a proselytizer, rather than a dedicated philosopher? What do we want in science—to conquer or to convince? . . . In the *Reich* of truth, no authority rules . . . [and] the names of famous men can never serve as protection against the skepticism of those who would test their doctrines. Here freedom and equality reign in the laws of research, proof, acceptance, and rejection. This empire need not fear that it will suffer from differences of opinion among its subjects. Rather, such differences enlarge and protect it by assuring that no one has limited elbowroom, that no symbolic books can survive, that disagreements can occur, and that the names of particular burghers and associations offer no protection and create no guarantee of infallibility.

No longer was it the antiphlogistonists who had to defend themselves against charges of divisiveness. Now, confronted by defeat, it was Gren who felt obliged to defend phlogistic advocacy by appealing to the antiauthoritarian ideals of science.[40]

Months passed, then in October Gren learned that van Mons had an ingenious counterargument concerning the reduction experiment. At glowing heat, van Mons suggested, the red oxide of mercury quickly gave off much of its oxygen, turning into a blackish-red oxide of mercury. Hence, when this new compound was put into a pneumatic apparatus, it had little oxygen left to yield. Gren, seeing that van Mons's suggestion reconciled the contradictory experimental results, readily embraced it. Soon he was writing his ally Westrumb to announce that he had abandoned his phlogistic system. He had been persuaded by van Mons that mercury calx was indeed a compound containing the base of vital air. Moreover, he had been persuaded by Göttling and Trommsdorff that burning phosphorus could exhaust a vessel of its vital air. Accepting all the salient facts of Lavoisier's system, he now endorsed the compromise theory of the Silesian technical chemist, J. B. Richter. That is, except for retaining phlogiston as the matter of light in order to explain the radiation accompanying many chemical reactions, he accepted Lavoisier's theory.[41]

40. Van Mons, letter to Gren (24 June 1793), *Journal der Physik*, 7 (1793), 343–347; and Gren, letter to van Mons (16 July 1793), *Journal der Physik*, 7 (1793), 348–352. For the date of Gren's letter, see the *Chemische Annalen*, no. 1 (1794), 116.
41. Van Mons, letter to Gren (8 or 12 October 1793), *Journal der Physik*, 8 (1793), 3–13, reprinted in *Chemische Annalen*, no. 1 (1794), 116–128; Gren, letter to Westrumb, *Chemische Annalen*, no. 2 (1793), 341–345; "An das chemische Publikum" (22 November 1793), *Allgemeine Literatur-Zeitung: Intelligenzblatt* (11 December 1793), 1072; letter to Hecker (4 December 1793), *Journal der Erfundungen*, 2, no. 5 (1794), 132–133; and letter to van Mons (12

Unlike Gren, Westrumb reacted directly to the experiments in Berlin. He vented his feelings in the *Chemische Annalen*. Sarcastically approving of Hermbstaedt's publication in "several learned dailies," he hoped that

> the public will show me the justice and respect that it has hitherto granted every discoverer—that it will believe in my statements until the contrary has been proven by physicists and chemists who bring somewhat cooler blood to their work than seems to flow in the veins of the antiphlogistonists.

He also rehearsed his procedures and results, implying that the calx he had sent out in sealed bottles the previous December was not appropriate for a truly decisive test. Then, unwilling to struggle further, Westrumb retreated into silence. A few months later, upon learning of Gren's adoption of Richter's system, he confessed in notes to his former ally's letter that he too had changed his mind about the reduction of mercury calx. Some of Wiegleb's recent experiments as well as some experiments that he had done at Gmelin's behest had led him, like van Mons, to recognize the difference between bright-red calx and blackish-red calx. Nonetheless, he was unwilling to adopt Richter's system. Instead he preferred to revert to his old theory that every calx was a compound of an earth and water, the basis of vital air. These notes were virtually Westrumb's last contribution to pure chemistry. Feeling that he had been "treated like a dumb school kid," he was happy to focus on applied chemistry and leave "learned disputes" for others.[42]

Like Westrumb, Trommsdorff reacted to the experiments in Berlin with open hostility. In the *Journal der Physik*, he conceded that Hermbstaedt's experimental procedures, though liable to objection, could probably not be discredited. Then, wondering how Gren, Westrumb, and he had gone astray, he queried,

> Can all three of us have been mistaken? So often and grossly mistaken? Or do people credit us with so little love of truth that we would deny facts out of prejudice for an accepted hypothesis? But really! The thought is shameful.

His recent results agreed with those obtained by Westrumb. However, he was not going to comply with the "fashion" of naming his witnesses, which made

---

December 1793), *Journal der Physik*, 8 (1794), 14–18. For Göttling's and his student A. N. Scherer's experiments with phosphorous, see their untitled notices, *Allgemeine Literatur-Zeitung: Intelligenzblatt* (27 November 1793), 1023–1024; (21 December 1793), 1112. For Richter's system, see his *Ueber die neueren Gegenstände der Chymie*, vol. III (Breslau: Korn, 1793). In discussing the reduction experiment, Richter indicated that he found Hermbstaedt's support for Lavoisier's reliability convincing (pp. x–xi, 56–58, 64–65). For a biographical profile of Richter, see Appendix I.

42. Westrumb, "Ein paar Worte die Reduktion des Quecksilbers betreffend," *Chemische Annalen*, no. 1 (1793), 401–404; footnotes to Gren's letter announcing his conversion to Richter's system, *Chemische Annalen*, no. 2 (1793), 341, 344–345; and letter to Oersted (18 August 1801), in *Correspondence de H. C. Oersted avec divers savants*, ed. M. C. Harding, vol. II (Copenhagen: Aeschehoug, 1920), pp. 586–588. This letter reveals that Westrumb was still performing the reduction experiment for anybody willing to observe it.

it seem that "no chemist trusts another anymore." Soon afterwards, in a note announcing his article, Trommsdorff insisted that he was not impressed by Hermbstaedt's thirteen witnesses. Someday, as had already happened to so many, his "reason" might be "captivated by faith." But he would "never swear to the new flag."[43]

Months later when Trommsdorff learned of Gren's shift to Richter's system, he sent Crell his "last statement concerning the phlogistic and antiphlogistic systems." He declared himself "completely neutral," implying that it was too early for anyone to commit himself to one theory or the other. Many experiments were still needed. He respected all who did good work, regardless of their views. Humbled by the year's developments, Trommsdorff retreated to empiricism rather than throw himself "into the arms of the antiphlogistonists."[44]

Thus, one result of the debate over the reduction experiment was that three of the most active phlogistic spokesmen in Germany stopped opposing Lavoisier's system in the fall of 1793. Gren adopted a compromise theory that rested on most of Lavoisier's facts and doctrines. Westrumb turned from pure to applied chemistry. Trommsdorff renounced theorizing for empiricism. Their defection was a serious loss for the phlogistic camp.

The second and more important result of the debate was that it discredited the phlogistic spokesmen and their cause. This effect is evident, for instance, in the successive reactions to the debate by G. F. Hildebrandt, young professor of medicine and chemistry in Brunswick, and later in Erlangen. From July 1792 through January 1793, he was inclined to trust Westrumb's account of the reduction experiment. However, Hermbstaedt's experiments during January and February 1793 appeared to fulfill all the conditions specified by Gren and Westrumb. Puzzled, he observed in March 1793 that "these three are all men who know how to work in chemistry. One would not lightly mistrust their results if their disagreement did not make it necessary to believe that one or the other was mistaken." The experiments carried out in Berlin on 3 April 1793 ended his puzzlement. In June Hildebrandt sent a note off to the *Intelligenzblatt* in which he promised to show that "Hermbstaedt's remarkable confirmation of the evolution of fire air (oxygen) from red mercury calx still does not suffice to overthrow the phlogistic system." After further reflection, however, he transferred his allegiance to the antiphlogistic theory, giving Hermbstaedt's experiments much of the credit for his change.[45]

---

43. Trommsdorff, "Noch einige Versuche mit dem für sich verkalkte Quecksilber, in Hinsicht auf die Entbindung der Lebensluft daraus," *Journal der Physik*, 7 (1793), 332–337; and "Erklärung," *Medicinisch-chirurgische Zeitung*, no. 3 (8 July 1793), 64.

44. Trommsdorff, "Letzte Erklärung wegen der phlogistische und antiphlogistische Systeme," *Chemische Annalen*, no. 2 (1793), 334–341, republished with slight revisions in his new *Journal der Pharmacie*, 1, no. 2 (1794), 103–108. Unlike Westrumb, Trommsdorff later admitted that the antiphlogistonists had succeeded in establishing the veracity of Lavoisier's account of the experiment, thereby doing much to promote their cause. See his *Versuch einer allgemeinen Geschichte der Chemie*, facsimile of the 1806 ed., pt. 3 (Leipzig: Zentral-Antiquariat, 1965), pp. 111–113.

45. Hildebrandt, *Chemische und mineralogische Geschichte des Quecksilbers* (Brunswick:

## END OF THE REVOLUTION (FALL 1793-1796)

The defeat of Westrumb, Gren, and Trommsdorff marked the turning point in the German antiphlogistic revolution. Only a few veterans continued to resist the new doctrines. When Crell, their leader, learned of Gren's adoption of Richter's compromise theory, he threw together a confused article that argued simultaneously for an eclectic system and against most antiphlogistic tenets. Soon afterwards, in the annual foreword to the *Chemische Annalen*, he reviewed the "struggle between the phlogistic and antiphlogistic chemistry." He conceded that during 1793 "the number of powerful friends" of Lavoisier's system had increased considerably, "many defenders of the phlogistic system had begun to waver," and "others have even been convinced to go over to the opposition." He implicitly denied Hermbstaedt any credit for this turn of affairs by attributing the trend to Göttling's and Trommsdorff's recent experiments with phosphorus. Looking ahead, he hoped that the two systems and their respective supporters would enjoy a "peaceful coexistence."[46] Crell must have realized that this was wishful thinking. He and his friends were soon publishing new critiques of Lavoisier's system. Crell and Gmelin led the way in 1795; Wiegleb joined them the following year.[47]

Phlogistic advocacy failed to retard, much less reverse antiphlogistic consolidation. Most German chemists felt that the time for controversy had passed. For instance in 1795, A. N. Scherer, who had just begun teaching chemistry in Jena, suggested that the term "antiphlogistic" be dropped because it was "spiteful."[48] That same year Lampadius, who had succeeded C. E. Gellert as professor of chemistry in Freiberg, decided to base his lectures on Lavoisier's theory. Eagerness to be in step with prevailing opinion evidently motivated his decision, for he was still a phlogistonist.[49] This consideration certainly underlay Gren's and Trommsdorff's decisions of 1796 to publish their texts according to the new system.[50]

---

Schulbuchhandlung, 1793), pp. ix-x, 70-74; "Vergleichende Uebersicht des phlogistischen und des antiphlogistischen Systems," *Chemische Annalen*, no. 1 (1793), 559; untitled notice (17 June 1793), *Allgemeine Literatur-Zeitung: Intelligenzblatt* (20 July 1793), 568; and "Etwas über die Entbindung der Feuerluft aus Metallkalken," *Chemische Annalen*, no. 2 (1793), 24-30; no. 1 (1794), 210-212. For a biographical profile of Hildebrandt, see Appendix I.

46. Crell, "Einige Bemerkungen über das phlogistische und antiphlogistische System," *Chemische Annalen*, no. 2 (1793), 346-352, 406-423; and "Vorbericht" (30 December 1793), *Chemische Annalen*, no. 2 (1793), iii-v. Around the same time as he wrote this foreword, Crell urged that the debate continue until a truly decisive experiment could be found. See Crell, footnote, *Chemische Annalen*, no. 1 (1794), 36-37.

47. Crell, "Ueber Sauerstoff und Säure," *Chemische Annalen*, no. 1 (1795), 227-243; Gmelin, "Winke an seine Zeitgenossen, den Streit über den Brennstoff betreffend," *Chemische Annalen*, no. 1 (1795), 287-302, 391-409; and Wiegleb, "Ueber die Entstehung und Natur der sogenannten Stick- oder azotische Luft, und die daraus gezogenen Folgen," *Chemische Annalen*, no. 2 (1796), 467-493.

48. Scherer, *Grundzüge der neuern chemischen Theorie* (Jena: Göpferdt, 1795), p. xii. For a profile of A. N. Scherer, who was not related to J. A. Scherer in Vienna, see Appendix I.

49. Lampadius, *Sammlung practisch-chemischer Abhandlungen*, vol. I (Dresden: Walther, 1795), pp. 134-140.

50. In the preface to his text, Gren presented the leading tenets of both systems, then announced that, though he still believed in phlogiston (as the matter of light), he would base his

To hold out for phlogiston was to lay oneself open to charges of obstructing progress. The dogged phlogistonist Gmelin first objected that the antiphlogistic chemistry had become so "modish" that many were adopting it so as not to seem "old-fashioned" and hence unworthy of respect. Soon he was complaining that the "excited and fanatical friends of the new chemistry" treated him and his allies as "obstinate," "vain," "malodorous," "stone-blind defenders of the old system."[51] And Crell lamented that, despite "resounding speeches for tolerance on every occasion from every mouth," only "one chemical faith" was permitted. "Dissenters," it seemed to him, were "relentlessly persecuted, not with fire and sword, but with bitterness, derision, and pitying scorn."[52] Indirectly confirming such complaints, Hermbstaedt boasted a few years later that

> it was not without trepidation that I dared to provide my German countrymen with a translation of [Lavoisier's *Traité*] in 1792. . . . In those days it was a crime if one did not trail after the field marshals of phlogiston in leading strings. One was abused without mercy if one . . . did not constrain one's spirit in chains. Nevertheless, I did not shrink from the threat of crucifixion; my translation appeared on German soil; my annotations buttressed the new principles advanced in the book; and on account of my cogency, I like to think, the importance of the new system was soon generally recognized in Germany. . . . Now, eleven years later, . . . even the most zealous phlogistonists have sworn allegiance to the antiphlogistonists' flag; . . . an example that truth always asserts her rights.[53]

A new orthodoxy prevailed within the German chemical community. Indeed, Crell's persistent use of the *Chemische Annalen* on behalf of the phlogistic cause was evidently a major reason for the founding of a rival periodical for chemistry by young Scherer in 1798.[54] Despite Scherer's personal foibles,

---

text on "the so-called antiphlogistic system." See his *Grundriss der Chemie*, vol. I (Halle: Waisenhaus, 1796), pp. vii–x. Trommsdorff presented the "new system," not as a "blind adherent," but rather out of the belief that it had the "most probability" of being correct. See his *Lehrbuch der pharmaceutischen Experimentalchemie nach dem neuen System* (Altona: Verlagsgesellschaft, 1796), p. xi. Elsewhere he expounded on the pedagogical reasons for teaching chemistry according to the new system. See his description of his text in the *Journal der Pharmacie*, 4, no. 1 (1797), 341–344.

51. Gmelin, letter to Crell, *Chemische Annalen*, no. 1 (1796), 255; and letters to C. E. Moll (14 May 1797, 25 August 1798, 17 March 1799, 5 October 1801), in C. E. Moll, *Mittheilungen aus seinem Briefwechsel*, vol. I (Augsburg: Volckhart, 1829), pp. 220–233.

52. Crell, "Vorbericht," *Chemische Annalen*, no. 2 (1798), xiii–xiv.

53. A. L. Lavoisier, *System der antiphlogistischen Chemie*, trans. S. F. Hermbstaedt, 2d ed. (Berlin and Stettin: Nicolai, 1803), p. iii.

54. Upon first announcing his intention to found a "new periodical work for chemistry," Scherer claimed that he got the idea from "French and English chemists." See his untitled notice (Edinburgh, November 1797), *Allgemeine Literatur-Zeitung: Intelligenzblatt* (23 December 1797), 1384. Later, when giving a full outline of his plans for the journal, he claimed that a truly comprehensive periodical was not available. See his "Ankündigung eines allgemeinen Journals der Chemie, herausgegeben von Dr. Alex. Nic. Scherer, Leipzig bey

the *Allgemeines Journal der Chemie* rapidly supplanted Crell's journal as the German chemical community's forum.[55] The German antiphlogistic revolution, which had begun in 1789 and whose central episode was the mercury calx debate of 1792/93, ended with the triumph of Lavoisier's theory in 1795/96 and, as it were, the forced retirement of the community's convener soon afterwards.

The picture of the German chemical community that emerges from the debate over the reduction experiment corroborates that educed earlier from the *Chemische Annalen* (see Chapter VI). As before, we find that the community was concentrated in Protestant Germany. It was phlogistonists there who, undaunted by the authority of Lavoisier and his growing band of followers, dared to challenge his reliability. And it was antiphlogistonists there who, rather than comporting themselves with the arrogance of certain winners, shouldered the responsibility of answering Lavoisier's critics. By contrast, not a single phlogistonist or antiphlogistonist in Catholic Germany bothered to enter the lists. The debate, it is clear, occurred partly because chemists in Protestant Germany regarded one another as the primary arbiters of chemical truth.

Also as before, we find evidence in the debate of a periphery-center structure in the German chemical community. The available sources indicate the names of almost forty friends of chemistry and chemists who became involved in the dispute between June 1792 and October 1793. Many served as witnesses and/or laboratory assistants. Some corresponded with spokesmen who represented their viewpoints. Some published a note or two bearing on the debate. Some conducted independent trials of the reduction experiment. And a few acted as champions for the rival camps. Not surprisingly, the more demanding the role, the greater was the likelihood that recognized chemists played the part. Only one of the twenty men to serve solely as witnesses—Karsten—was among Crell's contributors between 1784 and 1789. By contrast, all five of the chief contestants published in the *Chemische Annalen* during the 1780s. And four—Gren, Hermbstaedt, Klaproth, and Westrumb—were among the community's leaders as of 1789. The social structure of the German chemical community was clearly reflected in the debate.

Besides corroborating prior impressions that the German chemical community was concentrated in Protestant Germany and structured on a periphery-center pattern, the debate supports the suggestion that German chemists'

---

Breitkopf und Härtel" (Weimar, June 1798), *Allgemeine Literatur-Zeitung: Intelligenzblatt* (8 August 1798), 945–948. Later yet, Scherer complained, without specifically mentioning Crell's name, that Crell's journal was a mere newspaper, chaotic, and partial to its editor. See his foreword (November 1798), *Allgemeines Journal der Chemie*, 1 (1798), iv–v.

55. Crell continued to edit the *Chemische Annalen* until 1804, then agreed to close his journal and become a member of the editorial board of the *Neues allgemeines Journal der Chemie*. For details, see the biographical profiles of Crell, Scherer, and A. F. Gehlen in Appendix I.

backgrounds and occupations predisposed them to empiricism (see Chapter IV). Their empirical orientation was manifest in the debate's subject. By treating the reduction of mercuric oxide as a crucial experiment that could determine the direction of chemistry, German chemists drastically simplified the problem of identifying the most promising theory. Granted, this experiment had an important place in Lavoisier's case against phlogiston and in his arguments for his own system. Still, most European phlogistonists found ways to explain away Lavoisier's account of his findings. Leading German phlogistonists were virtually unique in searching, or acquiescing in the search, for evidence that Lavoisier had botched the reduction experiment. Once Gren and Westrumb committed themselves to opposing Lavoisier's theory by questioning its empirical underpinnings, the leading German antiphlogistonists readily agreed that Lavoisier's reliability was the central issue. In agreeing to limit the controversy in this way, they were probably mindful of their community's antipathy to theory. And they were definitely expressing their own propensity for "facts, not *Raisonnement*."

The debate's rhetoric also manifested the German chemical community's empiricism. Throughout the dispute, the rival spokesmen repeatedly appealed to empirical norms as a means of reinforcing their own credibility and undermining their adversaries' claims. To judge from this rhetoric, the German chemist was expected by his peers to experiment, not just cerebrate. When experimenting, he was expected to start with pure reagents and clean apparatus and to carry through his work without appreciable contamination or loss. As an experiment proceeded, he was expected to observe all significant details with cold-blooded impartiality. And in reporting his results, he was expected to resist all temptation to rectify his data according to preconceived theory. In short, members of the German chemical community expected one another to be diligent, clean, adroit, detached, and scrupulous experimenters. Even when they were in profound disagreement over theoretical matters, they could agree on this standard of performance.

Of course, espousing this standard was not the same as living up to it. Gren was not experimenting. Westrumb and others must have bollixed the reduction of mercuric oxide. Phlogistonists and antiphlogistonists alike issued biased reports about their procedures and results. Such behavior naturally exacerbated what was already an intense contest for the community's allegiance. In doubt about one another's detachment, the rival spokesmen went outside the community in hopes of finding impartiality. They mustered witnesses for their experiments, appealed to physicists for arbitration, and played to the *Intelligenzblatt*'s readership. Only as the phlogistonists sensed that the verdict was coming in against them did they proclaim the impropriety of looking to outsiders for assurance and support. By then, however, they and their cause had been discredited.

Interacting as peers and yet divided into warring sects, German chemists acquired full self-awareness as a German chemical community during the antiphlogistic revolution. Afterwards they quickly closed ranks, achieving greater social solidarity than ever before. Indeed, the few remaining phlogistonists, among them Crell, Westrumb, and Wiegleb, were virtually ostracized. Its formation complete, the young community was strong enough to dispense with all those who persisted in opposing the new consensus.

# 9 CONCLUSION

Social support for chemistry, the essential precondition for the creation of the German chemical community, increased markedly in Germany after 1720. Respect for the science grew. Patronage for its teaching expanded. Contributors to its advancement multiplied. This growth derived primarily from a new enthusiasm in Germany for health and prosperity. As educated and powerful Germans became more interested in material progress, they became more receptive and responsive to claims that chemistry was a broadly useful science deserving of their support.

By the 1780s chemistry was a *Lieblingswissenschaft*. To judge from subscription lists (our best indicator of moral support), men working in many occupations and living across Germany thought well enough of chemistry to have their names publicly associated with it. Yet, evidently reflecting opportunities for learning about the science, chemistry's public consisted largely of physicians and pharmacists. To judge from salaried chemical positions (our best indicator of material support), the professors and officials who managed higher education believed that administrators as well as physicians should know some chemistry. Yet, acting mainly within preexistent institutions, these men channeled most funds for chemical instruction into medical schools. To judge from the better-known German chemists and Crell's German contributors (our best indicators of manpower support), respect and patronage for chemistry gave rise to increasing participation in the science. By the 1780s some twenty Germans had an ongoing involvement in chemical research and another one to two hundred were occasionally carrying out original investigations. Those choosing to cultivate chemistry were, for the most part, men whose educations and occupations directed them toward the science's uses. They had begun learning it as medical students or as pharmaceutical apprentices. They were making part or all of their livings as physicians, as apothecaries, as technical chemists, and as teachers of chemistry whose students were primarily interested in the science's applications. Rooted, or mired if you will, in their culture's utilitarian support for chemistry, they were predisposed to value techniques and facts.

In 1778, taking advantage of German moral, material, and manpower support for chemistry, Crell founded a chemical periodical. During the next decade he resolutely developed it into a flourishing monthly. His early foreign travels must have helped prepare him for this venture, both by reinforcing his identification with Germany and by alerting him to French and especially British activity in the science. But it was his experiences after returning home that were crucial. His slow start in academia put him on the lookout for opportunities to make a name for himself. His service as professor of chemistry in Brunswick improved his knowledge of the science and its prospects. And his contacts with Baldinger, editor of the *Magazin vor Aerzte*, showed him how to go about editing a journal. Of course Crell was not the only German who was eager for renown, confident about chemistry's outlook, and familiar with Baldinger's example. In fact Weber and Göttling also started chemical journals around 1780. Still, first off the mark and more than astute in exploiting his advantage, it was Crell who got the satisfaction of establishing a forum for German chemistry.

Crell's forum opened the way to social solidarity for chemists in Germany. Thanks to the prior emergence of support for chemistry, many had already specialized sufficiently to place a value on an expert audience. Moreover, thanks to the growing popularity of cultural nationalism, many had already started thinking of themselves, not simply as chemists, but as "German chemists." As these men read and published in the *Chemische Annalen*, they easily began interacting as a national discipline-oriented community. Most, to judge from the patterns of referencing in Crell's journal and participation in the antiphlogistic revolution, looked to a few productive chemists for empirical information and theoretical guidance. These visible chemists, with the notable exception of Achard, comprised the community's central leadership group. That is, whether cooperating or competing, Crell, Gmelin, Gren, Hermbstaedt, Klaproth, Westrumb, and Wiegleb determined the direction of German chemistry between the mid-1780s and mid-1790s.

The importance of these disciplinary leaders is evident throughout Germany's antiphlogistic revolution. Initially, prominent chemists there rejected Lavoisier as a French fashionmonger. They were confident that, with appropriate modifications, existing theory would take account of the new findings from gas chemistry. However in 1789, with the publication of the *Traité* and especially the conversions of Hermbstaedt and others in Germany, leading German chemists began to see that Lavoisier's system would have to be discredited, or embraced. From the outset, Crell, Westrumb, and Wiegleb reacted to the appearance of antiphlogistonists within the community's ranks as criticism of their leadership. Nonetheless, Hermbstaedt and his allies continued to press their case and win new recruits, including Klaproth in the spring of 1792. The chief phlogistonists, joined by Gren, desperately counterattacked. Reverting to habits of thought fostered by chemistry's applications, they questioned the experimental abilities of Lavoisier and, by implication, all his German supporters. They did so by challenging the reliability of Lavoisier's account of the reduction of mercuric oxide, thereby elevating this experiment to the status of a crucial experiment. For a time the phlogistonists dared to hope for vindication. Then in April 1793, Hermbstaedt, aided by Klaproth, established to the community's satisfaction that it was the phlogistonists, not Lavoisier, whose skills were deficient. While Gren soon reconciled himself to the new orthodoxy and Westrumb withdrew from the fray, Crell, Wiegleb, and Gmelin continued to argue for the phlogistic cause. However, German chemists were no longer interested in polemics. Convinced that the revolution was over, they rapidly closed ranks behind Lavoisier's champions. It was time, they thought, to move forward, even though this meant parting company with their original leaders.

## IMPLICATIONS

The history of the formation of the German chemical community has its own intrinsic interest. I for one am intrigued by the number and variety of Germans who joined chemistry's audience during the late Enlightenment, by the increasing dedication with which chemists approached their science, by the ingenious tactics with which Crell developed the first periodical for chemistry, by the nationalistic rhetoric with which the phlogistonists advocated their cause, and by the rapidity with which Hermbstaedt and his allies achieved hegemony after April 1793. Yet this monograph is also of interest for the light that it sheds on larger topics—the growth of science in eighteenth-century Germany, the emergence of national discipline-oriented communities, and the role of social factors in scientific revolutions.

During the eighteenth century, the present study suggests, Germans became increasingly supportive of the natural sciences. They did so in response much less to foreign trends than to domestic developments. To be sure, Germans were not indifferent to the concurrent flowering of esteem, patronage, and enthusiasm for natural knowledge elsewhere in Europe. Indeed, emulating the French, Frederick II of Prussia and two other German princes established academies with salaried positions. But to judge from the case of chemistry, these rulers' hopes for glory motivated only a fraction, albeit a striking fraction, of the new moral, material, and manpower support provided the natural sciences in eighteenth-century Germany. Widespread desire for moral uplift, though it did not come into play in our case, and burgeoning zeal for material progress evidently motivated far more of this new support. The mundane motivation, we have seen, benefited the natural sciences by inclining educated and powerful Germans to presume that scientific knowledge is useful knowledge. Significantly, this presumption tended to engender support for the natural sciences along disciplinal lines. The men most likely to step forward as each discipline's patrons, spectators, and participants were, to judge from the case of chemistry, that discipline's actual and anticipated beneficiaries. As the new utilitarianism was motivating fresh support for the republic of letters' scientific wing, it was also laying the basis for this republic's fragmentation into discipline-oriented communities.

In Germany the intelligentsia's dissolution into specialized communities may well have been inaugurated by the formation of the German chemical community. Yet the success of German periodicals for astronomy, botany, physics, and physiology suggests that these sciences also became the foci of discipline-oriented scientific communities in the late eighteenth century. Thanks to extensive social support, specialized periodicals, and cultural nationalism, German science had reached a significant juncture. Hitherto, patrons and users had exercised a large influence on the research of Germans seeking to advance the natural sciences. Now salient figures in the new discipline-oriented communities began, as we saw in the German antiphlogistic revolution, to play the major role in setting priorities. This shift was of great potential significance. Disciplinary leaders were advantageously situated to promote more rigorous standards, to encourage attention to basic issues, and to represent disciplinary interests in higher councils. They were in an excellent position both to foster and to benefit from developments which, between the 1790s and 1870s, transformed German universities into centers for advanced instruction, specialized research, and predictable careers. They were, in brief, ready to professionalize their disciplines.

Besides setting the stage for the eventual professionalization of the sciences, the formation of national discipline-oriented communities resulted in a quickening of the pace of scientific change. In particular, major theoretical changes no longer took so many decades as to vitiate, by my way of thinking, the treatment of them as "revolutions." Once scientists have coalesced into a disciplinary community, such changes can proceed so quickly as to approximate revolutions in politics. Like a revolution in a polity, a revolution in a scientific community is, to judge from the German antiphlogistic revolution, influenced from the outset by the community's structure and outlook. New facts or theories do not in themselves have the power to provoke a sense of crisis about a discipline's direction. Rather it evidently requires the agitation of one or more dissidents within the community's leadership group to invest novel information with such power. At first the dissidents find that marginal figures in the community, whether beginners or mature workers in neighboring fields, are more susceptible to recruitment than their colleagues. Within a few years, however, they are so successful in winning recruits or, perhaps with a crucial experiment, in discrediting (not necessarily disproving) their rivals, that everyone senses that the revolution has passed its turning point. At this juncture, the pressures for consensus in the community evidently become irresistible. Its members rapidly consolidate behind the revolutionary leaders, spurning all those who, like Crell, Wiegleb, and Gmelin, seek to prolong debate. Scientific revolutions are, in short, social as well as intellectual upheavals.

## EPILOGUE

During the two decades following the antiphlogistic revolution, conditions were not favorable for chemical research or recruitment in Germany. Military engagements, political realignments, and economic dislocations crowded one after the other. German chemists found it difficult to maintain their concentration and obtain the wherewithal for their inquiries. They often had to stand by as youths whom they might otherwise have won for chemistry were caught in the flow of events. To make matters worse, ideological developments were rapidly undermining the intelligentsia's moral support for the science. Opposition to the Enlightenment, which had been growing even before the French Revolution, fed first on fear of Jacobinism, then on resistance to Napoleonic imperialism. Young intellectuals who could not abide the cant and compromises of enlightened absolutism joined with conservatives in denouncing the Enlightenment's advocates. They succeeded in branding their foes as misguided partisans of an alien ideology, or, more kindly, as stodgy relics of a bygone era. As disenchantment with the Enlightenment grew, chemistry lost its status as a *Lieblingswissenschaft*. To the new generation of intellectuals, the science's reputation for usefulness seemed banal. Beauty, not utility, was their inspiration. They were also skeptical of the chemist's belief that natural secrets could be fathomed solely by analysis. In their opinion no single discipline held the key to nature.[1]

Though belonging to the preceding generation, the physicist G. C. Lichtenberg understood and sympathized with this new viewpoint. He crafted it into a fictitious "dream," which he published in 1794. As the dream opened, Lichtenberg was high above the earth, facing an old man who inspired his devotion and trust. The ancient gave him a small sphere, asked him to determine its nature in a nearby laboratory, then disappeared. Lichtenberg wiped the sphere clean, tested its electrical and magnetic properties, and measured its hardness and specific weight. Then he analyzed it, finding three different earths, iron, salt, and one unknown substance. Just as he finished, the ancient reappeared, studied the results with a smile, then announced that the sphere was really the Earth. Astounded, Lichtenberg asked what had happened to the oceans, only to learn that they were on his dust cloth. The elder went on to recount the consequences of Lichtenberg's various tests, then gave him a bag, told him to test its contents chemically, and again vanished. Thinking that the bag might contain the sun or a planet, Lichtenberg resolved to proceed with greater care. But when he opened the bag, he found an old book in an incomprehensible script. All he could read was the title page: "Test this chemically, my son, and tell me what you find." Lichtenberg wondered how he could determine the meaning of a book by chemical means. A chemical analysis would yield no more than rags and ink. Suddenly, the light dawned on him. He shouted out, "I understand, I understand, Immortal Being; oh forgive me; I accept your good reproof!"[2]

Though confronted by such reproofs as well as daily tribulations, the situation of German chemists was not entirely bleak during this unsettled era. The bureaucrats, an increasingly powerful breed, continued to value and patronize chemistry on account of its broad utility. They made substantial improvements in material support for the science in schools and academies in Berlin,

---

1. For an illuminating discussion of the rebellion against the Enlightenment, see H. Brunschwig, *La Crise de l'état prussien à la fin du xviii$^e$ siècle et la genèse de la mentalité romantique* (Paris: PUF, 1947). Although the romantics and idealists rejected the Enlightenment's utilitarian approach to chemistry, they seem to have had no qualms about seizing upon a few of the science's concepts for their own projects. See P. Kapitza, "Die frühromantische Theorie der Mischung," *Münchener Germanistische Beiträge*, 4 (1968); and D. von Engelhardt, *Hegel und die Chemie: Studie zur Philosophie und Wissenschaft um 1800* (Wiesbaden: Pressler, 1975).

2. For this "dream," see *Georg Christoph Lichtenberg: Schriften und Briefe*, ed. W. Promies, vol. III (Munich: Hanser, 1972–1974), pp. 108–111, K47–K49.

Erlangen, Freiberg, Göttingen, Jena, Kiel, Marburg, Munich, and Prague. Indeed, these improvements more than offset the reduction in the number of salaried chemical positions that resulted from the closing of many small universities.[3]

Besides the bureaucrats' reassurance, German chemists were also able, because they were bound together in a self-aware community, to give one another valuable encouragement. Whenever they acknowledged indebtedness to prior work, they reinforced their colleagues' desire to go on with their research. Whenever they achieved something significant, they animated their colleagues' spirit of emulation. Those in favorable positions could go beyond such informal encouragement. They frequently arranged for fellow chemists to be named to scientific societies or awarded honorary doctorates. And, thanks to increasing deference to the community's judgments of quality, they occasionally helped place younger colleagues in administrative, teaching, or academic jobs.[4]

Drawing strength from one another, German chemists strove to cope with the difficult situations that confronted them. Some, to be sure, were distracted, even overwhelmed, by military, political, and economic developments. Others responded to the intelligentsia's new disdain for utilitarianism and specialization by devoting less attention to chemistry. But most German chemists, though they were no longer so willing to devote their own time to useful applications, remained convinced that they were deserving of support on account of their science's potential for improving health and increasing production. Moreover, though they tempered their claims, they remained convinced that chemistry was an indispensable key to nature's secrets.[5]

These convictions served German chemists well as they transformed their discipline-oriented community into a profession during the decades following the reestablishment of peace in 1815. Liebig, his allies, and his successors were able to capitalize, without corrosive cynicism, on chemistry's reputation for usefulness. They harnessed the energies of upwardly mobile youths who were willing to gamble that a knowledge of the science would prove valuable. They cultivated the support of officials who envisioned how chemistry

    3. Considerable material bearing on patronage for chemical instruction and research between 1795 and 1815 appears in Appendices I and II. Also see the *Berlinisches Jahrbuch für die Pharmacie*, 17 (1816), 310–313; and G. Lockemann, "Der chemische Unterricht an den deutschen Universitäten im ersten Viertel des neunzehnten Jahrhunderts," in *Studien zur Geschichte der Chemie, Festgabe Edmund O. v. Lippman*, ed. J. Ruska (Berlin: Springer, 1927), pp. 148–158.
    4. For numerous examples of such mutual reinforcement, see the chemists born after 1760 in Appendix I. Also see W. Prandtl, *Deutsche Chemiker in der ersten Hälfte des neunzehnten Jahrhunderts* (Weinheim: Verlag Chemie, 1956).
    5. The development of chemistry in Germany between 1795 and 1815 has only recently begun to receive serious attention. For studies illuminating how German chemists, especially younger chemists influenced by *Naturphilosophie*, developed plant chemistry and responded to chemical atomism, see R. Löw, "Pflanzenchemie zwischen Lavoisier und Liebig," *Münchener Hochschulschriften: Reihe Naturwissenschaften*, 1 (1977); Löw, "The Progress of Organic Chemistry during the Period of German Romantic Naturphilosophie (1795–1825)," *Ambix*, 27 (1980), 1–10; and A. J. Rocke, "The Reception of Chemical Atomism in Germany," *Isis*, 70 (1979), 519–536.

could promote health and production and realized how an active professor of chemistry could enhance a university's renown and augment its attendance. Drawing upon the first group for recruits and the second group for funds, they revolutionized chemical education by bringing ever larger numbers of students into ever bigger laboratories to learn the latest techniques and theories. In doing so, they made chemistry into a profession with regular careers by 1870 and prepared the way for the internationalization of this profession by 1900.[6]

Even as chemistry was becoming a profession, it was branching into narrower specialty-oriented communities. The first such communities focused on organic, inorganic, physiological, and physical chemistry. In this century the process of branching has continued, evidently keeping pace with the rapid growth in the population of research chemists. Today, to judge from the titles of periodicals, the chemical profession contains several subprofessions and scores of specialty-oriented communities. In the light of these developments, the formation of the German chemical community in the late eighteenth century appears as an important beginning of a recurrent process that has helped shape modern science.

6. B. H. Gustin, "The Emergence of the German Chemical Profession 1790–1867" (Diss., University of Chicago, 1975); O. Krätz, "Der Chemiker in den Gründerjahren," in *Der Chemiker im Wandel der Zeiten*, ed. E. Schmauderer (Weinheim: Verlag Chemie, 1973), pp. 259–284; and H.-W. Schütt, "Zum Berufsbild des Chemikers im Wilhelminischen Zeitalter," in ibid., pp. 285–309.

# Appendix I

# BIOGRAPHICAL PROFILES

This appendix presents profiles of sixty-five men born between 1635 and 1775 and living between 1700 and 1800 who made their names as chemists while residing in Germany. I have selected most of them on the basis of their contemporary renown, as revealed by lists of contributors to chemistry drawn from the work of eighty-two observers of the eighteenth and early nineteenth centuries. To be precise, fifty-eight of the sixty-five men were selected because five or more of these observers identified them as meritorious chemists as indicated on Table A1. The remaining seven men were chosen for a variety of reasons: Scopoli, Jacquin, and Ingen-Housz, because they were the best-known foreign-born chemists active in Germany; Born, because he played a leading role in the antiphlogistic revolution in Hapsburg Germany; and Wurzer, Link, and Pfaff, because, after becoming professors, they received two or more calls to German university chairs involving responsibility for chemistry. The biographical profiles supplement the text in two important ways. First, whenever one of the sixty-five chemists is mentioned in the text, the

TABLE A1: Recognition of German Chemists as Meritorious Contributors to Chemistry

| Number of recognizers | Chemists born before 1676 | Chemists born 1676–1700 | Chemists born 1700–1725 | Chemists born 1726–1750 | Chemists born 1751–1775 |
|---|---|---|---|---|---|
| 5–8: | Dippel<br>J. Hoffmann<br>Schlüter | Eller<br>Gericke | Andreae<br>Cartheuser<br>Ludolf<br>Mangold | Bucholtz<br>Cartheuser<br>Heyer<br>Weber<br>Wenzel<br>Winterl | Bucholz<br>Gehlen<br>Girtanner<br>Hildebrandt<br>Lampadius<br>Rose<br>Scherer<br>Suckow |
| 9–16: | Bohn<br>Wedel | Henckel<br>Juncker<br>Rothe<br>Teichmeyer | Gellert<br>Lehmann<br>Meyer<br>Spielmann<br>Vogel | Gerhard<br>Gmelin<br>Hagen<br>Leonhardi<br>Meyer<br>Poerner<br>Weigel | Achard<br>Göttling<br>Richter<br>Trommsdorff |
| 17–32: | F. Hoffmann<br>Kunckel | Neumann<br>Pott | Cramer<br>Marggraf | Crell<br>Klaproth<br>Wiegleb | Gren<br>Hermbstaedt<br>Westrumb |
| 33+: | Stahl | — | — | — | — |

reader can turn to his profile for background information. Second, because the profiles are presented in the order of the chemists' births, the reader can consult them serially to get a detailed sense for trends and contexts. Besides supplementing the text, the profiles may serve as starting points for further work. They may stimulate fresh monographs on important neglected figures such as Kunckel, Pott, Marggraf, Cramer, and Westrumb. And they may facilitate needed studies of the Ludolf-Mangold quarrel, the Pott-Eller feud, and the pinguic acid controversy.

The profiles follow a set format: identification, background, education, career, state honors, major scientific societies and honorary degrees, informal recognition, bibliography.

*IDENTIFICATION:* Each profile begins with the chemist's name, in the form used most commonly in his publications, except that Latin forms have been germanized. Unused baptismal names appear in parentheses. Immediately below the chemist's name are the years and places of his birth and death. Here and elsewhere in the profiles, I follow custom in anglicizing German place-names (e.g., Vienna) and using pre-Soviet forms for Russian and East European place-names (e.g., St. Petersburg). I indicate the approximate location of villages and burgs by giving the name of a nearby large town or the territory.

*BACKGROUND:* Following the identification, each profile presents the chemist's father and his religion. I give the father's name, vital data, education, and career, using in this and subsequent sections the following abbreviations:

Mag. Phil., Ph.D. = Master of Philosophy, Doctor of Philosophy; degrees granted by university philosophical faculties upon satisfactory completion of the required examinations and disputations and payment of the specified fees.

Bac. Med., M.D. = Baccalaureate of Medicine, Doctor of Medicine; degrees granted by university medical faculties upon satisfactory completion of the required examinations and disputations and payment of the specified fees. Candidates for the M.D. were usually expected to defend, but not necessarily to write, an inaugural dissertation.

PD = Privatdozent or private lecturer; a teacher who had obtained the faculty's permission to lecture but, usually, was still awaiting an official appointment and salary.

AOP = Extraordinary professor; a professor who, though he had an official appointment, usually did not receive any salary or share in the student fees for matriculation or degrees.

OP = Ordinary professor; a professor who typically received a salary and shared in student fees.

The names of relatives, teachers, colleagues, and friends who have profiles in this appendix appear in italics. In giving the religion, I often rely on such indirect evidence as the chemists' Christian names or the prevailing denomination in his birthplace.

*EDUCATION:* After the background, each profile recounts the chemist's education, emphasizing his training in chemistry and his contacts with chemists.

*CAREER:* Turning from the youth to the mature man, each profile next describes the chemist's career. I summarize all available information on his primary position and auxiliary posts, his applications for jobs and rejections of offers, his successful ventures and unrealized schemes, his income and assets, his moves and travels. When the chemist was employed in two or more institutions at the same time, I deal with the institutions in the order in which he was hired, tracing his career in a given institution before proceeding to the next. In naming students, subordinates, colleagues, correspondents, and friends, I only indicate chemists and patrons of chemistry. Though chemists with M.D.'s often took on private patients to supplement their income, I only list a private medical practice when this was a man's sole means of support. And though chemists often received royalties, I only report on publishing when the chemist edited a periodical dealing with chemistry.

*STATE HONORS:* As a means of indicating official recognition of achievement, each profile lists the chemist's honorific titles in chronological order.

*MAJOR SCIENTIFIC SOCIETIES AND HONORARY DEGREES:* As a means of indicating general intellectual recognition of achievement, each profile lists scientific society memberships and honorary degrees in chronological order. In giving memberships, I restrict consideration to ten major societies, using the following abbreviations:

Berlin = Society of Sciences (founded in 1700), which became the Academy of Sciences and Fine Literature in 1744. For membership, see K.-R. Biermann and G. Dunken, *Biographischer Index der Mitglieder: Deutsche Akademie der Wissenschaften zu Berlin* (Berlin: Akademie Verlag, 1960).

Berlin NF = Society of Scientific Friends (founded in 1773). For membership, see *Verzeichnis der Mitglieder der Gesellschaft Naturforschender Freunde seit ihrem Bestehen 1773–1907* (Berlin: Starcke, 1907).

Erfurt = Academy of Useful Sciences (founded in 1754). For membership, see the articles by R. Thiele and D. Oergel in Akademie gemeinnütziger Wissenschaften zu Erfurt: *Jahrbücher*, N.F. 28 (1902), 1–46; N.F. 30 (1904), 1–224.

Göttingen = Society of Sciences (founded in 1751). For membership, see M. Arnim, *Mitglieder-Verzeichnisse der Gesellschaft der Wissenschaften zu Göttingen 1751–1927* (Göttingen: Dieterich, 1928).

Leopoldina = Collegium Naturae Curiosorum (founded in 1652), which became the Imperial Leopoldine Academy in 1687 and the Imperial Leopoldine-Caroline Academy in the eighteenth century. Its seat was at the residence of the president until 1879. For membership, see J. D. F. Neigebaur, *Geschichte der kaiserlichen Leopoldino-Carolinischen Deutschen Akademie der Naturforscher während des zweiten Jahrhunderts ihres Bestehens* (Jena: Frommann, 1860), pp. 185–314.

London = Royal Society (founded in 1662). For membership, see *The Rec-*

*ord of the Royal Society of London for the Promotion of Natural Knowledge*, 4th ed. (London: Royal Society, 1940), pp. 517–566.

Munich = Academy of Sciences (founded in 1759). For membership, see *Geist und Gestalt: Biographische Beiträge zur Geschichte der Bayerischen Akademie der Wissenschaften*, supp. vol., pt. 1, *Gesamtverzeichnis der Mitglieder . . . 1759–1959*, ed. U. Thürauf (Munich: Beck, 1963).

Paris = Academy of Sciences (founded in 1666), which became the Institut in 1795. For membership, see *Index biographique des membres et correspondants de l'Académie des sciences du 22 décembre 1666 au 15 novembre 1954* (Paris: Gauthier-Villars, 1954).

St. Petersburg = Academy of Sciences (founded in 1725). For membership from 1725 to 1803, see *Procès-verbaux des séances de l'Académie impériale des sciences depuis sa fondation jusqu'a 1803*, 4 vols. (St. Petersburg: Akademiia nauk, 1897–1911). For membership from 1827 to 1847, see *Recueil des actes de la séance publique de l'Académie impériale des sciences de St. Petersbourg* 1–20 (1827–1848). For men who became members after 1803 but died before 1827, I have relied on Dr. B. V. Levshin, Director of the Academy's Archives in Moscow.

Stockholm = Academy of Sciences (founded in 1739). For membership, see E. W. Dahlgren, *Kungl. Svenska Vetenskapsakademien: Personförteckningar, 1739–1915* (Stockholm: Almqvist-Wiksells, 1915).

*INFORMAL RECOGNITION:* As a means of indicating recognition of achievement within chemistry, each profile provides a chronological listing of observers who identified the man as a meritorious chemist. In cases where an observer identified a given chemist as meritorious on various occasions, I only report the observer's earliest identification. I use the following abbreviations for my sources:

1710?: *Rothe* = Rothe, G. *Gründliche Anleitung zur Chymie*. 3d ed. Leipzig: Eysseln, 1727, p. 10.

1718: *Stahl* = Stahl, G. E. *Zufällige Gedancken . . . über . . . Sulphure*. Halle: Wäysenhaus, 1718, pp. 13–19, 55.

1729: *Teichmeyer* = Teichmeyer, H. F. *Institutiones Chemiae*. Jena: Bielck, 1729, p. [v].

1732: Boerhaave = Boerhaave, H. *Elementa Chemiae*. Vol. I. Leiden: Severin, 1732, pp. 27–29.

1736: Burghart = Burghart, G. H. *Zum allgemeinen Gebrauch Wohleingerichtete Destillier-Kunst*. Breslau: Korn, 1736, pp. 43–45.

1737?: *Neumann* = Neumann, C. *Chemiae Medicae*. Edited by C. H. Kessel. Vol. I, pt. 1. Züllichau: Waysenhaus, 1749, pp. 32–50.

1740: Kestner = Kestner, C. W. *Medicinisches Gelehrten-Lexikon*. Jena: Meyer, 1740, pp. 256, 451, 466, 478, 578, 585, 709, 723, 806, 813, 866.

1741: Shaw = Boerhaave, H. *A New Method of Chemistry*. Edited by P. Shaw. 2d ed. Vol. I. London: Longman, 1741, p. 63.

1742: Anon. = *Grosses vollständiges Universal-Lexicon*. Edited by J. H. Zedler. Vol. XXXIV. Leipzig: Zedler, 1742, pp. 1109.

1748: *Mangold* = Mangold, C. A. *Chymische Erfahrungen.* Erfurt: Nonne, 1748, p. [i].
1749: *Juncker* = Juncker, J. *Conspectus Chemiae.* 3d ed. Vol. I. Halle: Waysenhaus, 1749, pp. 28–40.
1750: *Gellert* = Gellert, C. E. *Anfangsgründe zur Metallurgischen Chimie.* Leipzig: Wendler, 1751, p. [ix].
1752: *Börner* = Börner, F. *Nachrichten von den vornehmsten Lebensumständen und Schriften jeztlebender berühmter Aerzte und Naturforscher in und um Deutschland.* Vol. II. Wolfenbüttel: Meissner, 1752, p. 487.
1753: *Vogel* = *Göttingische Anzeigen von gelehrten Sachen* (26 November 1753), 1284.
1756?: *Cullen* = Cullen, W. Manuscript of his chemical lectures (ca. 1756). National Library of Medicine, Bethesda, Maryland.
1758?: *Cullen* = *An Eighteenth Century Lectureship in Chemistry.* Edited by A. Kent. Glasgow: Jackson, 1950, p. 27. [Note: A. Donovan informs me that this study is based on a Cullen manuscript of ca. 1758, not 1756.]
1758: *Hoffmann* = Hoffmann, G. A. *Unterricht in der Chymie.* Reprint of the 1758 ed. Gotha: Ettinger, 1774, p. 14.
1759: *Wallerius* = Wallerius, J. G. *Chemia Physica.* Vol. I. Stockholm: Salvi, 1759, p. 19.
1760: *Wallerius* = Wallerius, J. G. *Chemiae Physicae.* Vol. I. Stockholm: Salvi, 1760, pp. 35–40.
1766: *Lehmann* = See Freyberg, B. von. "Johann Gottlob Lehmann." *Erlanger Forshungen: Reihe B: Naturwissenschaften,* 1 (1955), 59.
1767: *Baldinger* = Baldinger, E. G. *Ehrengedächtnis des Herrn Christoph Andreas Mangold.* Jena: Cuno, 1767, p. 30.
1767: *Rose* = *Allgemeine deutsche Bibliothek,* 5, no. 1 (1767), 220.
1767: *Zückert* = *Allgemeine deutsche Bibliothek,* 4, no. 2 (1767), 299.
1768: *Westfeld* = *Allgemeine deutsche Bibliothek,* 7, no. 1 (1768), 297.
1769: *Suckow* = Suckow, L. J. D. *Entwurf einer physischen Scheidekunst.* Frankfurt a. M. and Leipzig: Göbhardt, 1769, p. 7.
1770: *Madihn* = Lange, J. J. *Grundlegung zu einer chemischen Erkenntnis der Körper.* Edited by J. J. Madihn. Halle: Curt, 1770, pp. 7–9.
1770: *Wiegleb* = Wiegleb, J. C. *Vertheidigung der Meyerischen Lehre vom Acido Pingui.* Altenburg: Richter, 1770, pp. [i–iii].
1771, 1773: *Cartheuser* = Cartheuser, F. A. *Mineralogische Abhandlungen.* 2 vols. Giessen: Krieger, 1771–1773, I: 46, 180, 182; II: 188.
1773: *Gerhard* = Gerhard, C. A. *Beiträge zur Chymie.* Vol. I. Berlin: Himburg, 1773, pp. [iii], 8.
1773: *Weber* = Weber, J. A. *Monath-Schrift von nüzlichen and neuen Erfahrungen.* Tubingen, 1773, p. 4.
1774: *Wiegleb* = *Allgemeine deutsche Bibliothek,* 21 (1774), 579.
1775: *Suckow* = Suckow, G. A. *Von dem Nuzzen der Chymie.* Mannheim and Lautern: Oekonomische Gesellschaft, 1775, pp. [1–7].
1777: *Black* = Black, J. "A Course of Chemical Lectures," Vol. I., p. 41. Lane Medical Library, Stanford University.

1777: Menn = Menn, J. G. *Rede von der Nothwendigkeit der Chemie.* Cologne: Universitätsdruckerey, 1777, p. 4.
1777: Raspe = Born, I. *Travels through the Bannat of Temeswar.* Translated by R. E. Raspe. London: Miller, 1777, p. xv.
1777: *Scopoli* = Scopoli, J. A. *Fundamenta Chemiae.* Prague: Gerle, 1777, p. 13.
1778: *Hagen* = Hagen, H. *Abhandlungen Chemischen und Physikalischen Inhalts.* Edited by K. G. Hagen. Königsberg: Hartung, 1778, p. xi.
1780: Anon. = *Nürnbergische gelehrte Zeitung* (1780), 15.
1780: *Gmelin* = Gmelin, J. F. *Einleitung in die Chemie.* Nuremberg: Raspe, 1780, pp. 12–13.
1780: *Weigel* = Wallerius, J. G. *Der Physischen Chemie.* Edited by C. E. Weigel. 2d German ed. Vol. I. Leipzig: Crusius, 1780, pp. 38–41.
1781: *Crell* = *Die neuesten Entdeckungen in der Chemie*, 1 (1781), [ii].
1781: *Wiegleb* = Wiegleb, J. C. *Handbuch der allgemeinen Chemie.* Vol. I. Berlin and Stettin: Nicolai, 1781, pp. 110–113.
1784: Hahnemann = See Lippmann, E. O. von. *Beiträge zur Geschichte der Naturwissenschaften und der Technik.* Vol. II. Weinheim: Verlag Chemie, 1952, p. 298.
1784: Scherf = *Archiv der medizinischen Polizey und der gemeinnützigen Arzneikunde*, 2 (1784), 210.
1784: *Suckow* = Suckow, G. A. *Anfangsgründe der ökonomischen und technischen Chymie.* Leipzig: Weidmann, 1784, p. 5.
1785, 1786: *Crell* = *Chemische Annalen*, no. 1 (1785), 345–346; no. 2 (1785), 535; no. 1 (1786), 192.
1786: Beseke = Beseke, J. M. G. *Entwurf eines Systems der transzendentellen Chemie.* Leipzig: Müller, 1787, pp. xi–xii.
1786: Blumenbach = Blumenbach, J. F. *Introductio in historiam medicinae litterariam.* Göttingen: Dieterich, 1786, pp. 284–290, 331–333, 414–417.
1786: Guyton de Morveau = *Chemische Annalen*, no. 2 (1786), 137.
1786: *Hermbstaedt* = *Neue Beiträge zur Natur- und Arznei-Wissenschaft*, 3 (1786), 21.
1786: *Wiegleb* = Wiegleb, J. C. *Handbuch der allgemeinen Chemie.* 2d ed. Vol. I. Berlin and Stettin: Nicolai, 1786, pp. 130–133.
1787: Fuchs = Fuchs, G. F. C. *Chemischer Lehrbegrif nach Spielmanns Grundsätzen.* Leipzig: Kummer, 1787, p. iv.
1787: *Gmelin* = *Chemische Annalen*, no. 1 (1787), 396.
1787: *Gren* = Gren, F. A. C. *Systematisches Handbuch der gesammten Chemie.* Vol. I. Halle: Waisenhaus, 1787, pp. 10–29.
1788: *Leonhardi* = Macquer, P. J. *Chymisches Wörterbuch.* Edited by J. G. Leonhardi. 2d German ed. of 2d French ed. Vol. I. Leipzig: Weidmann, 1788, pp. 657–662.
1788: *Westrumb* = *Chemische Annalen*, no. 2 (1788), 307; and Westrumb, J. F. *Kleine physikalisch-chemische Abhandlungen.* Vol. II, pt. 2. Leipzig: Müller, 1788, p. 223.

1789: *Gmelin* = Gmelin, J. F. *Grundriss der allgemeinen Chemie.* Göttingen: Vandenhoeck and Ruprecht, 1789, pp. xxvii–xxviii.
1789: Hopson = Wiegleb, J. C. *A General System of Chemistry.* Edited by C. R. Hopson. London: Robinson, 1789, p. i.
1789: *Westrumb* = Westrumb, J. F. *Kleine physikalisch-chemische Abhandlungen.* Vol. III, pt. 1. Leipzig: Müller, 1789, pp. 400, 410.
1790: *Hermbstaedt* = *Bibliothek der neuesten physisch-chemischen . . . Literatur,* 3; no. 1 (1790), 109.
1790: *Leonhardi* = Macquer, P. J. *Chymisches Wörterbuch.* Edited by J. G. Leonhardi. 2d German ed. of 2d French ed. Vol. VI. Leipzig: Weidmann, 1790, p. 372.
1791: *Crell* = *Annali di Chimica,* 3 (1791), 47.
1791: Denina = Denina, C. *La Prusse littéraire sous Frederic II.* Vol. III. Berlin: Rottmann, 1791, pp. 31–32.
1791: *Göttling* = *Almanach oder Taschenbuch für Scheidekünstler und Apotheker,* 12 (1791), unpaginated calendar following the preface.
1791: Hahnemann = *Bibliothek der neuesten physisch-chemischen . . . Literatur,* 3, no. 2 (1791), vi.
1791: Nicolai = König, A. B. *Versuch einer Historische Schilderung . . . Berlin.* Vol. V, pt. 2. Berlin: Pauli, 1799, p. 167.
1792: Humboldt = *Die Jugendbriefe Alexander von Humboldts 1787–1799.* Edited by I. Jahn and F. G. Lange. Berlin: Akademie Verlag, 1973, p. 184.
1792: *Klaproth* = Akademie der Wissenschaften, Berlin: *Mémoires,* (1792–1793), 18.
1793: *Richter* = Richter, J. B. *Ueber die neuern Gegenstände der Chymie.* Vol. III. Breslau and Hirschberg: Korn, 1793, p. [i].
1793: Scherer = Scherer, J. B. A. *Beweis, dass Johann Mayow vor Hundert Jahren den Grund zur antiphlogistischen Chimie und Physiologie gelegt hat.* Vienna: Wappler, 1793, p. 132; and *Ueber das Einathmen der Lebensluft.* Vienna: Stahel, 1793, p. 28.
1793: *Wurzer* = Wurzer, F. *Rede über die vornehmsten Schicksale der Chemie.* Bonn: Abshoven, 1793.
1794: Baldinger = *Medicinisches und physisches Journal,* no. 33 (1794), 52.
1796?: Black = Black, J. *Lectures on the Elements of Chemistry.* Edited by J. Robison. Vol. I. Edinburgh: Mondell, 1803, p. 549.
1796: Fourcroy = *Encyclopédie méthodique: chimie, pharmacie, et métallurgie.* Vol. III. Paris: Agasse, 1796–1797, pp. 332, 711–714.
1796: *Girtanner* = Ibid., p. 617.
1796: *Link* = Link, H. F. *Beyträge zur Physik und Chemie.* Vol. II. Rostock and Leipzig: Stiller, 1796, pp. 6, 15, 20.
1796: S. in M. = *Journal der Pharmacie,* 3 (1796), 313.
1796: Van Mons = *Encyclopédie méthodique: chimie, pharmacie et métallurgie.* Vol. III. Paris: Agasse, 1796–1797, pp. 613–617.
1796: *Wiegleb* = Wiegleb, J. C. *Handbuch der allgemeinen Chemie.* 3d ed.

Vol. I. Berlin and Stettin: Nicolai, 1796, pp. 111–116.
1797: Freund = *Berlinisches Jahrbuch der Pharmazie*, 3 (1797), 49.
1797: Mann = See *The Banks Letters*. Edited by W. Dawson. London: British Museum, 1958, p. 575.
1798: *Bucholz* = Bucholz, C. F. *Beyträge zur Erweiterung und Berichtigung der Chemie*. Vol. I. Erfurt: Beyer and Maring, 1799, p. 6.
1798: Frank = *Berlinisches Jahrbuch der Pharmazie*, 4 (1798), 17–18, 27–28, 30, 32–33.
1798 . . . 1803: *Scherer* = Frontis-portraits and biographies selected by A. N. Scherer, editor for the *Allgemeines Journal der Chemie*, 1–10 (1798–1803).
1800: Anon. = Gren, F. A. C. *Principles of Modern Chemistry*. Vol. I. London: Cadell and Davies, 1800, p. 12.
1802: *Göttling* = Göttling, J. F. A. *Praktische Anleitung zur prüfenden und zerlegenden Chemie*. Jena: Mauke, 1802, p. vii.
1802: Salzmann = Salzmann, C. G. *Denkwürdigkeiten aus dem Leben ausgezeichneter Teutschen des achtzehnten Jahrhunderts*. Schnepfenthal: Erziehungsanstalt, 1802, p. 789.
1803: Ed. Board = Editorial board for the *Neues allgemeines Journal der Chemie*, 1–2 (1803), title pages. [This editorial board was probably put together by Hermbstaedt and Klaproth.]
1804 . . . 1810: *Gehlen* = New editorial board members and biographies selected by A. F. Gehlen, editor for the *Neues allgemeines Journal der Chemie*, 3–6 (1804–1805) and the *Journal für die Chemie und Physik*, 1–9 (1806–1810).
1804: *Gmelin* = Gmelin, J. F. *Grundriss der allgemeinen Chemie*. 2d ed. Göttingen: Vandenhoeck and Ruprecht, 1804, pp. 17–19.
1804: Kastner = *Correspondance de H. C. Oersted*. Edited by M. C. Harding. Vol. II. Copenhagen: Ascheoug, 1920, pp. 420–421.
1806: *Trommsdorff* = Trommsdorff, J. B. *Versuch einer allgemeinen Geschichte der Chemie*. Facsimile of the 1806 ed. 3 vols. Leipzig: Zentral-Antiquariat, 1965, frontis-portraits; II: vi–vii; III: 122.
1807: Kastner = Kastner, C. W. G. *Grundriss der Chemie*. Vol. I. Heidelberg: Mohr and Zimmer, 1807, pp. 5–8.
1808: *Göttling* = Göttling, J. F. A. *Elementarbuch der chemischen Experimentirkunst*. Vol. I. Jena: Seidler, 1808, p. viii.
1809: *Bucholz* = Gren, F. A. C. *Grundriss der Chemie*. Edited by C. F. Bucholz. 3d ed. Vol. I. Halle and Berlin: Waisenhaus, 1809, p. 11.
1811: Hausmann = *Annalen der Physik*, 38 (1811), 1.
1811 . . . 1820: Schweigger = New editorial board members and biographies selected by J. S. C. Schweigger, editor for the *Journal für die Chemie und Physik*, 1–28 (1811–1820).
1811: Sertürner = See Akademie gemeinnütziger Wissenschaften zu Erfurt: *Jahrbücher*, N.F. 55 (1941), 155.
1814: Kastner = Kastner, C. W. G. *Einleitung in die neuere Chemie*. Halle and Berlin: Waisenhaus, 1814, p. 11.

1814: *Wurzer* = Wurzer, F. *Handbuch der populären Chemie.* 2d ed. Leipzig: Barth, 1814, pp. 12–13.

1816: *Bucholz* = *Almanach oder Taschenbuch für Scheidekünstler und Apotheker*, 38 (1817), [xiii–xxv].

1817: *Bucholz* = Gren, F. A. C. *Grundriss der Chemie.* Edited by C. F. Bucholz. 4th ed. Vol. I. Halle and Berlin: Waisenhaus, 1818, pp. 9–12.

1817: *Döbereiner* = *Journal für die Chemie und Physik*, 23 (1817), 82.

1818: Hohnbaum = Deutsche Akademie der Naturforscher, Leopoldina: *Verhandlungen*, 1 (1818), 52–53.

1818: *Trommsdorff* = *Neues Journal der Pharmacie*, 2, no. 2 (1818), 529–530.

1819: Brandes = *Almanach oder Taschenbuch für Scheidekünstler und Apotheker*, 40 (1819), 359.

1819: Du Mênil = *Journal für die Chemie und Physik*, 27 (1819), 48.

1821: *Pfaff* = Pfaff, C. H. *Handbuch der analytischen Chemie für Chemiker, Staatsärzte, Apotheker, Oekonomen und Bergwerks-Kundige.* Vol. I. Altona: Hammrich, 1821, pp. vi, 8–15.

1822: *Lampadius* = Lampadius, W. A. *Grundriss des Systems der Chemie.* Freiberg: Craz and Gerlach, 1822, p. xxxvii.

BIBLIOGRAPHY: Each profile ends with a bibliography and, usually, cross-references to other profiles in Appendix I and/or institutional histories in Appendix II. For the sake of brevity, I use the following abbreviations:

ADB = *Allgemeine deutsche Biographie.* 56 vols. Leipzig: Duncker and Humblot, 1875–1919.

AEWK = *Allgemeine Encyclopädie der Wissenschaften und Künste.* 167 vols. Leipzig: Gleditsch, then Brockhaus, 1818–1889.

ALZ: IB = *Allgemeine Literatur-Zeitung: Intelligenzblatt* (1788–1800).

Baldinger = Baldinger, E. G. *Biographien jetztlebender Aerzte und Naturforscher in und ausser Deutschland.* Jena: Hartung, 1768–1772.

Bergman = "Torbern Bergman's Foreign Correspondence," edited by G. Carlid and J. Nordström. *Lychnos-Bibliotek*, 23, no. 1 (1965).

Berzelius = "Brevväxling mellan Berzelius och Eilhard Mitscherlich (1819–1847)." In *Jac. Berzelius Bref*, edited by H. G. Söderbaum, vol. VI, pt. 1. Uppsala: Almqvist and Wiksell, 1932.

Biereye = Biereye, J. "Erfurt in seinen berühmten Persönlichkeiten." Akademie gemeinnütziger Wissenschaften zu Erfurt: *Sonderschriften*, 11 (1937).

Börner = Börner, F. *Nachrichten von den vornehmsten Lebensumständen und Schriften jeztlebender berühmter Aerzte und Naturforscher in und um Deutschland.* 3 vols. Wolfenbüttel: Meissner, 1749–1756.

Callisen = Callisen, A. C. P. *Medicinisches Schriftsteller-Lexicon der jetzt lebenden Aerzte, Wundärzte, Geburtshelfer, Apotheker, und Naturforscher aller gebildeten Völker.* 33 vols. Copenhagen: K. taubstummen Institute, 1830–1845.

DAB = "Deutsche Apotheker-Biographie," edited by W.-H. Hagen and H.-

D. Schwartz, 2 vols. Internationale Gesellschaft für Geschichte der Pharmazie: *Veröffentlichungen*, 43 (1975) and 46 (1978).

Dreyhaupt = Dreyhaupt, J. C. von. *Beschreibung des . . . Saalkreises*. Vol. II. Halle: Schneider, 1750.

DSB = *Dictionary of Scientific Biography*. Edited by C. C. Gillispie et al. 16 vols. New York: Scribner's, 1970–1980.

Elwert = Elwert, J. K. P. *Nachrichten von dem Leben und den Schriften jeztlebender teutscher Aerzte, Wundärzte, Thierärzte, Apotheker und Naturforscher*. Hildesheim: Gerstenberg, 1799.

Euler = "Die Berliner und die Petersburger Akademie der Wissenschaften im Briefwechsel Leonhard Eulers," edited by A. P. Juskevic and E. Winter. *Quellen und Studien zur Geschichte Osteuropas*, 3, nos. 1–3 (1959–1976).

Forster = *George Forsters Werke: Sämtliche Schriften, Tagebücher, Briefe*. Vol. XII: *Tagebücher*, edited by B. Leuschner; Vol. XIII: *Briefe bis 1783*, edited by S. Scheibe; Vol. XIV: *Briefe 1784–Juni 1787*, edited by B. Leuschner; Vol. XVI: *Briefe 1790 bis 1791*, edited by B. Leuschner and S. Scheibe. Berlin: Akademie Verlag, 1973–1980.

Friedrich = *Friedrichs des Grossen Korrespondenz mit Aerzten*. Edited by G. L. Mamlock. Facsimile of the 1907 ed. Wiesbaden: Sändig, 1966.

Goethe & Carl August = *Briefwechsel des Herzogs-Grossherzogs Carl August mit Goethe*. Edited by H. Wahl. Vol. I. Berlin: Mittler, 1914.

Goethe & Voigt = "Goethes Briefwechsel mit Christian Gottlob Voigt," edited H. Tummler. Goethe-Gesellschaft, Weimar: *Schriften*, 53–54 (1949–1951).

Goetten = Goetten, G. W. *Das Jetztlebende Gelehrte Europa*. Vol. II. Brunswick and Hildescheim: Schröder, 1736.

Gossmann = Gossmann, H. "Das Collegium Pharmaceuticum Norimbergense und sein Auffluss auf das Nürnbergische Medizinalwesen." *Quellen und Studien zur Geschichte der Pharmazie*, 9 (1966).

Humboldt = *Die Jugendbriefe Alexander von Humboldts (1787–1799)*. Edited by I. Jahn and F. G. Lange. Berlin: Akademie Verlag, 1973.

Joecher = Joecher, C. G., et al. *Allgemeines Gelehrten-Lexicon*. Facsimile of the 1750–1877 ed. 4 vols. plus 7 supp. vols. Hildesheim: Olms, 1960–1961.

Justi = Justi, K. W. *Grundlage zu einer hessischen Gelehrten-, Schriftsteller- und Künstler-Geschichte*. Marburg: Garthe, 1831.

König = König, A. B. *Versuch einer Historische Schilderung der Hauptveränderungen, der Religion, Sitten, Gewohnheiten, Künste, Wissenschaften, etc. der Residenzstadt Berlin seit den ältesten Zeiten, bis zum Jahre 1786*. Vol. V, pts. 1–2, Berlin: Pauli, 1798–1799.

Leiden = *Album studiosorum Academiae lugduno batavae MDLXXV–MDCCLXXV*. Edited by W. N. Du Rieu. The Hague: Nijhoff, 1875.

Lenz = Lenz, M. *Geschichte der Königlichen Friedrich-Wilhelms-Universität zu Berlin*. 4 vols. Halle: Waisenhaus, 1910–1918.

Lichtenberg = *Lichtenbergs Briefe*. Edited by A. Leitzmann and

C. Schüddekpof. 3 vols. Leipzig: Dieterich, 1901–1904.
Luca = Luca, I de. *Das gelehrte Oesterreich*. 2 vols. Vienna: Gehlen, then Trattner, 1776–1778.
Moll = Moll, C. E. von. *Mittheilungen aus seinem Briefwechsel*. 3 vols. Augsburg: Volckhart, 1829–1834.
NDB = *Neue Deutsche Biographie*. Vols. I–XII. Berlin: Duncker and Humblot, 1953–1980.
Oersted = *Correspondance de H. C. Oersted avec divers savants*. Edited by M. C. Harding. 2 vols. Copenhagen: Aschehoug, 1920.
Partington = Partington, J. R. *A History of Chemistry*. Vols. II–III. London: MacMillan, 1961–1962.
Pohl = Pohl, D. *Zur Geschichte der pharmazeutischen Privatinstitute in Deutschland von 1779 bis 1873*. Diss., Marburg University; Marburg: Mauersberger, 1972.
Pr. Hdbh. = *Handbuch über den königlich preussischen Hof und Staat*. 54 vols. Berlin: Decker or Unger, 1795–1848.
Rotermund = Rotermund, H. W. *Das gelehrte Hannover*. 2 vols. Bremen: Schünemann, 1823.
St. Petersburg Correspondence = "Uchenaia korrespondentsiia Akademii nauk XVIII veka, 1766–1782," edited by I. I. Livbenko. Akademiia nauk SSSR, Arkhiv: *Trudy Arkhiva*, 2 (1937).
St. Petersburg Procès-verbaux = *Procès-verbaux des séances de l'Académie impériale des sciences depuis sa foundation jusqu'a 1803*. 4 vols. St. Petersburg: Akademiia nauk, 1897–1911.
Schlichtegroll = Schlichtegroll, F. *Nekrolog, enthalt. Nachrichten von den Leben merkwürd. verstorbener Deutschen*. 22 vols. Gotha: Perthes, 1790–1800.
Strieder = Strieder, F. W., et al. *Grundlage zu einer hessischen Gelehrten und Schriftsteller Geschichte*. 18 vols. Göttingen: Barmeier, then Kassel: Cramer, 1781–1819.
Wurzbach = Wurzbach, C. J. *Biographisches Lexikon des Kaiserthums Oesterreich*. 60 vols. Vienna: Hof- und Staatsdruckerei, 1856–1891.
Zedler = *Grosses vollständiges Universal-Lexicon*. Edited by J. H. Zedler. Facsimile of the 1730–1754 ed. 68 vols. Graz: Akad. Druck- und Verlagsanstalt, 1961–1964.

The bibliographies include all available autobiographies and all pertinent biographies in three modern biographical dictionaries (DAB, DSB, and NDB). Otherwise, they are limited to the particular materials from which the profiles were constructed. Moreover, when the same information is to be found in more than one place, they give precedence to modern studies over earlier publications and to published works over archival sources.

# PROFILES

**KUNCKEL, Johann**
1638?, Plön, Holstein / 1703, Bernau?, Brandenburg

**Background** Son of a glassmaker and alchemist, probably Jürgen Kunckel, supervisor of the glassworks at Rixdorf during the 1620s, at Wittenberg from 1630? to 1637?, at Ascheberg near Plön from 1637 to 1642, and at Depenau from 1642. Lutheran. **Education** In addition to receiving instruction from his father, he apparently received pharmaceutical training in Hamburg.

**Career**

1658?–1660?: Alchemist in service of Duke Franz Carl of Saxe-Lauenburg (d. 1660).

1661?–1663?: Probably worked in an apothecary shop in Hamburg. He married there in 1662.

1663: Tried to establish a new apothecary shop in Eckernförde, but gave up after the owner of the existing shop objected to the town council.

1663?–1665?: Alchemist and apothecary in service of Duke Julius Heinrich of Saxe-Lauenburg (d. 1665), who resided at Ratzeburg and Schlackenwerth in Bohemia.

1665?–1667?: Probably lived in Hamburg. During this period he seems to have visited Holland and learned the techniques of Italian glassmaking.

1667?–1676: Alchemist in service of Saxon Elector Johann Georg II. From 1670 he was director of the Elector's alchemical laboratories in Dresden and Annaberg. In 1676 he was dismissed for failing to achieve positive results.

1676: Visited Hamburg, where the alchemist H. Brand showed him phosphorus.

1676: Matriculated at Wittenburg University, giving Plön as his birthplace.

1676–1678: Taught chemistry at Wittenberg University on a private basis. W. Homberg, who later achieved fame as a chemist in France, may have been among his students. He soon found that earning a living as a private teacher was a "sour piece of bread." Although he was reinstated as an alchemist to the Saxon Elector in 1677, he continued teaching because he was unable to secure payment of his salary (1,000 Rtl).

1678: Visited Elector Frederick William in Berlin twice, the second time to give advice about the reliability of another alchemist's claims. He made such a good impression that the Elector decided to hire him as a glass chemist.

1678–1691: Glass chemist (salary: 500 Tl) in Berlin.

    1678–1688: Enjoyed the patronage and encouragement of Elector Frederick William and leading figures in the Court. In 1679 he leased the Elector's glassworks at Drewitz near Potsdam. In 1684 he and Privy Finance Councillor D. von Knyphausen established a cobalt-blue works near Wernigerode. And in 1685, after being granted title to properties in the Havel, he founded a new glassworks on the Pfaueninsel. All in all he received 26,749 Tl plus these properties from Frederick William.

    1688–1691: Fell from favor upon the accession of Elector Frederick II. He was required to report on the uses to which he had put the prior elector's funds. Failing to account for more than 13,744 Tl, he was ordered to return 8,000 Tl during the next four years. Shortly before this decision, his glassworks on the Pfaueninsel burned down. To meet his obligations he sold his interest in the cobalt-blue works in 1690 and relinquished his lease on the Drewitz glassworks in 1691.

1692: Fled Berlin because he could not pay his debt to the Elector. He was captured and jailed but influential friends soon secured his release.

1693: Visited Stockholm to give advice on mining matters. While there, he was ennobled, possibly in partial payment for an alchemical secret.

1694–1703: Resided on an estate at Dreissighufen (about 30 miles northeast of Berlin) that he acquired through trade of his properties in the Havel. During this period he was still at work on the manuscript of his "Laboratorium Chymicum" which appeared posthumously in 1716.

1694: Borrowed 1,000 Tl from the Berlin Apothecary F. Zorn, whom he probably met in Hamburg ca. 1691.

1695–1696: Visited Sweden to advise on copper smelting techniques. His proposals did not survive the experimental stage.

1702: Visited Stockholm.

1703: Died while traveling, probably at Bernau about ten miles from his estate, rather than at Pernau in Livonia as is usually claimed. He had debts of 6,300 Rtl.

**State Honors** 1658?: Saxe-Lauenburg Chamberlain. 1670?: Saxon Privy Chamberlain. 1678: Brandenburg Privy Chamberlain. 1693: Swedish Mining Councillor. 1693: Swedish Ennoblement as "von Lövenstjern." **Major Scientific Societies** 1693: Leopoldina. 1699: Paris (corresponding member). **Informal Recognition** 1710?: *Rothe*. 1718: *Stahl*. 1732: Boerhaave. 1736: Burghart. 1737?: *Neumann*. 1740: Kestner. 1749: *Juncker*. 1758?: Cullen. 1759: Wallerius. 1766: *Lehmann*. 1770: Madihn. 1777: Black. 1777: Raspe. 1777: *Scopoli*. 1780: *Gmelin*. 1781: *Wiegleb*. 1786: Blumenbach. 1787: *Gren*. 1789: Hopson. 1791: *Göttling*. 1798: Frank. 1806: *Trommsdorff*. 1807: Kastner.

**Bibliography**
Bölsche, L. "Johann Kunckel der Glasmacher und Alchymist des Grossen Kurfürsten." Verein für die Geschichte Berlins: *Zeitschrift*, 26 (1909), 163–169, 184–194.
DSB, VII: 524–526.
Heine, A. "Johan Kunchel von Löwenstern." *Tidsskrift for Industri*, 13 (1912), 127–160.
Hucke, K. "Glasmacherei." In *Gottorfer Kultur im Jahrhundert der Universitätsgrundüng*, edited by E. Schlee. Schleswig: Schleswig-Holsteinisches Landesmuseum, 1965, pp. 428–434.
Lindroth, S. *Gruvbrytning och kopparhantering vid Stora Kopparberget*. Vol. II. Uppsala: Almqvist and Wiksell, 1955, pp. 293–302, 451–542.
Maurach, H. "Johann Kunckel (1630–1703)." Deutsches Museum, Munich: *Abhandlungen und Berichte*, 5 (1933), 31–63.
Rau, H. G. "Johann Kunckel, Geheimer Kammerdiener des Grossen Kurfüsten, und sein Glaslaboratorium auf der Pfaueninsel in Berlin." *Medizinhistorisches Journal*, 11 (1976), 129–148.
Stein, G. "Johann Kunckels 'Ars Vitraria' als Synthese internationalen Glaswissens im 17. Jahrhundert." Deutsche Akademie der Naturforscher, Leopoldina: *Nova Acta*, 16 (1954), 479–489.
Stürzbecher, M. *Berlins alte Apotheken*. Berlin: Hessling, 1965, pp. 35, 37.

**BOHN, Johann**
1640, Leipzig / 1718, Leipzig

**Background** Son of Johann Bohn (?, Nuremberg–1672, Leipzig), a wealthy merchant. Lutheran. **Education** At the age of three, he was matriculated in Leipzig University in anticipation of future attendance. He actually began his university studies in 1658, taking a bachelor of arts that same year. He then proceeded to Jena University where he studied with the chemist W. Rolfinck. A year or so later he returned to Leipzig University. He earned a Mag. Phil. and bachelor of medicine there in 1661 and obtained his medical license in 1663. Then he embarked on a study tour through Denmark, Holland, England, France, and Switzerland, returning home in 1665. He earned his M.D. the next year.

**Career**
1666–1718: PD, then OP in Leipzig University's Medical Faculty. Among his students were G. *Rothe* (1700–1704; colleague: 1709–1710) and H. F. *Teichmeyer* (1702–1703).
1666–1668: PD.
1668–1691: OP of anatomy and surgery.
1691–1718: OP of therapy.
1690–1718: Leipzig Town Physician.
1690–1718: Member of the Lesser, then from 1691, the Greater Prince's Council.

**State Honors** None. **Major Scientific Societies** None. **Informal Recognition** 1710?: *Rothe*. 1718: *Stahl*. 1732: Boerhaave. 1736: Burghart. 1737?: *Neumann*. 1749: *Juncker*. 1756?: Cullen. 1760: Wallerius. 1770: Madihn. 1777: *Scopoli*. 1781: *Wiegleb*. 1787: *Gren*. 1791: *Göttling*. 1806: *Trommsdorff*.

**Bibliography**
DSB, II: 237–238.
Rabl, D. "Geschichte der Anatomie an der Universität Leipzig." *Studien zur Geschichte der Medizin*, 7 (1909), 39–41.
Zedler, IV: 436; supp. IV: 95–96.
Also see Leipzig University (*Matrikel*, II: 39).

**WEDEL, Georg Wolfgang**
1645, Golssen, Lower Lusatia / 1721, Jena

**Background** Son of Johann George Wedel (?, Eschenbach, Palatinate–1665, Spremberg), who entered Wittenberg University in 1631, earned his Mag. Phil. there in 1633, served as pastor in Golssen until 1648, and ended his career as pastor and church inspector in Spremberg. Lutheran. **Education** In 1656, after preliminary instruction from his father who wanted him to become a physician, he journeyed by way of Leipzig University to Schulpforta. At the Prince's School there, he did so well in ancient languages and natural philosophy that he decided to become a professor. In 1662 he proceeded to Jena University where he first studied mathematics and natural philosophy, particularly with E. Weigel, and then took up medicine. His favorite teacher was the chemist W. Rolfinck, with whom he lived for a time. In 1667 he practiced medicine in Landsberg for three months, traveled in Silesia and Saxony, and then earned his medical license at Jena University with a disputation on nocturnal pollutions.

**Career**
1667: PD in Jena University's Medical Faculty.
1667–1672: District Physician in Gotha.
1669: Received his M.D. at Jena University.
1672: Studied in Holland for a few months by way of preparing for a chair in Jena University's Medical Faculty.
1672–1721: OP in Jena University's Medical Faculty. Among his students were *F. Hoffmann* (1678–1681; colleague: 1681–1682), *G. E. Stahl* (1679–1683; colleague: 1683–1687), *J. F. Henckel* (1698–1705?), *H. F. Teichmeyer* (1703–1707; colleague: 1707–1721), and *J. T. Eller* (1709–1711).
1672–1673: OP designate.
1673: OP of anatomy, surgery, and botany.
1673–1719: OP of theoretical medicine (salary, 1684: increased from 170 fl to 200 fl).
1719–1721: OP of practical medicine and chemistry (salary: 300 fl).
1679–1683: Personal Physician to the Duke of Saxe-Weimar (salary: 100 Tl).
1698?–1721: Estate-owner at Schwartza near Jena.

**State Honors** 1685: Ducal Saxon Councillor and Physician. 1694: Imperial Ennoblement as Palatine Count. 1705: Saxe-Weimar Titular Councillor. 1716: Imperial Councillor. 1718: Saxe-Weimar Court Councillor. 1721: Electoral Mainz Councillor and Physician. **Major Scientific Societies** 1672: Leopoldina. 1706: Berlin. **Informal Recognition** 1710?: *Rothe*. 1729: *Teichmeyer*. 1736: *Burghart*. 1737?: *Neumann*. 1742: *Anon.* 1760: *Wallerius*. 1787: *Gren*. 1791: *Göttling*. 1806: *Trommsdorff*.

**Bibliography**
Dittrich, J. (Clergyman, Golssen). Letter concerning Wedel's family (11 March 1970).
DSB, XIV: 212–213.
Staatsarchiv, Weimar. Letter bearing on Wedel's relations with the Saxe-Weimar government (11 November 1970).
Wedel, G. W. Autobiography (1672). Archiv, Deutsche Akademie der Naturforscher (Leopoldina), Halle.
Zedler, LIII: 1804–1820.
Also see Jena University (Chemnitius, pp. 63–70; Giese and Hagen, pp. 174–179), Leipzig University (*Matrikel*, II: 484), and Wittenberg University ("Matrikel," 14: 356).

### HOFFMANN, Johann Moritz
1653, Altdorf / 1727, Ansbach

**Background** Son of Mortiz Hoffmann (1621/22, Fürstenwalde, Brandenburg–1698, Altdorf), who studied at the universities in Frankfurt a. d. Oder, Padua, and Altdorf (M.D., 1643) and served as PD, AOP, then from 1649 OP of medicine at Altdorf University. Lutheran. **Education** After instruction from private tutors and at the town school, he enrolled at Altdorf University in 1669. He earned a Mag. Phil. there in 1671, then studied medicine at Frankfurt a. d. Oder and Padua universities. In Padua he learned chemistry from J. Barner. In 1673 he toured Italy, meeting A. Kircher in Rome. He then returned to Altdorf where he earned his M.D. in 1675 with a dissertation on convulsive ailments.

**Career**
1675–1713: PD, AOP, then OP in Altdorf University's Medical Faculty.
　1675–1677: PD.
　1677–1681: AOP of anatomy and chemistry.
　1681: OP of anatomy.
　1682–1703: OP of anatomy and chemistry.
　1703–1709: OP of theoretical medicine, botany, and chemistry.

1710–1713: OP of practical medicine, botany, and chemistry.
1695–1727: Personal Physician to the Margrave of Ansbach.
1695–1713: Resided in Altdorf. He accompanied the Margrave and members of his family to Italy (1695–1696, 1701), Bad Ems (1701), and Hanover (1705).
1713–1727: Resided in Ansbach.

**State Honors** 1695: Ansbach Councillor. 1704: Sulzbach Councillor and Physician. 1706: Bayreuth Councillor and Physician. 1709: Culmbach Councillor and Physician. **Major Scientific Societies** 1684: Leopoldina. **Informal Recognition** 1736: Burghart. 1760: Wallerius. 1787: *Gren*. 1791: *Göttling*. 1796: Fourcroy.

**Bibliography**
Hoffmann, J. M. Autobiography (ca. 1706). Archiv, Deutsche Akademie der Naturforscher (Leopoldina), Halle.
Zedler, XIII: 452–453.
Also see Altdorf University (Flessa, pp. 25–28, 33–35; Sauer-Haeberlein, pp. 57–72).

## STAHL, Georg Ernst
1659, Ansbach / 1734, Berlin

**Background** Son of Johann Lorentz Stahl (1620, Ansbach–1689/99, Ansbach), who was Secretary of the Ansbach Court Council, of the Consistory from 1664, and of the Divorce Court from 1672. Lutheran (probably became a Pietist). **Education** Though he matriculated at Altdorf University in 1670, he actually attended the Latin School in Ansbach. In 1675, while still in school, he immersed himself in chemistry after reading a manuscript of J. Barner's lectures. The following year he learned enameling from a journeyman goldsmith. And in early 1679 he read *J. Kunckel's* book on chemical principles. Consequently, he was well versed in chemistry when he entered Jena University to study medicine in the spring of 1679. While there he probably studied chemistry with both *G. W. Wedel* and *F. Hoffmann*. In 1683 he earned his M.D. under R. W. Krause with a dissertation on intestinal diseases.

**Career**
1683–1687: PD at Jena University. Among his colleagues was *Wedel*. He taught chemistry and conducted much chemical research in this period.
1687–1696: Court and Personal Physician to the Duke of Saxe-Weimar (salary, 1687?–1696: 200 fl). From 1694 he was no longer in residence.
1694–1715: OP at Halle University. His appointment was probably arranged by *Hoffmann*, with whom he later quarreled. Among his students were *J. Juncker* (1698–1702), *G. Rothe* (1704–1708), *J. H. Pott* (1709–1715?), and *J. F. Henckel* (1711).
1694–1715: OP of physiology, pathology, hygiene, materia medica, and botany in the Medical Faculty (salary, 1694–1707: 200 Rtl; 1707–1715: 300 Rtl).
1709–1715: OP of natural philosophy in the Philosophical Faculty (salary: 100 Rtl).
1712–1715: Royal Physician to the Prussian King (salary: 500 Tl). He served in Berlin only when needed.
1715–1734: First Royal Physician in Berlin (salary, 1716: 1,800 Tl plus firewood and fodder for four horses). President of Berlin's Medical Board, which in 1725 became the Higher Medical Board for the whole of Prussia. Among his colleagues on the Board were *C. Neumann* (1724–1734) and *J. T. Eller* (1725–1734).
1719–1734: Member of the new Sanitation Board.
1726: Journeyed to St. Petersburg to attend the ailing Czar Peter, but arrived too late to do any good.

**State Honors** 1712?: Prussian Court Councillor. **Major Scientific Societies** 1700: Leopoldina. ca. 1715: Berlin (declined membership). **Informal Recognition** 1710?: *Rothe*. 1732: Boerhaave. 1736: Burghart. 1737?: *Neumann*. 1740: Kestner. 1749: *Juncker*. 1750: *Gellert*. 1752: Börner. 1753: *Vogel*. 1756?: Cullen. 1758: Hoffmann. 1759: Wallerius. 1766: *Lehmann*. 1767: Zückert. 1769: Suckow. 1770: Madihn. 1771: *Cartheuser*. 1773: *Weber*. 1777: Black. 1777: Menn. 1777: Raspe. 1777: *Scopoli*. 1780: *Gmelin*. 1781: *Wiegleb*. 1785: *Crell*. 1787: *Gren*. 1788: *Leonhardi*. 1789: Hopson. 1789: *Westrumb*. 1791: *Göttling*. 1796: Fourcroy. 1798: Frank. 1802: Salzmann. 1804: Kastner. 1806: *Trommsdorff*. 1814: *Wurzer*. 1821: *Pfaff*.

**Bibliography**
Altmann, E. "Christian Friedrich Richter (1676–1711): Arzt, Apotheker und Liederdichter des Halleschen Pietismus." *Arbeiten zur Geschichte des Pietismus*, 7 (1972), 21–22, 30, 70–74, 82–83, 110–111, 127.

*Die Behördenorganisation und die allgemeine Staatsverwaltung Preussens im 18. Jahrhundert.* Edited by G. Schmoller, O. Krauske, and V. Loewe. Vols. II–III. Berlin: Parey, 1898–1901; II: 278–280, 297; III: 192.

Dreyhaupt, 724–726.

DSB, XII: 599–606.

Fasch, A. H. *S. P. D. Lecturis!* Jena, 1684.

*Materialy dlia istorii Imperatorskoi akademii nauk.* Vol. VI. St. Petersburg: Akademiia nauk, 1890, pp. 90–91.

Partington, II: 653–686.

Staatsarchiv, Weimar. Letter bearing on Stahl's salary while employed by the Duke of Saxe-Weimer (3 May 1972).

Strube, I. "Der Beitrag George Ernst Stahls (1659–1734) zur Entwicklung der Chemie." Diss., Leipzig University, 1960.

Stürzbecher, M. "Friedrich Hoffmann und George Ernst Stahl als Leibärzte in Berlin." *Forschung, Praxis, Fortbildung,* 17 (1966), 535–537.

Vogtherr, F. *Geschichte der Stadt Ansbach.* Ansbach: Brügel, 1927, p. 81.

Zaunick, R. "Von den Vor- und Nachfahren Georg Ernst Stahls." *Sudhoffs Klassiker der Medizin,* 36 (1961), 76–88.

Also see Altdorf University (*Matrikel*, I: 365), Berlin Academy (Harnack, I: 204, 229–230), Berlin Medical-Surgical College (Stürzbecher, p. 772), Halle University (Kaiser and Krosch, 14: 43, 361–363; Schrader, II: 427–428, 441, 453), and Jena University (*Matrikel*, II: 774).

## HOFFMANN, Friedrich
1660, Halle / 1742, Halle

**Background** Son of Friedrich Hoffmann (1626, Halle–1675, Halle), who studied with the chemist W. Rolfinck and others at Jena University, received an M.D. there in 1650, and later became Personal Physician to the Governor of Magdeburg, Town Physician in Halle, wealthy shareholder in the Halle saltworks, and a minor chemist. Lutheran. **Education** At the age of thirteen, after schooling from his parents and tutors, he entered Halle's gymnasium. There he gave particular attention to mathematics. Meanwhile he began learning anatomy and chemistry from his father. Although he was orphaned in 1675, he stayed in school, finishing with an essay on "the world" three years later. He then journeyed to Jena University to study medicine with *G. W. Wedel*. Besides his formal studies, he pursued chemistry in his lodgings. His chemical reputation was soon so great that his fellow students, perhaps including *G. E. Stahl*, sought out his instruction. In 1680 he visited Erfurt University to take C. Cramer's renowned chemistry course. The following year he earned his M.D. under *Wedel* at Jena University with a dissertation on suicide.

**Career**

1681–1682: PD at Jena University.

1682–1687: Practiced medicine in Minden where his brother-in-law, J. M. von Unverfährt, was Governor. In 1684 he spent several months in Holland and England extending his medical and natural knowledge. In London he met R. Boyle.

1685–1687: Regimental Physician, then Court and Provincial Physician for the Principality of Minden and the County of Ravensburg.

1687–1693: Visited Holland with his brother-in-law, who had become Governor of Halberstadt, then in early 1688 settled in Halberstadt as District Physician (salary: 200 Tl).

1693–1709: OP of medicine at Halle University, then from 1694 OP of practical medicine, anatomy, surgery, and chemistry in the Medical Faculty (salary: 200 Rtl) and of natural philosophy in the Philosophical Faculty (salary: 100 Rtl). Among his colleagues was *Stahl* (1694–1709) and among his students, *J. Juncker* (1698–1702).

1700: Turned down a call to succeed H. Meibom in Helmstedt University's Medical Faculty.

1702: Treated the Elector of Mainz and visited the Landgraves of Hesse-Darmstadt and Hesse-Cassel.

1703: Turned down a call to succeed Royal Physician Weiss in Berlin, but nevertheless was appointed Prussian Court Councillor.

1704: Made his first visit to the spas in Carlsbad.

1705: Attended the funeral of the Prussian Queen in Berlin and the jubilee of Frankfurt a. d. Oder University.

1707: Treated King Frederick I at Carlsbad.

1708: Purchased a house in Halle (cost: 4,300 Tl).

1709–1712: Royal Physician in Berlin and Ordinary Member of the Berlin Society of Sciences. Eventually falling into disfavor, he returned to Halle University where he had prudently retained a claim on his chair in the Medical Faculty.

1712–1742: OP in Halle University's Medical Faculty (salary: 200 Rtl until 1740 when he had his salary transferred to his son). Among his associates during this period were *Stahl* (colleague: 1712–1715), *Juncker* (colleague: 1718–1742), *J. H. Pott* (student: 1712–1715; colleague: 1719–1720), *P. Gericke* (student: 1716–1717; colleague: 1724–1730), *J. F. Cartheuser* (student: 1729–1731; colleague: 1731–1740), and *A. S. Marggraf* (student: 1733–1734). He was responsible for practical medicine, anatomy, surgery, and chemistry until 1718 when the expansion of the salaried faculty rendered the statutory distribution of subjects obsolete.

1715?–1742?: Manufactured and sold various patent medicines, especially Hoffmann's drops.
1721: Treated the Empress at Carlsbad.
1732: Treated Emperor Charles VI at Carlsbad.
1734–1735: Treated King Frederick William in Potsdam.
1736: Turned down a call to succeed C. T. Behrens as Ducal Physician in Wolfenbüttel.
1742: Died with assets between 70,000 and 100,000 Rtl, including an estate at Schwertz that he had purchased the year before for 32,000 Rtl.

**State Honors** 1694: Electoral Brandenburg Physician. 1703: Prussian Court Councillor. 1727: Imperial Ennoblement as Palatine Count. 1735: Prussian Privy Councillor. **Major Scientific Societies** 1696: Leopoldina. 1701: Berlin. 1720: London. 1734: St. Petersburg. **Informal Recognition** 1732: Boerhaave. 1736: Burghart. 1737?: *Neumann.* 1749: *Juncker.* 1752: Börner. 1753: *Vogel.* 1756?: Cullen. 1758: Hoffmann. 1759: Wallerius. 1766: *Lehmann.* 1777: Black. 1777: Raspe. 1777: Scopoli. 1780: *Gmelin.* 1781: *Wiegleb.* 1784: Suckow. 1785: *Crell.* 1787: *Gren.* 1789: Hopson. 1791: *Göttling.* 1796: Fourcroy. 1798: Frank. 1806: *Trommsdorff.* 1814: *Wurzer.*

**Bibliography**
Dreyhaupt, 635–640.
DSB, VI: 458–461.
Goetten, 96–149.
Kaiser, W., and W. Piechocki. "Anfänge einer pharmazeutischen Industrie in Halle und ihre Begründer." *Münchener Medizinische Wochenschrift,* 109 (1967), 1746–1747.
NDB, IX: 416–419.
Partington, II: 690–700.
Piechocki, W. "Das Testament des halleschen Klinikers Friedrich Hoffmann des Jüngeren (1660–1742)." *Acta Historica Leopoldina,* 2 (1965), 107–144.
Rothschuh, K. E. "Studien zu Friedrich Hoffmann (1660–1742)." *Sudhoffs Archiv,* 60 (1976), 163–193, 235–270.
Schulze, J. H. "Lebenslauf D. Friedrich Hofmanns." In F. Hoffmann, *Vernünftige und gründliche Abhandlung von den fürnehmsten Kinderkrankheiten.* Frankfurt a. M. and Leipzig: Möller, 1741, pp. 223–292.
Also see Stahl (Stürzbecher), Berlin Academy (Harnack, I: 146, 170–184), and Halle University (Kaiser and Krosch, 13: 827; 14: 43–44; Schrader, II: 427, 441, 453).

**DIPPEL, Johann Conrad**
1673, Frankenstein near Darmstadt / 1734, Wittgenstein near Berleburg
**Background** Son of Johann Philip Dippel (1636, Rodheim near Giessen–1704, Nieder-Ramstadt near Darmstadt), who, after studying at Wittenberg University in the mid-1650s, was a tutor in Zwingenberg, then a parson in Nieder-Beerbach and, from 1678, in Nieder-Ramstadt. Lutheran (became an outspoken Pietist). **Education** In 1691, having attended Darmstadt's *Pädagogium,* he entered Giessen University with the intention of earning three doctorates. Initially he gave his attention to theology, taking every opportunity to defend orthodoxy and criticize Pietism. After two years, he qualified for his M.A. by participating in a disputation on nihilism and paying fees of 200 fl. Then, continuing his theological reading, he served as a tutor in the Odenwald. However, in 1695, troubled with doubts, he returned to Giessen University in the hopes of getting a professorship in the Philosophical Faculty. Rejected on account of his extreme skepticism, he proceeded to Strasbourg University. In addition to studying theology and medicine there, he taught a private course on chiromancy and unsuccessfully sought chairs in philosophy and theology. Soon after publishing a theological disputation in 1696, he fled Strasbourg to avoid arrest for dueling and debts. He ended up at his father's home in Nieder-Ramstadt where he spent several months reading and writing.

### Career

1697–1699: Resided in Giessen. During this period he evidently gave private lectures, began learning alchemy, and published several theological polemics. In 1699 he was ordered to return home on account of his heretical publications.

1699–1704: Resided with his father in Nieder-Ramstadt. While there he continued publishing theological tracts in defiance of government edicts. He also pursued alchemical investigations on an estate purchased with 50,000 fl provided by various patrons including, perhaps, Landgrave Ernst Ludwig. In 1704 his father petitioned unsuccessfully for his appointment as the next parson in Nieder-Ramstadt.

1704–1707: Resided in Berlin. He was the protégé of Count A. von Wittgenstein, who supplied him with ample funds for alchemical investigations (including 1,000 fl per year for rent). In 1705 he served as a referee when the alchemist, D. M. Cajetano, did experiments for King Frederick I. Two years later he fled Berlin to avoid imprisonment for writing in protest of the Swedish monarch's attempt to suppress Pietism.

1707–1714: Resided near Utrecht in Holland. He continued his writing and alchemical work, apparently supporting himself by practicing medicine. In 1711 he earned an M.D. at Leiden University with a dissertation that included many of his chemical findings. Around 1714 he gave private instruction to J. Juncker, whom he probably met in Giessen in 1695.

1714–1719: Resided in Altona where he practiced medicine and continued his writing. Initially favored by the Danish authorities, he eventually was jailed.

1719–1726: Prisoner on Bornholn Island. He owed his release to the intercession of the Danish Queen.

1726–1728: Traveled through Scandinavia, practicing medicine in aristocratic circles in Copenhagen, Stockholm, and other cities.

1728–1729: Returned to Germany and settled in Liebenburg near Goslar. He practiced medicine and experimented there until he was forced to leave by state edict.

1729–1734: Resided in Berleburg, Count C. von Wittgenstein's haven for harrassed Pietists. He continued his theological, medical, and alchemical work there until his death. In 1732 he sold "a particular chemical secret" to Landgrave Ernst Ludwig for a note of 100,000 Rtl.

**State Honors** 1707: Danish Chancellery Councillor (the title was revoked in 1719 when he was imprisoned). 1732: Hesse-Darmstadt Court Councillor. **Major Scientific Societies** None. **Informal Recognition** 1740: Kestner. 1753: *Vogel*. 1786: Blumenbach. 1791: *Göttling*. 1802: Salzmann.

### Bibliography

Bender, W. *Johann Konrad Dippel: Der Freigeist aus dem Pietismus*. Bonn: Weber, 1882.

Buchner, K. "Johann Konrad Dippel." *Historisches Taschenbuch*, 3d series, 9 (1858), 207–355.

NDB, III: 737–738.

Strieder, III: 89–111.

Voss, K. L. "Christianus Democritus: Das Menschenbild bei Johann Conrad Dippel." *Zeitschrift für Religions- und Geistesgeschichte: Beihefte*, 12 (1970).

Also see Wittenberg University ("Matrikel," 14: 554).

### SCHLÜTER, Christoph Andreas

1673, Goslar / 1744?, ?

**Background** Son of Heinrich Zacharias Schlüter (1644, Goslar–1698?, Goslar?), who served as Smelting Comptroller in the Lower Harz for twenty-eight years. Lutheran. **Education** He apparently learned mining, smelting, and related subjects from his father and other relatives. It was probably during his youth that he visited the main smelting works of the Harz, Saxony, and Bohemia.

### Career

1698?–1744: Hanover-Brunswick Mining Official in the Lower Harz.

1702: Married a daughter of Goslar's Mayor and Town Physician, J. G. Trumph, (1644, Goslar–1711, Goslar), who wrote two chemical works as a medical student of W. Rolfinck at Jena University in the late 1660s.

1703: Became a member of Goslar's Merchant Guild.

1724: Promoted from Smelting Comptroller to Comptroller.

**State Honors** None. **Major Scientific Societies** None. **Informal Recognition** 1760: Wallerius. 1767: Rose. 1777: Raspe. 1787: *Gren*. 1790: *Hermbstaedt*. 1791: *Göttling*. 1796: Fourcroy.

## Bibliography

Asch (Niedersächsisches Staatsarchiv, Hanover). Letter revealing that Schlüter was listed as *Zehntner* in the Hanoverian *Staatskalender* until 1744 and that negotiations bearing on a proposed sale of his former collections to the Elector began in 1750 (29 December 1975).

Grüning (Stadtkirchenamt, Goslar). Letter reporting that Goslar's church records indicate that H. Z. Schlüter was baptised on 9 April 1644, married on 26 April 1669, and possibly buried on 2 March 1698, and that C. A. Schlüter was born on 19 March 1673 and married on 14 November 1702 (2 December 1975).

Hillebrand (Stadtarchiv, Goslar). Letters bearing on the careers of C. A. Schlüter and J. G. Trumph (7 November 1975; 6 February 1976).

König, J. (Niedersächsisches Staatsarchiv, Wolfenbüttel). Letter reporting that the last of Schlüter's reports to the Brunswick authorities was dated Goslar, 1 February 1742 (14 January 1976).

Schlüter, C. A. *Gründlicher Unterricht von Hütte-Werken*. Brunswick: Meyer, 1738, unpaginated foreword.

Zedler, XLV: 1280.

## HENCKEL, Johann Friedrich
1678, Merseburg / 1744, Freiberg

**Background** Son of Johann Andreas Henckel (?, Ilmenau–1722, Merseburg) who studied under the chemist W. Rolfinck and others at Jena University from 1663, earned his M.D. there in 1669, and then became Town Physician in Merseburg. Lutheran (probably became a Pietist).   **Education** After attending Merseburg's Cathedral School from 1685 to 1694 and an unknown gymnasium, he entered Jena University in 1698. He began studying theology but soon switched to medicine. Among his teachers was his father's old friend, *G. W. Wedel.*

## Career
1706?–1711: Served in Dresden as an assistant to *G. E. Stahl*'s former student, the physician and chemist E. P. Meuder. Later practiced medicine there on his own.

1711: Went to Halle University where he earned an M.D. under H. Henrici with a dissertation on tonics.

1712–1730: Practiced medicine in Freiberg.

1714?–1730: Taught private courses for "lovers of chemistry" in Freiberg. *C. Neumann* was probably among his students by 1719.

1718–1730?: District Physician.

1721–1730: Town Physician (salary: 53 fl).

1721?–1729: Mine and Smelting Works Physician.

1725: Nominated by the philosopher C. Wolff for the chair of chemistry in the new St. Petersburg Academy, but nothing came of this.

1730–1732: Personal Physician to Count von Wackerbarth in Dresden.

1732?: Received a call, possibly to St. Petersburg, offering a salary of 800 Tl plus extras.

1732–1744: Saxon Mining Councillor in Dresden, then Freiberg (salary: 600 Tl). Upon moving to Freiberg in 1733, he was provided with 200 Tl for a laboratory and an annual budget of 200 Tl plus 4 wagon loads of charcoal. He was responsible for investigating Saxony's mineral wealth.

1733?–1744: Taught private courses on metallurgical chemistry for which Saxon students paid 100 Tl and foreign students as much as 330 Tl. He took only six students at a time. Among his pupils were *A. S. Marggraf* (1734–1735), M. V. Lomonosov (1739–1740), *J. R. Spielmann* (1742), and F. A. von Heynitz (1743–1744).

**State Honors** None.   **Major Scientific Societies** 1726: Berlin. 1728: Leopoldina.   **Informal Recognition** 1749: *Juncker.* 1750: *Gellert.* 1753: *Vogel.* 1756?: *Cullen.* 1759: *Wallerius.* 1767: *Rose.* 1770: *Madihn.* 1777: *Scopoli.* 1780: *Gmelin.* 1781: *Wiegleb.* 1784: *Suckow.* 1787: *Gren.* 1788: *Leonhardi.* 1791: *Göttling.* 1796: Fourcroy. 1821: *Pfaff.*

## Bibliography
*Briefe von Christian Wolff aus den Jahren 1719–1753: Ein Beitrag zur Geschichte der Kaiserlichen Akademie der Wissenschaften zu St. Petersburg.* Facsimile of the 1860 ed. Hildesheim: Olms, 1971, p. 67.

DSB, VI: 259–260; VIII: 468.

Herrmann, W. "Bergrat Henckel: Ein Wegbereiter der Bergakademie." *Freiberger Forschungshefte: Kultur und Technik*, D37 (1962).

*Mineralogische, Chemische und Alchymistische Briefe von reisenden und anderen Gelehrten*

*an den ehemaligen Chursächsischen Bergrath J. F. Henkel.* 3 vols. Dresden: Walther, 1794–1795. (Cross-referenced as *Min. Briefe.*).
NDB, VII: 515–516.
Zedler, XX: 1441.
Also see Halle University (Kaiser and Krosch, 15: 1021) and Jena University (*Matrikel*, II: 383).

**JUNCKER, Johann**
1679, Londorf near Giessen / 1759, Halle
**Background** Son of Johann Ludwig Juncker (d. after 1706), a well-off tenant farmer in Londorf until 1696 when he was forced by war to migrate to Franconia, then the Rhenish Palatinate. Lutheran (became a Pietist). **Education** After attending school in nearby Allendorf and in Obernhof near Wetterau, he spent four years in Giessen's *Pädagogium* which was headed by the Pietist J. H. May. Around 1695 he began attending philosophical lectures at Giessen University. In 1696 he proceeded to Marburg University. Soon after arriving, he had to interrupt his studies to help his parents escape the ravages of war. Once the worst danger was past, he enrolled at Leipzig University. Dissatisfied there, he soon switched to Halle University to study with the Pietist A. H. Francke. He remained in Halle until 1702, giving most of his attention to ancient languages and theology. However, he also pursued medicine, taking courses from *F. Hoffmann* and *G. E. Stahl*. He probably would have gone into medicine at this time had his parents and friends not dissuaded him.
**Career**
1701–1707: Taught in Francke's *Pädagogium* in Halle for a year, in Lemgo for another year, in the new convent school in Schaaken for four years, and then in the orphanage in Pyrmont for a few weeks. While in Schaaken, he studied medicine with a Dr. Westhof. He also fell in love with Abbess Charlotte Sophia, sister of the Count of Waldeck and Pyrmont. In May 1707 this relationship was consummated in a secret wedding at the residence of the Countess von Perleburg. Two months later Juncker wrote his patron Francke from Erfurt explaining the situation. Rather than being offended by the mismatch, Francke evidently offered him a teaching position in the *Pädagogium* in Halle. While in Halle, Juncker seems to have secured his parents' permission to pursue medicine. Then he apparently returned to Erfurt where, rather than attending lectures, he studied *Stahl*'s medical and chemical works in private.
1708–1716: Practiced medicine first in the environs of Schwarzenau in the county of Wittgenstein, then from 1711 in Ibbenbüren in the county of Lingen. Around 1714 he went to Holland to learn more medicine. In addition to visiting various universities, he studied with *J. C. Dippel*, whom he had probably met at Giessen in 1695.
1717–1759: Ordinary Physician for the *Pädagogium* and Orphanage in Halle. Among his subordinates in the Orphanage was the journeyman pharmacist *J. F. Meyer* (1728–1729).
1718–1759: PD, then from 1729 unsalaried OP of clinical medicine in Halle University's Medical Faculty (1743–1759: shared in the income from student fees). In order to qualify for lecturing, he earned his M.D. under M. Alberti with a dissertation on the medical means of stimulating natural functions. Among his associates were *Hoffmann* (colleague: 1718–1742), *J. H. Pott* (colleague: 1719–1720), *P. Gericke* (colleague: 1724–1730), *J. A. Cramer*(?) (student: 1728–1729), *J. F. Cartheuser* (student: 1729–1731; colleague: 1731–1740), and *A. S. Marggraf* (student: 1733–1734).
**State Honors** 1730: Danish Court Councillor. **Major Scientific Societies** 1754: Erfurt (?). **Informal Recognition** 1736: Burghart. 1737?: *Neumann*. 1741: Shaw. 1752: Börner. 1758?: Cullen. 1759: Wallerius. 1770: Madihn. 1777: Raspe. 1781: *Wiegleb*. 1787: Gren. 1791: *Göttling*.
**Bibliography**
ADB, XIV: 692.
*Commentarii de Rebus in Scientia Naturali et Medicina Gestis*, 9 (1760), 350–361.
DSB, VII: 188.
Joecher, supp. II: 2347–2349.
Kaiser, W. "Die medizinische Studienreform des frühen 18. Jahrhunderts: Zur 300. Wiederkehr des Geburtstages von Johann Juncker (1679–1759)." *Zeitschrift für die gesamte innere Medizin und ihre Grenzgebiete*, 34 (1979), 340–348.
Korn (Hessisches Staatsarchiv, Marburg). Letter revealing that Juncker matriculated at Marburg University from Rabenau near Londorf on 30 October 1696 (31 August 1972).
NDB, X: 661–662.

Petry, A. (Clergyman, Rabenau-Londorf). Letter confirming Juncker's birth date as 23 December 1679 (29 January 1973).
Also see Halle University (Kaiser and Krosch, 13: 827; 14: 1–6, 396–416; 15: 258–261; Kaiser and Krosch, "Collegium," pp. 33–64; "Matrikel," p. 250), and Leipzig University (*Matrikel*, II: 207).

## ROTHE, Gottfried
1679, Lissa near Görlitz / 1710, Leipzig

**Background** Son of Aegidius Rothe (1646, Kamenz, Upper Lusatia–1711, Lissa), who, after attending schools in Halle and Torgau as well as Wittenberg University (Mag. Phil., 1669), taught natural philosophy at Wittenberg for a few years and then went to Lissa as a parson in 1676. Lutheran (probably Pietist). **Education** From 1700 to 1704 he studied medicine at Leipzig University, giving special attention to chemistry. Among his teachers was *J. Bohn*. Then he proceeded to Halle University, where he worked closely with *G. E. Stahl*. In 1708 he earned a medical license under *Stahl* with a dissertation on metallic salts. Returning to Leipzig, he received his M.D. there on the occasion of the three-hundredth anniversary of the university in late 1709.

**Career**
1709–1710: Practiced medicine in Leipzig and gave private lectures on chemistry (these were published posthumously in 1718 and subsequently issued in many editions and translations).
**State Honors** None. **Major Scientific Societies** None. **Informal Recognition** 1736: Burghart. 1737?: *Neumann*. 1740: Kestner. 1752: Börner. 1758: Hoffmann. 1760: Wallerius. 1770: Madihn. 1787: *Gren*. 1791: *Göttling*.

**Bibliography**
Partington, II: 687–688.
Wenzel, P. (Ratsarchiv, Görlitz). Letter concerning Rothe's ancestors (3 July 1972).
Zedler, XXXII: 1133.
Also see Henckel (*Min. Briefe*, III: 40), Halle University (Kaiser and Krosch, 15: 1017; "Matrikel," p. 368), and Leipzig University (*Matrikel*, II: 369–370).

## NEUMANN, Caspar
1683, Züllichau, Brandenburg / 1737, Berlin

**Background** Son of George Neumann (1647?, ?–1695, Züllichau), merchant and organist in Züllichau. Lutheran (probably became a Pietist). **Education** He attended the local Latin school until his father's death thwarted plans to send him into the clergy. He was then apprenticed to his godfather, Senator J. Romcke, owner of Zullichau's Löwen-Apotheke. He learned so quickly that in 1698, three years before completing his apprenticeship, he was put in charge of an apothecary shop, brewery, and distillery owned by Romcke in nearby Unruhstadt. In 1704 war forced him to flee to Berlin where he served briefly as a journeyman in C. Schmedicke's Apotheke zum Schwarzen Adler. From 1705 to 1711, he served as a journeyman to the Royal Prussian Traveling Apothecary Conradi. During this period he accompanied King Frederick I on his travels through Germany and Holland, often playing the clavier for him. Around 1710 he declined an offer to go to St. Petersburg as Court Apothecary (salary: 600 rubles plus extras). Using this call for leverage, he arranged, with the help of *F. Hoffmann*(?), to pursue formal studies at royal expense. He first visited the Harz mining district, then Holland, where he heard H. Boerhaave lecture on chemistry, and finally, England. On Frederick's death in early 1713, Neumann's funds were cut off, leaving him stranded in London.

**Career**
1713–1716: Assistant in London to the wealthy Dutch surgeon A. Cyprian, who spent £1000 annually on chemical experiments. In his spare time he offered courses on chemistry and made friends with pharmacists and chemists, including A. G. Hanckewitz. He was doing so well that in 1715 he decided against returning to Prussia to become Royal Field Apothecary.
1716: Traveled in King George I's entourage to Hanover. While visiting Berlin to settle his affairs, he was persuaded by *G. E. Stahl* to reenter Prussian service.
1717–1719: Took a study tour to England, France, and Italy at royal expense. In Paris he spent much time with E. F. and C. J. Geoffroy, became acquainted with R. A. F. de Réaumer, and taught a chemistry course. J. T. *Eller* may have been among his students. Later, in Freiberg, Neumann probably studied briefly with *J. F. Henckel*.
1719–1737: Court Apothecary in Berlin (salary: 200 Tl; budget around 10,000 Tl). Among his apprentices was *A. S. Marggraf* (1726–1731).

1721-1737: Ordinary Member of the Berlin Society of Sciences. Among his colleagues were *J. H. Pott* (1722-1737) and *Eller* (1725-1737).
1731-1737: Treasurer (salary, July 1731-1737: 200 Rtl).
1723-1737: Professor of "practical" chemistry in Berlin's new Medical-Surgical College (salary, 1724: 50 Rtl; 1725-1737: 220 Rtl). Among his colleagues were *Pott* (1723-1737) and *Eller* (1725-1733).
1724-1737: Member for pharmacy of the Berlin Medical Board, which in 1725 became the Higher Medical Board. Among his colleagues were *Stahl* (1724-1734) and *Eller* (1725-1737).
1727: Traveled to Vienna, visiting *Henckel* in Freiberg on his way back.
1736-1737: Founded and ran a private pharmaceutical school.

**State Honors** 1733: Prussian Court Councillor. **Major Scientific Societies and Honorary Degrees** 1721: Berlin. 1725: London. 1727: Honorary M.D. from Halle University. 1728: Leopoldina. **Informal Recognition** 1736: Burghart. 1740: Kestner. 1741: Shaw. 1742: Anon. 1749: *Juncker*. 1752: *Börner*. 1753: *Vogel*. 1756?: Cullen. 1758: Hoffmann. 1759: Wallerius. 1766: *Lehmann*. 1767: Zückert. 1769: Suckow. 1770: Madihn. 1777: Menn. 1777: Raspe. 1777: *Scopoli*. 1780: *Gmelin*. 1781: *Wiegleb*. 1784: *Suckow*. 1786: Blumenbach. 1787: *Gren*. 1789: Hopson. 1791: *Göttling*. 1791: Nicolai. 1796: Fourcroy. 1802: Salzmann. 1806: *Trommsdorff*. 1807: Kastner. 1814: *Wurzer*. 1821: *Pfaff*.

**Bibliography**
DAB, II: 465-467.
Dann, G. E. (Professor, Dransfeld). Letter bearing on Neumann's master, Schmedicke (17 April 1974).
DSB, X: 25-26.
Exner, A. *Der Hofapotheker Caspar Neumann (1683-1737)*. Diss.; Berlin University; Berlin: Triltsch, 1938.
Gossmann, 118, 189-192.
Hörmann, J. "Die Königliche Hofapotheke in Berlin 1598-1898." *Hohenzollern-Jahrbuch*, 2 (1898), 220-222.
Mayer, H. "Caspar Neumanns Lebensbeschreibung." *Apotheker-Zeitung*, 49 (1934), 590-595.
Partington, II: 702-706.
Pott, J. H. *Chymische Untersuchungen welche fürnemlich von der Lithogeognosia . . . handeln*. Potsdam: Voss, 1746, p. 2.
Stürzbecher, M. "Caspar Neumann." *Berliner Medizin*, 7 (1957), 156-158.
Also see Berlin Academy (Harnack, I: 226, 228, 230; Kirsten) and Berlin Medical-Surgical College (Dorwart, 265-266; Lehmann, 55-57, 65-70).

**TEICHMEYER, Hermann Friedrich**
1685, Münden, Hanover / 1744, Jena

**Background** Son of Hermann Theodor Teichmeyer (?, Münden-1703, Münden), who studied at Jena University with the chemist W. Rolfinck, earned his M.D. at Erfurt University in 1670 and gave private lectures there for a few years, and then served in Münden as Town Physician (1673-1703), as a Town Councilman (1684-1696), and as Commissioner of the Merchant's Guild (1697-1703). Lutheran. **Education** After attending schools in Münden and Altenburg, he entered Leipzig University in 1702. Among his teachers during his year studying the philosophical and medical sciences there was *J. Bohn*. Following his father's death, he proceeded to Jena University. He studied with *G. W. Wedel* and others there for several years, earning his Ph.D. in 1707 and his M.D. under *Wedel* in 1708 with a dissertation on the medical uses of cubeb.

**Career**
1707-1744: PD, AOP, and OP at Jena University. Among his colleagues was *Wedel* (1707-1721) and among his students were *J. T. Eller* (1709-1711), *J. F. Cartheuser* (ca. 1724), *H. Ludolf* (1737-1739), and *C. A. Mangold* (1737-1740?).
1707-1716: PD.
1717-1727: OP of natural philosophy in the Philosophical Faculty and from 1719, AOP in the Medical Faculty.
1727-1744: OP of anatomy, surgery, and botany in the Medical Faculty (salary, 1744: 210 Rtl plus goods).
1711-1728: District Physician for Jena and Physician to the Count of Hatzfield-Gleichen.
1720?-1744?: Produced and sold secret remedies in his laboratory.

1729?–1744: Owned estates at Wenigen-Jena and Camsdorf.
1733: Nominated, as second choice after his colleague G. E. Hamberger, for a chair of chemistry in Göttingen University by Hanoverian Court Physician P. G. Werlhof, but nothing came of this.

**State Honors** 1729?: Saxe-Eisenach Court Councillor and Physician. **Major Scientific Societies** 1725: Berlin. 1731: Leopoldina. **Informal Recognition** 1736: Burghart. 1737?: *Neumann.* 1749: *Juncker.* 1752: Börner. 1753: *Vogel.* 1758: Hoffmann. 1760: Wallerius. 1781: Wiegleb. 1787: Gren. 1791: Göttling.

**Bibliography**
Börner, I: 474–475.
Brethauer, K. (Educator, Münden). "Aus der Galerie berühmter Mündener." *Göttingen Tageblatt* (15/16 February 1969).
———. Letter concerning Teichmeyer's family (11 March 1970).
Zedler, XLII: 606–611.
Also see Erfurt University ("Matrikel," 10: 125), Göttingen University (Rössler, pt. B, pp. 299–300), and Jena University (Chemnitius, pp. 63–70; Giese and Hagen, pp. 201–209; *Matrikel*, II: 807; "Modell-Buch").

### ELLER, Johann Theodor
1689, Plötzkau near Bernburg / 1760, Berlin

**Background** Son of Jobst Hermann Eller von Brockhausen, a military officer, then a justice of the peace and innkeeper in Plötzkau, and finally a resident of Erxleben near Haldensleben. His mother was from the nobility. Lutheran. **Education** After being tutored at home, he attended the gymnasium in Quedlinburg. In 1709 he entered Jena University with the intention of studying law. Soon, however, he switched into medicine and attended the courses of *G. W. Wedel*, *H. F. Teichmeyer*, and others. In 1711 he journeyed to Leiden by way of Halle, where he probably met *G. E. Stahl*. For the next few years he studied in Leiden, probably taking chemistry from H. Boerhaave. He also served as prosecutor to the anatomist F. Ruysch in Amsterdam. In 1716 he took his M.D. at Leiden University, then returned to Germany to study chemistry and metallurgy in the Saxon and Harz mining districts. In Freiberg he may have become acquainted with *J. F. Henckel*. From the German mining districts, he went to Paris where he pursued chemistry with J. Grosse, S. Boulduc, L. Lemery, and probably *C. Neumann*. In 1719 he visited Leiden and in 1720 proceeded to London. There he met all the leading scientists, including I. Newton, and pursued chemistry in A. G. Hanckewitz's laboratory for over a year. He then returned to Germany.

**Career**
1721–1723: Physician to the Prince of Anhalt-Bernburg. He was the first in the principality to inoculate against smallpox. He also invented a patent medicine for arthritis that Bernburg's apothecaries subsequently sold under the rubric "Eller's Drops."
1723?: Gave a course on anatomy to the military surgeons in Magdeburg.
1724–1760: Royal Physician in Berlin. Among his colleagues was *Stahl* (1724–1734), who probably arranged his appointment.
1724–1735: Field Physician.
1727–1735: Medical Director in Berlin's Charité Hospital.
1730?–1760: Dean of the Higher Medical Board and Member of the Sanitation Board.
1735–1760: First Royal Physician and General Staff Physician.
1741–1760: Voting member of the Directory for the Poor.
1725–1760: Professor in, then Director of Berlin's Medical-Surgical College.
   1725–1733: Professor of therapy, pathology, and physiology (salary: 200 Rtl). Among his colleagues were *Neumann* and *J. H. Pott*. From 1727, he gave clinical instruction in Berlin's Charité Hospital.
   1733–1760: Director of the College (salary: 200 Rtl). Among his protégés was *C. A. Gerhard* (1755–1760).
1725–1760: Ordinary Member of the Berlin Society of Sciences, which in 1744 became the Berlin Academy. Among his colleagues were *Neumann* (1725–1737), *Pott* (1725–1760), *A. S. Marggraf* (1738–1760), and *J. G. Lehmann* (1756–1760).
   1736–1743: Co-Director of the Berlin Society (salary: none).
   1744–1760: Director of the Berlin Academy's Physical Class (salary, 1744–1748: none; 1748–1760: 100 Rtl).

1735: Investigated the newly popular mineral springs at Freienwalde and helped devise a plan of development.
1740–1741: Worked briefly with *Pott, J. A. Cramer*, and others trying to develop porcelain for King Frederick II.
1760: Bequeathed 3,112 Tl to the Berlin Academy. The annual interest of 100 Tl was to be used for prizes for essays on scientific and agricultural questions.

**State Honors** 1724: Prussian Court Councillor. 1755: Prussian Privy Court Councillor. **Major Scientific Societies** 1725: Berlin. 1738: Leopoldina. **Informal Recognition** 1759: Wallerius. 1770: Madihn. 1780: *Gmelin*. 1791: *Göttling*. 1802: Salzmann. 1806: *Trommsdorff*.

**Bibliography**
Diepgen, P., and E. Heischkel. *Die Medizin an der Berliner Charité bis zur Gründung der Universität*. Berlin: Springer, 1935, pp. 4–6, 14–15.
DSB, VI: 352–353.
Eller, H. C., to S. Formey. Letter bearing on Formey's éloge and Eller's will (24 November 1760). Staatsbibliothek Preussischer Kulturbesitz, West Berlin (Darmst. F la 1746: Eller).
Formey, S. "Éloge de Monsieur Eller." Akademie der Wissenschaften, Berlin: *Histoire* (1761), 498–509.
Friedrich, 41–52, 119.
Joecher, supp. III: 870–871.
Kaiser, W. "Johann Theodor Eller (1689 bis 1760)." *Zahn-, Mund- und Kieferheilkunde*, 64 (1976), 277–287.
Leiden, 821, 830, 866.
NDB, IV: 456–457.
Partington, II: 715–717.
Pott, J. H. *Animadversiones physico chemicae circa varias hypotheses et experimenta D. Dr. et Consiliar. Elleri*. Berlin, 1756, p. 66.
Also see Marggraf (Formey, p. 71), Berlin Academy (Harnack, I: 226, II: 227–228; Kirsten; Müller, pp. 91–92), Berlin Medical-Surgical College ("Matrikel," 11: 129–130; Stürzbecher, pp. 774–775), and Jena University (*Matrikel*, II: 235).

**POTT, Johann Heinrich**
1692, Halberstadt / 1777, Berlin

**Background** Son of Johann Andreas Pott (1622, Halberstadt—1729, Halberstadt), attorney, canon, and eventually Prussian Councillor. Lutheran (Pietist). **Education** After attending the Cathedral School in Halberstadt, he went to Francke's *Pädagogium* in Halle. In 1709 he entered Halle University to study theology. Soon, however, he switched to medicine so that he could pursue his interest in chemistry with *G. E. Stahl*. Around 1713 he went to Klein-Oehrner (Mansfeld), where he learned assaying from Mining Master Lages. Then, after traveling in the German mining districts, he returned to Halle University. In 1716 he earned an M.D. under *F. Hoffmann* with a dissertation on metallic sulfurs.

**Career**
1716–1719: Practiced medicine in Halberstadt.
1719–1720: PD at Halle University. Among his colleagues were *Hoffmann* and *J. Juncker*.
1720–1722: Went to Berlin, where he practiced medicine, pursued experimental chemistry under the guidance of *C. Neumann*, and sought patronage with the help of influential relatives and his former teacher *Stahl*.
1722–1760: Ordinary Member of the Berlin Society of Sciences, which in 1744 became the Berlin Academy (salary, 1722–1744: none; 1744–1754: 150 Rtl; 1755–1760: none). Among his colleagues were *Neumann* (1722–1737), *J. T. Eller* (1725–1760), *A. S. Marggraf* (1738–1760), and *J. G. Lehmann* (1756–1760).
  1754: His salary as academician was added to his salary as professor when *Marggraf* became director of the Academy's new chemical laboratory, probably so that there would not be two academicians for chemistry.
  1760: He formally withdrew from active membership when *Marggraf* was named Director of the Physical Class as *Eller*'s successor.
1723–1777: Professor in Berlin's new Medical-Surgical College. Among his colleagues were *Neumann* (1723–1737) and *Eller* (1725–1733). Among his students were *Marggraf* (ca. 1730), *J. R. Spielmann* (1741–1742), *J. G. R. Andreae* (1744), *R. A. Vogel*(?) (1747), *F. A.*

*Cartheuser* (1752–1753), *C. A. Gerhard*(?) (mid-1750s), and *J. C. F. Meyer* (1760–1761).
1723–1737: Professor of "theoretical" or "rational pharmaceutical" chemistry (salary, 1724–1725: 100 Rtl; 1726–1737: 200 Rtl).
1737–1777: Professor of chemistry (salary, 1738–1754: 400 Rtl; 1755–1776: 550 Rtl). In 1770 he stopped lecturing on account of bad eyesight.
ca. 1740: Visited *J. F. Henckel* in Freiberg.
1740–1746?: Attempted, at the behest of King Frederick II, to develop a satisfactory porcelain. Initially he worked with *Eller* and *J. A. Cramer*. But from 1741 he worked with less distinguished men. In 1745–1746 he established a porcelain works in Freienwalde at royal expense (1,200 Tl). However, its product was unsatisfactory. All in all, he carried out more than 30,000 unsuccessful experiments (cost: 6,000 Tl).
Before 1756: Received "advantageous" calls to Utrecht, St. Petersburg (1754?) and other places.

**State Honors** None. **Major Scientific Societies** 1722: Berlin. **Informal Recognition** 1748: *Mangold*. 1749: *Juncker*. 1750: *Gellert*. 1752: *Börner*. 1753: *Vogel*. 1756?: *Cullen*. 1759: *Wallerius*. 1767: *Baldinger*. 1769: *Zückert*. 1769: *Suckow*. 1770: *Madihn*. 1773: *Gerhard*. 1774: *Wiegleb*. 1775: *Suckow*. 1777: *Black*. 1777: *Menn*. 1777: *Raspe*. 1777: *Scopoli*. 1780: *Gmelin*. 1785: *Crell*. 1786: *Blumenbach*. 1787: *Gren*. 1788: *Leonhardi*. 1789: *Hopson*. 1791: *Göttling*. 1791: *Nicolai*. 1796: *Fourcroy*. 1802: *Salzmann*. 1806: *Trommsdorff*. 1807: *Kastner*. 1814: *Wurzer*. 1821: *Pfaff*.

**Bibliography**
Bensch, A. D. *Die Entwicklung der Berliner Porzellanindustrie unter Friedrich dem Grossen*. Berlin: Heymann, 1928, pp. 22–23.
Börner, II: 485–494.
Cassebaum, H. "Die Stellung der Braunstein-Untersuchungen von J. H. Pott (1692–1777) in der Geschichte des Mangans." *Sudhoffs Archiv*, 63 (1979), 136–153.
DSB, XI: 109.
[Formey, S.] "Éloge de M. Pott." Akademie der Wissenschaften, Berlin: *Nouveaux Mémoires* (1777), 55–66.
Friedrich, 45, 51–52.
Gossmann, 119, 191, 193, 197, 199.
[Lehmann, J., et al.] *Kurtze Untersuchung der wahren Ursachen, welche den Professor . . . Pott verleitet*. [Berlin], 1756.
König, pt. 2: 246–249.
Meyer-Erlach, G. "Die Pott in Halberstadt." *Familiengeschichtliche Blätter*, 38 (1940), 63–64.
Pott, R. *Johann Heinrich Pott*. Jena: Dufft, 1876.
Schulze, J. "Die ersten Versuche der Porzellanfabrikation in Brandenburg." *Forschungen zur brandenburgischen und preussischen Geschichte*, 4 (1935), 149–153.
Zedler, XXVIII: 1911–1912.
Also see Lehmann (Freyberg, pp. 45, 75–81, 136), Marggraf (Formey, p. 71), Berlin Academy (Harnack, I: 232; Kirsten), and Berlin Medical-Surgical College (Dorwart, p. 265; Lehmann, pp. 57–58, 65–71).

**GERICKE, Peter**
1693, Stendal / 1750, Helmstedt

**Background** Son of Andreas Gericke (d. ca. 1722, Stendal), a brewer. Lutheran. **Education** In 1713, after attending the local town school and Berlin's Joachimsthal Gymnasium, he enrolled at Jena University to study theology. Two years later he returned home to secure parental approval of his decision to become a physician. Early in 1716 he entered Halle University with free board. There he soon became close to C. Wolff, the philosopher, and *F. Hoffmann*. He did so well that *Hoffmann* recommended that he be sent abroad by the Prussian government to continue his medical studies. Nothing came of this, so in 1718 he proceeded by way of Leipzig to Altdorf University. There he studied with L. Heister and J. J. Baier (chemistry), earning his M.D. in 1721 with a dissertation on the study of novelties in medicine.

**Career**
1721: PD at Altdorf University.
1722–1723: Began a trip to Holland and England, but the death of his father obliged him to return to Stendal. He practiced medicine there briefly.

1724–1730: AOP in Halle University's Medical and Philosophical Faculties. His appointment was arranged by M. L. von Printzen, von Scharden, and *Hoffmann*, all of whom cleared him of the taint of his friendship with Wolff, who had been exiled. Among his colleagues were *Hoffmann* and *J. Juncker* and among his students were *J. A. Cramer* (1728–1729) and *J. F. Cartheuser* (1729–1730).

1730–1750: OP of anatomy, pharmacy, and chemistry, then in 1741 of theoretical medicine, materia medica, and chemistry in Helmstedt University's Medical Faculty (salary, 1730–1741: 300 Rtl; 1741–1747: 375 Rtl; 1747–1750: 475 Rtl). His appointment was arranged by his former teacher Heister, who had first tried to get P. G. Werlhof for the job. Among his students was *Cramer* (1734–1735). Between 1731 and 1745 he made several fruitless petitions to the government for a laboratory. In 1737 he was Helmstedt University's delegate to Göttingen University's opening ceremonies.

**State Honors** 1744: Brunswick Court Councillor and Personal Physician.   **Major Scientific Societies** 1731: Berlin.   **Informal Recognition** 1749: *Juncker*. 1752: Börner. 1760: Wallerius. 1781: *Wiegleb*. 1787: *Gren*. 1791: *Göttling*.

**Bibliography**
Goetten, 345–346.
"Leben des Herrn Hofrath Gericke." *Braunschweigische Anzeigen* (1751), 441–445.
Niedersächsisches Staatsarchiv, Wolfenbüttel (37 Alt 442 and 644).
Also see Halle University (Friedländer, pp. 20–21; Kaiser and Krosch, 14: 368–370; "Matrikel," p. 171).

### CARTHEUSER, Johann Friedrich
1704, Hayn, Stollberg / 1777, Frankfurt a. d. Oder

**Background** Son of Valentin Heinrich Cartheuser (1670, Hayn–1726, Hayn), who studied at Jena University, then became a parson in Hayn. Lutheran.   **Education** After completing the gymnasium in Quedlinburg, he went to Jena University to study medicine. Among his teachers there was *H. F. Teichmeyer*. He did so well that he received permission to practice medicine in Thuringia. However, finding the physician's life not to his liking, he set his sights on becoming a professor. In 1729 he proceeded to Halle University, where he probably heard the lectures of *F. Hoffmann, P. Gericke,* and *J. Juncker*. In 1731 he earned his M.D. under M. Alberti with a dissertation on asthma.

**Career**
1731–1740: PD at Halle University. Among his colleagues were *Hoffmann* and *Juncker*. During this period he took several trips to investigate mineral springs and mines. Upon his departure in 1740, he sold his house for 1,300 Tl.
1740–1777: OP in Frankfurt a.d. Oder University's Medical Faculty. His initial appointment (to a supernumerary chair) was begrudged by many at the university, so he had to appeal to the Berlin authorities to secure his travel costs, his share of student fees, and his allotments of wood and grain. Soon, however, he had earned the confidence of his colleagues. Among his students were his son *F. A. Cartheuser* (colleague: 1754–1766) and *C. A. Gerhard* (1757–1760).
1740–1744: OP of chemistry, pharmacy, and materia medica.
1744–1759: OP of chemistry, pharmacy, materia medica, anatomy, and botany (salary, 1744: 419 Tl).
1759–1777: OP of pathology and therapy.

**State Honors** None.   **Major Scientific Societies** 1755: Erfurt. 1758: Berlin.   **Informal Recognition** 1753: *Vogel*. 1760: Wallerius. 1787: *Gren*. 1791: *Göttling*. 1802: Salzmann. 1807: Kastner.

**Bibliography**
Börner, I: 248–258.
Bornhak, C. *Geschichte der preussischen Universitätsverwaltung bis 1810*. Berlin: Reimer, 1900, pp. 68, 115, 139.
Dreyhaupt, 602.
Joecher, supp. II: 730–731.
Lans, J. (Clergyman, Hayn). Letter dealing with Cartheuser's family (16 March 1970).
Also see Jena University (*Matrikel*, II: 97) and Halle University (Conrad, p. 56; Kaiser and Krosch, 14: 373–376).

## MEYER, Johann Friedrich
1705, Osnabrück / 1765, Osnabrück

**Background** Son of Johann Andreas Meyer (1670, Weissensee–1714, Osnabrück), who earned his M.D. at Erfurt University in 1702, then became a physician in Osnabrück. Lutheran (Pietist). **Education** Destined for the clergy, he entered the local gymnasium at the age of seven. However, on account of his father's early death, he became an apprentice in Osnabrück's Hirsch-Apotheke, which was owned by his maternal grandmother. During his apprenticeship he studied a manuscript of A. Q. Rivinus's chemical lectures. In 1726 he went to Leipzig in search of a journeyman's position. Being unsuccessful, he spent six months reading, botanizing, and experimenting in the laboratory of the apothecary-scientist, J. H. Lincke. Then he went to Nordhausen where he found a position in the Ratsapotheke. He resigned two months later on account of poor health and toured the Harz mining district. Then he served as a journeyman pharmacist in Trier for a year and in Halle's Orphanage for two years. In Halle he probably became acquainted with *J. Juncker*, the physician at the Orphanage.

**Career**
1730–1765: Administrator, then from 1737, owner of the Hirsch-Apotheke in Osnabrück. He corresponded with *J. G. R. Andreae, J. H. Pott, A. S. Marggraf,* and *J. C. Wiegleb*.

**State Honors** None. **Major Scientific Societies** None. **Informal Recognition** 1767: Zückert. 1770: *Wiegleb.* 1775: *Suckow.* 1777: Raspe. 1780: *Gmelin.* 1786: *Hermbstaedt.* 1787: *Gren.* 1788: *Leonhardi.* 1791: *Göttling.* 1807: Kastner.

**Bibliography**
Baldinger, E. G. "Vorrede worinnen Herrn Meyers Leben erzählt und von dessen Verdiensten gehandelt wird." In J. C. Wiegleb, *Kleine chymische Abhandlungen.* Langensalza: Martin, 1767, pp. 3–28.
DAB, II: 432.
DSB, IX: 346–347.
Tschanter, K., and W. Schneider. "Johann Friedrich Meyer." *Die pharmazeutische Industrie,* 17 (1955), 450–452.
Zedler, XVII: 1307–1309.

## LUDOLF, Hieronymus
1708, Erfurt / 1764, Erfurt

**Background** Son of Hieronymus Ludolf (1679, Erfurt–1728, Erfurt), who studied under *G. W. Wedel* at Jena University, earned his M.D. at Erfurt University in 1703, and served as OP of mathematics and medicine at Erfurt University from 1717 to 1728 (responsible for chemistry 1719–1727). Lutheran (Pietist). **Education** After instruction at home and Erfurt's Barfüsser School, he entered Erfurt's Town Council Gymnasium in 1720. A few years later he began studying at Erfurt University. Although he focused on philosophy and law, he also attended his father's courses on chemistry and anatomy. In 1728 he proceeded to Jena University to study law. Soon afterwards his father's death obliged him to seek funds for continuing his education from rich relatives in Copenhagen. Having no luck, he returned to Erfurt. The following year he went to Wetzlar where he served as an amanuensis to a highly placed uncle. In 1731, perceiving that his chances of promotion were small, he resumed his law studies at Jena University. Three years later, having come under the influence of Pietism, he went to Berleburg as a tutor. He stayed until 1737, even though his poverty made it necessary for him to serve as a watchmaker's assistant. Once again he returned to Jena University, this time to study medicine. Among his teachers was *H. F. Teichmeyer.* He also worked with the alchemists R. J. F. Schmid and Baron von Blaka. Finally in 1739, he returned to Erfurt University and earned his M.D. with a dissertation on vitriolic acid.

**Career**
1739–1764: PD, AOP, then OP at Erfurt University. Among his students and colleagues were *R. A. Vogel* (student?: 1740–1744; colleague: 1748–1752) and his cousin *C. A. Mangold* (colleague: 1745, 1748–1753). Around 1743 he constructed a large laboratory. For the next few years, he ran a chemical boarding school, charging students 100 Rtl per year for room, board, and instruction. He also produced patent medicines.
1739–1740: PD.
1740–1745: AOP in the Philosophical Faculty.
1745–1764: OP of mathematics in the Philosophical Faculty and of chemistry in the Medical

Faculty (salary, 1745–56: none; 1756–1763: 44 Rtl plus goods; 1764: 91 Rtl plus goods). From 1753 to 1764 he was on leave in Mainz.
1745–1753: Town Physician in Erfurt (salary: 40 Rtl plus goods).
1753–1764: Electoral Physician in Mainz.
1755–1764: AOP of chemistry in Mainz University's Medical Faculty.
1764: Returned to Erfurt and died soon afterwards.
**State Honors** 1753: Electoral Mainz Court Councillor. 1756?: Imperial Ennoblement as Palatine Count. **Scientific Societies** 1754: Erfurt. **Informal Recognition** 1759: Wallerius. 1777: *Scopoli*. 1784: Hahnemann. 1787: *Gren*. 1791: *Göttling*.
**Bibliography**
Börner, I: 826–843.
Also see Erfurt University (Hampel, pp. 21, 23, 40–43, 51–60, 80–93; Loth, p. 220) and Mainz University (*Hofkalender* for 1754–1764).

**MARGGRAF, Andreas Sigismund**
1709, Berlin / 1782, Berlin

**Background** Son of Henning Christian Marggraf (1680, Neuhausen–1754, Berlin), who settled in Berlin as a grocer, leased the Ratsapotheke there in 1707, acquired the Apotheke zum Bären in 1720, and later engaged in some chemical research and corresponded with *J. F. Henckel*. Lutheran. **Education** After schooling, he became an apprentice in his father's shop. In 1726 he transferred to Berlin's Court Apothecary Shop where he received instruction from *C. Neumann*. He also attended *J. H. Pott*'s lectures at the Medical-Surgical College. Upon completing his apprenticeship in 1731, he served as a journeyman in Rössler's shop in Frankfurt a. M. for a year, then in J. J. Spielmann's Hirsch-Apotheke in Strasbourg for another year. In the spring of 1733, he proceeded to Halle University where, besides studying medicine, he took chemistry from *F. Hoffmann, J. Juncker*, J. H. Schulze, and J. J. Lange. In the summer of 1734, he went on to Freiberg. There he pursued chemistry under *Henckel* and assaying under Süssmilch. Finally, in the spring of 1735, he returned to Berlin by way of the Harz mining district.

**Career**
1735–1753: Administrator of his father's Apotheke zum Bären until 1753 when it was sold to his brother-in-law (price: 7,000 Tl).
1737: Declined a call to Brunswick to head the Ducal Apothecary Shop and serve as an advisor on mining matters.
1738–1782: Ordinary Member of the Berlin Society of Sciences, which in 1744 became the Berlin Academy (salary, 1738–1743: none; 1744–1746: 100 Rtl; 1746–1770: 350 Rtl; 1770–1782: 400 Rtl). Among his colleagues were *Pott* (1738–1760), *J. T. Eller* (1738–1760), *J. G. Lehmann* (1756–1761), *C. A. Gerhard* (1768–1782), and *F. C. Achard* (1776–1782).
1753–1782: Director of the Academy's new Chemical Laboratory (salary: 250 Rtl plus free residence in the laboratory).
1760–1782: Director of the Academy's Physical Class (salary, 1762–1782: 200 Rtl).
1740s–1770s: Gave private courses on chemistry in Berlin. Among his students were *J. R. Spielmann* (1741–1742), *Gerhard* (late 1750s), *J. C. F. Meyer* (ca. 1760), *Achard* (ca. 1770), and *M. H. Klaproth* (1770s).
1782: Upon his death, his mineral collection was offered for sale at 200 louis d'or.
**State Honors** None. **Major Scientific Societies** 1738: Berlin. 1757: Erfurt. 1776: St. Petersburg. 1777: Paris (one of six foreign associates). **Informal Recognition** 1748: *Mangold*. 1750: *Gellert*. 1753: *Vogel*. 1756?: Cullen. 1759: Wallerius. 1767: Baldinger. 1769: Suckow. 1770: Madihn. 1770: *Wiegleb*. 1771: Cartheuser. 1773: *Gerhard*. 1775: *Suckow*. 1777: Black. 1777: Menn. 1777: Raspe. 1777: *Scopoli*. 1778: *Hagen*. 1780: *Gmelin*. 1781: *Crell*. 1786: Blumenbach. 1787: Fuchs. 1787: *Gren*. 1789: Hopson. 1791: Denina. 1791: *Göttling*. 1791: Nicolai. 1796: Fourcroy. 1802: Salzmann. 1807: Kastner. 1814: *Wurzer*. 1818: *Trommsdorff*. 1821: *Pfaff*.
**Bibliography:**
Baldinger, pt. 1: 87–98.
Bernoulli, J. Notice of sale. *Gothaische gelehrte Zeitung* (30 November 1782), 805–806.
Boswell, J. *Boswell on the Grand Tour: Germany and Switzerland, 1764*. Edited by F. A. Pottle. New York: McGraw-Hill, 1928, p. 94.

Condorcet, M. "Éloge de M. Margraaf." Academie des Sciences, Paris: *Histoire* (1782), 122–130.
Crell, L. "Lebensgeschichte Andreas Sigismund Marggraf's," *Chemische Annalen*, no. 1 (1786), 181–192.
DAB, II: 401–402.
Dann, G. E. "Marggraf-Briefe." *Zur Geschichte der Pharmazie*, 20 (1968), 20–22.
DSB, IX: 104–107.
Formey, J. H. S. "Éloge de M. Marggraf." Akademie der Wissenschaften, Berlin: *Histoire* (1783), 63–72.
Grottkass, R. E. "Unbekannte Briefe und Einzelheiten über Marggraf und Achard." *Deutsche Zuckerindustrie*, 56 (1931), 604–606.
Partington, II: 723–729.
Pott, J. H. *Fortsetzung seiner physicalisch-chymischen Anmerkungen.* Berlin, 1756, p. 7.
Zaunick, R. "Andreas Sigismund Marggraf." *Zeitschrift für die Zückerindustrie*, 9 (1959), 71–75.
Also see Henckel (*Min. Briefe*, II: 341–359), Klaproth (Dann, pp. 33–35, 146, table VII), Kunckel (Stürzbecher, pp. 16–17), and Berlin Academy (Harnack, I: 244, 283, 353, 380, 382, 466, 487, 490–491; II: 276; Kirsten).

**CRAMER, Johann Andreas**
1710, Quedlinburg / 1777, Berggieshübel near Dresden

**Background** Son of Johann Andreas Cramer (d. 1718?, Quedlinburg), a merchant who served as Town Treasurer and who leased the Prince of Anhalt's iron works at Mägdesprung. Lutheran. **Education** As a boy he took such pleasure in accompanying his father on visits to mines and smelteries that he did poorly in school. Upon his father's death, he became the ward of his brother-in-law, C. G. Schwalbe, a physician who had studied chemistry with H. Boerhaave in Leiden and L. Lemery in Paris. In 1726 young Cramer was sent by way of Halle, where he matriculated for law, to Hamburg. However, after one semester at the Johanneum Gymnasium there, he was called back home because Schwalbe feared his growing enthusiasm for the life of the sea. He then went to Halle University. Interested in chemistry, he began studying medicine; but finding that anatomy made him sick, he switched to law. Nevertheless, he continued the study of chemistry, evidently taking the courses of *J. Juncker* and *P. Gericke*. In 1729 he proceeded to Helmstedt University where he spent a year or so. Then he tried his hand at practicing law in the mining town of Blankenburg. In his spare time he pursued chemistry and metallurgy so avidly that he came to be known as "black Cramer." In 1734 he returned to Helmstedt University, where, besides studying with *Gericke*, he set up his own laboratory. The following year he journeyed to Leiden and enrolled to study medicine. In addition to pursuing his studies, he worked as a chemical experimenter for I. Lawson, a rich English medical student. During the winter of 1737–1738 he gave a course on assaying to a private scientific club among whose members were Boerhaave, C. Linnaeus, and G. van Swieten.

**Career:**
1738–1740: Traveled to England where he visited mines and taught a private course on smelting in London.
1740–1742: Traveled around Germany in search of a satisfactory position. In late 1740 he visited Berlin, where he soon got involved in the porcelain-making venture of *J. T. Eller*, a friend of his guardian Schwalbe, and *J. H. Pott*. Frederick II promised *Eller* that Cramer could have a salary of 500 to 600 Rtl if he could produce good porcelain. Nothing came of this, so Cramer proceeded to Leipzig where he gave a private chemistry course in fall 1741. Around this time he also visited most of the Saxon mining towns.
1742–1743: Councillor in Brunswick with responsibility for investigating the smelting works in the Weser district (salary: 500 Tl). Upon completing the investigation, he considered returning to England. However, he decided to remain in the Duchy of Brunswick because his friend Cabinet Secretary H. B. Schrader succeeded in arranging a good position for him.
1744–1773: Treasury Councillor in Blankenburg. Primarily responsible for promoting mining and forestry in the district, he was apparently also expected to teach the mining sciences, including smelting and assaying. Among those who came to learn from him were *J. G. R. Andreae* (1746) and *C. E. Gellert* (1753). He enjoyed complete freedom from fiscal accountability until 1766 when his enemies persuaded the Duke to scrutinize his conduct. During the protracted investigation, he received job offers from the St. Petersburg Academy, which wanted him to succeed *J. G. Lehmann*, and the Prussian government, which valued his prior

advice for improving the mints, the copper smelteries at Rothenburg an der Saale, and the ironworks in Neumark. He declined both offers, thinking it would be ignoble to leave the duchy while his honor was in question. He was finally reinstated in 1769 after a hearing in which Hanoverian mining officials testified on his behalf. But four years later he fled the duchy when he was subjected to new accusations as a result of Schrader's death.

1774: Consultant to Count von Bolza in Berggieshübel near Dresden. They tried to revive the copper mines there.

1775–1776: Consultant to the Hapsburg Mining Administration concerning the smelting works in Hungary. Upon the completion of his report, he was asked to implement his suggestions. However, despite promises of a budget of 60,000 fl, a salary of 3 ducats a day, and a return of half the savings over the next twelve years, he decided to return to Saxony.

1777: Consultant in Dresden and Berggieshübel to the Saxon Mining Administration.

**State Honors** None. **Major Scientific Societies** None. (The St. Petersburg Academy offered him a chair in 1767, but did not make him a member.) **Informal Recognition** 1750: *Gellert*. 1752: Börner. 1756?: Cullen. 1758: Hoffmann. 1759: Wallerius. 1767: Rose. 1768: Westfield. 1769: Suckow. 1773: *Gerhard*. 1777: Raspe. 1780: *Gmelin*. 1784: *Suckow*. 1787: *Gren*. 1789: Hopson. 1790: *Hermbstaedt*. 1791: *Göttling*. 1796?: Black. 1807: Kastner.

**Bibliography**
Baldinger, pt. 1: 78.
Bernstorff, H. "Personal-Register der Verwaltungen des Domaniums und der Klostergüter des Herzogtums Braunschweig . . . (1662) bis 1883." Niedersächsisches Staatsarchiv, Wolfenbüttel.
Claus. "Zum Andencken Herrn Johann Andreas Cramer." *Chemische Annalen*, no. 2 (1786), 376–384.
Cramer, F. "Johann Andreas Cramer." *Harzboten* (1828), 194–204, 277–287.
DSB, XV: 93–94.
Friedrich, 45.
Knausche (Staatsarchiv, Hamburg). Letter concerning Cramer's enrollment in the Johanneum from the fall of 1726 to the spring of 1727 (9 February 1972).
Leiden, 821, 956.
Linnaeus, C. *A Selection of the Correspondence of Linnaeus and other Naturalists*. Edited by J. E. Smith. Vol. II. London: Longmans, 1821, pp. 171–185.
St. Petersburg Correspondence, 74, 78, 90.
St. Petersburg Procès-verbaux, II: 627.
Zedler, XXXV: 1804–1805.
Also see Henckel (*Min. Briefe*, I: 226), Pott (Bensch), Spielmann (Wittwer, p. 549), Freiberg Mining Academy (Weber, pp. 42–43), Halle University ("Matrikel," p. 99), and Helmstedt University ("Matrikel").

**GELLERT, Christlieb Ehregott**
1713, Hainichen, Saxony / 1795, Freiberg

**Background** Son of Christian Gellert (1672, Zeitz–1747, Hainichen), who earned a Mag. Phil. at Leipzig University (1693) and served as deacon, then parson (1705–1747) in Hainichen. Lutheran (Pietist). **Education** After attending the Latin school in Freiberg, he entered the St. Afra Prince's School in Meissen in 1729. Three years later he was expelled for striking a teacher. He spent a short time at home, then entered Leipzig University where he studied the philosophical sciences. In 1735, having been called to a teaching position in St. Petersburg, he took his Mag. Phil. at Wittenberg University because the fees were lower there than at Leipzig.

**Career**
1735–1742: Co-rector, then in 1740 Pro-rector of the St. Petersburg Academy's Lutheran Gymnasium (traveling money: 100 rubles; salary, 1735–1740: 350 rubles; 1740–1742: 400 rubles). At first he taught history, geography, and logic; later, logic, metaphysics, natural philosophy, and mathematics.

1736–1744: Adjunct member of the St. Petersburg Academy's Physical Class. He probably received intensive chemical instruction from *J. F. Henckel*'s student M. W. Lomonosov, who returned to Russia in 1741. At odds with the Academy's administrators, he resigned in 1744 and returned to Germany.

1744–1749: Tried to get a position in Berlin, where he probably met *J. H. Pott*, *A. S. Margraf*, and *J. T. Eller*, in Freiberg, where he would have succeeded *J. F. Henckel*, and in

Leipzig, where he was patronized by the former Russian Mine Director, Schönberg. In 1747, encouraged by a salary of 200 Tl, he began offering private courses in Leipzig University in the hope of proving his worth. However, in 1749 when the Philosophical Faculty denied his petition for a chair, he moved to Freiberg.

1749–1765: Taught metallurgical chemistry in Freiberg. His fees were high, e.g., 500 Tl for a course he gave five Italian aristocrats in 1750 and 400 Rtl for a course he gave B. von Heynitz in 1763.

1753: Visited the Harz mining district in order to study assaying with *J. A. Cramer*, whose text he had translated in 1746.

1753–1795: Saxon Mining Commission Councillor, then from 1782 Saxon Mining Councillor with residence in Freiberg (salary, 1753–1761: 500 Tl; 1762: 700 Tl; 1763–1795: 900 Tl).

1753–1762: He was responsible for testing mining machines and smelting processes and for analyzing minerals. Instead of using *Henckel*'s laboratory, which he found to be inadequate, he constructed one in his residence.

1762–1795: Chief Smelting Administrator. In 1766 he developed an efficient oven for smelting silver. In the late 1780s he developed the cold amalgamation process for extracting silver. Among his subordinates was *C. F. Wenzel* (1779–1793).

1765–1795: Professor of metallurgical chemistry in the new Freiberg Mining Academy (salary: 200 Tl). Among his students was his successor *W. A. Lampadius* (1794–1795).

**State Honors** None.  **Major Scientific Societies** 1736: St. Petersburg (adjunct).  **Informal Recognition** 1752: Börner. 1758: Hoffmann. 1759: Wallerius. 1769: Suckow. 1777: Raspe. 1777: *Scopoli*. 1781: *Wiegleb*. 1787: *Gren*. 1790: *Hermbstaedt*. 1791: *Göttling*. 1796: Fourcroy. 1807: Kastner.

**Bibliography**
ALZ: IB (12 August 1795), 708.
Becker, W., and A. Lange. *Christian Fürchtegott Gellert, Christlieb Ehregott Gellert: Gedenckschrift*. Hainichen: Stadtrat, [1965].
Čenakal, V. L., and J. C. Kopelvič. "Christlieb Ehregott Gellert in Petersburg." *Freiberger Forschungshefte: Kultur und Technik*, D46 (1964), 23–46.
Distel, T. "Schreiben des Freiberg Gellert vom Jahre 1747, mit wissenschaftlichen Beilagen." Freiberger Altertumsverein: *Mitteilungen*, 30 (1894), 108–109.
Euler, no. 3: 43–45, 93, 99, 105–106.
NDB, VI: 175–176.
Schlichtegroll, vol. II for 1795, pp. 382–391.
St. Petersburg Correspondence, 470–471.
Staatsarchiv, Dresden. Letter containing information on Gellert's salary as a mining official (8 May 1972).
Also see Lampadius (Seifert, pp. 17–24), Freiberg Mining Academy (Baumgärtel, pp. 82, 95; Weber, pp. 157–158), Leipzig University (*Matrikel*, II: 127; III: 107), and Wittenberg University ("Matrikel," 2: 174).

**LEHMANN, Johann Gottlob**
1719, Langenhennersdorf near Pirna / 1767, St. Petersburg

**Background** Son of Martin Gottlob Lehmann (d. 1729, Grünewalde), scion of a rich administrative family, lessee of an estate at Langenhennersdorf from 1718 to 1722, then owner of an estate at Eckersdorf (cost 10,000 Tl).  Lutheran (Pietist?).  **Education** After private instruction, he went to the Prince's School at Schulpforta in 1735. He soon returned home, however, on account of poor health. In 1738 he entered Leipzig University to study medicine. The following year he proceeded to Wittenberg University where he received instruction in chemistry and other medical subjects from A. Vater. He earned his M.D. there in 1741 with an anatomical dissertation.

**Career**
1741–1750: Practiced medicine in Dresden. In his spare time he visited the Harz, Saxon, and Bohemian mining districts. In 1747 he belonged to a small society that met weekly to discuss mathematics and natural philosophy. He described himself then as an "ongoing metallurgist." He probably met *C. E. Gellert*.

1750–1753: Went to Berlin, perhaps hoping to find work in Frederick II's as yet unsuccessful venture to manufacture porcelain. He soon sought out *J. H. Pott, A. S. Marggraf*, and *J. T. Eller*. Early in 1751 he was prospecting along the Oder River, possibly at the expense of the

Prussian government. Late that year he catalogued a coin collection in Stettin. During 1752–1753 he was prospecting in the Harz and Oder regions for the Prussian government.
1753–1754: Prussian Mine Director in Hasserode in the Harz. He probably visited *J. A. Cramer* in nearby Blanckenburg.
1754–1761: Prussian Mining Councillor in Silesia, then from 1756 in Berlin.
1756–1761: Extraordinary Member of the Berlin Academy (salary: none, but during this period he was provided with 302 Rtl for experiments and 1,490 Rtl for living expenses). Among his colleagues were *Pott* (1756–1760), *Eller* (1756–1760), and *Marggraf*, with whom he worked very closely.
1756?–1761: Gave private courses in mineralogy and assaying. *C. A. Gerhard* was probably among his students.
1761–1767: Professor of chemistry and director of the natural collections in the St. Petersburg Academy (salary: 1,000 rubles).
1765: Investigated the saltworks in "old Russia."

**State Honors** 1765: Russian Court Councillor. **Scientific Societies** 1754: Berlin. 1757: Erfurt. 1761: St. Petersburg. **Informal Recognition** 1759: Wallerius. 1768: Westfeld. 1771: *Cartheuser*. 1777: Raspe. 1777: *Scopoli*. 1780: *Gmelin*. 1781: *Wiegleb*. 1784: *Suckow*. 1787: *Gren*. 1791: *Göttling*. 1796: Fourcroy.

**Bibliography**
DSB, VIII: 146–148.
Euler, no. 3: 239.
Freyberg, B. von. "Johann Gottlob Lehmann (1719–1767): Ein Arzt, Chemiker, Metallurg, Bergmann, Mineraloge und grundlegender Geologe." *Erlanger Forschungen: Reihe B: Naturwissenschaften*, 1 (1955).
Prescher, H. "Johann Gottlob Lehmann." *Sächsische Heimatblätter*, 15 (1969), 274–277.

**MANGOLD, Christoph Andreas**
1719, Erfurt / 1767, Erfurt

**Background** Son of Balthasar Christian Mangold (1674,?–1738, Erfurt), deacon of the Barfüsser Church of Erfurt. Lutheran. **Education** After attending Erfurt's Barfüsser School and Town Council Gymnasium, he studied the philosophical and medical sciences at Erfurt University from 1734 to 1737. Then he proceeded to Jena University where, in conjunction with his medical studies, he took chemistry from G. E. Hamberger, *H. F. Teichmeyer*, and G. A. Fuchs. While there, he also worked with the alchemist R. J. F. Schmid. In the early 1740s, he returned to Erfurt University for further medical study.

**Career**
1745: AOP in Erfurt University's Philosophical Faculty. Among his colleagues was his cousin *H. Ludolf*.
1746–1748: Personal Physician to ailing Count G. A. von Gotter, a favorite of King Frederick II and owner of a luxurious estate two hours from Erfurt. During this period he accompanied Gotter to Berlin where he met *J. H. Pott* and *A. S. Marggraf*.
1748–1752: AOP in Erfurt University's Philosophical Faculty. Among his colleagues were *Ludolf* and *R. A. Vogel*.
1750: Nominated by A. von Haller as the best candidate for a salaried chair of chemistry in the new Göttingen Society of Sciences, but nothing came of this.
1751: Earned his M.D. at Erfurt University with a dissertation on the rules for perfecting medical practice.
1752–1754: Personal Physician to Count Gotter. They traveled by way of Strasbourg, where Mangold met *J. R. Spielmann*, to Montpellier. They spent a year there, then returned to Berlin where Mangold participated in two sessions of the Berlin Academy (November 1753, May 1754).
1754–1767: AOP, then OP at Erfurt University.
1754–1767: AOP and from 1763 OP in the Philosophical Faculty.
1755–1757: AOP of chemistry in the Medical Faculty.
1757–1767: OP of chemistry, botany, and from 1765 anatomy in the Medical Faculty (salary, 1757–1764: 150 Tl plus wood; 1765–1767: 194 Tl plus goods). In 1763 he and his colleague J. P. Nonne established a large laboratory in their residence.
1759?: Declined a call to Frankfurt a. d. Oder University.
1767: Nominated to succeed *J. G. Lehmann* in the St. Petersburg Academy by P. I. Kennedy of the Munich Academy, but nothing came of the recommendation.

**State Honors** None. **Major Scientific Societies** 1754: Erfurt (in charge of the "department of chemistry"). **Informal Recognition** 1753: *Vogel*. 1759: Wallerius. 1767: Baldinger. 1781: *Wiegleb*. 1787: *Gren*. 1791: *Göttling*.

**Bibliography**

Baldinger, E. G. *Ehrengedächtnis des Herrn Christoph Andreas Mangold*. Jena: Cuno, 1767.
Beck, A. *Graf Gustav Adolf von Gotter*. Gotha: Perthes, 1867, pp. 27, 76, 79, 83–85.
Joachim, J. "Die Anfänge der königlichen Sozietät der Wissenschaften zu Göttingen." Gesellschaft der Wissenschaften. Göttingen: Philologisch-Historische Klasse: *Abhandlungen*, 3 Folge: 19 (1936), 77.
St. Petersburg Correspondence, 77.
Schyra, B. "Christoph Andreas Mangold (1719–1767) und die Grundzüge seines Systems der Medizin." *Beiträge zur Geschichte der Universität Erfurt*, 7 (1960), 67–75.
———. "Das Reformwerk Johann Wilhelm Baumers und Christoph Andreas Mangolds . . . in Erfurt um die Mitte des 18. Jahrhunderts." Diss., Erfurt Medical Academy, 1959.
Zedler, XIX: 954.
Also see Berlin Academy (Kirsten) and Erfurt University (Hampel, pp. 21, 40–43; Loth, p. 222; Loth, "Entwicklung," p. 245).

## SPIELMANN, Jacob Reinbold
1722, Strasbourg / 1783, Strasbourg

**Background** Son of Johann Jacob Spielmann (1695, Strasbourg–1740, Strasbourg), who, after serving as a journeyman in Breslau (1715–1716), assumed control of the family's Hirsch-Apotheke in Strasbourg. Lutheran. **Education** As a youth he attended the local gymnasium and, in 1733, enjoyed fruitful contacts with *A. S. Marggraf*, who was serving in his father's shop. In 1734 his father forced him to leave school and begin an apprenticeship. Determined to be more than a pharmacist, he audited courses at the university whenever possible. In 1740 he left Strasbourg to serve as a journeyman in Nuremberg's Hospital Apothecary Shop. There his master J. A. Beurer, a former student of *C. Neumann* and *J. H. Pott*, taught him much about chemistry. In 1741 he proceeded by way of Erlangen, Frankfurt a. M., Erfurt, Leipzig (where he met *J. A. Cramer*), and Halle (where he met *F. Hoffmann*) to Berlin. He enrolled in the Medical-Surgical College as a medical student and took, among other things, *Pott*'s course on chemistry. He also pursued the science with his old friend *Marggraf*. In 1742 he went to Freiberg where he studied under *J. F. Henckel* for several months. Then he proceeded to Paris where he studied chemistry with J. Grosse and C. J. Geoffroy and became acquainted with R. A. F. de Réaumur. In 1743 he returned home and assumed control of the family apothecary shop.

**Career**

1743–1781: Owner of the Hirsch-Apotheke in Strasbourg. In 1781 he turned the shop over to his son, J. J. Spielmann, Jr. on the proviso that he have access to its laboratory.
1748: Earned his M.D. at Strasbourg University with a dissertation on the principle of salt.
1748–1783: PD, AOP, then OP at Strasbourg University. Among his students were *J. A. Weber*(?) (1758–1759) and *C. Girtanner* (1779–1780).
1749–1759: AOP in the Medical Faculty. In 1754 he qualified for the Ph.D. in anticipation of competing for the unsalaried chair of poetry, which he hoped would serve as a stepping stone to the next vacant chair in medicine.
1756–1759: OP of poetry in the Philosophical Faculty. His inaugural address argued for "the necessity of physicians reading the ancient poets." His courses emphasized the natural descriptions of the Greek and Latin authors, especially Lucretius.
1759–1783: OP of chemistry, botany, and materia medica in the Medical Faculty (salary, 1759–1771: 300 to 560 livres; 1771–1783: 600 livres plus goods).
1782: Visited Württemberg.

**State Honors** None. **Major Scientific Societies** 1758: Berlin. 1760: Leopoldina. 1763: St. Petersburg. 1772: Paris (corresponding member). 1774: Berlin NF. **Informal Recognition** 1767: Baldinger. 1769: Suckow. 1777: Raspe. 1777: *Scopoli*. 1780: *Weigel*. 1781: *Crell*. 1781: *Wiegleb*. 1786: *Hermbstaedt*. 1787: *Gren*. 1789: Hopson. 1791: *Göttling*.

**Bibliography**

Bachoffner, P. "Une correspondance inédite de J. R. Spielmann." Internationale Gesellschaft für Geschichte der Pharmazie: *Veröffentlichungen*, N.F. 28 (1966), 17–32.
Baldinger, pt. 1: 75–86.

Brachmann, W. "Johann Jacob Spielmann in Breslau," *Deutsche Apotheker-Zeitung*, 91 (1951), 218–219.
DAB, II: 644–645.
"Éloge de M. Spielmann." Société Royale de Médecine, Paris: *Histoire* (1782/83), 116–126.
Gossmann, 189–193, 196–197.
Grass, U. (Apothecary, Alpirsbach). Letter providing J. J. Spielmann's dates (14 October 1980).
Wittwer, P. L. "Lebensgeschichte Dr. Jac. Reinbold Spielmann, der Arzneygelahrtheit Prof. in Strassburg." *Chemische Annalen*, no. 1 (1784), 545–580.
Also see Berlin Medical-Surgical College ("Matrikel," 11: 134) and Strasbourg University (Berger-Levrault; Wieger, p. 56).

### SCOPOLI, Giovanni Antonio
1723, Cavalase, Trent / 1788, Pavia

**Background** Son of Francesco Antonio Scopoli, doctor of civil and canon law, "chancellor" of criminal affairs in Cavalese, and later steward and military commissioner to the Prince-Bishop of Trent. His mother was from the nobility. Catholic. **Education** After attending school in Trent and the gymnasium in Hall, he went to Innsbruck University around 1740. He studied medicine there, pursuing botany on his own. He evidently received an M.D. in 1743.

**Career**
1743–1754: Practiced medicine and pursued natural history in Venice, Trent, and Styria.
1754: Went to Vienna where he satisfied Vienna University's Medical Faculty that he was qualified for service as a government physician. He became the protégé of Maria Theresa's advisor G. van Swieten (*J. A. Cramer's* old friend).
1754–1769: Imperial Mining Physician at Idria (salary, 1754: 700 fl plus free residence). In 1755 he unsuccessfully sought the directorship of the new Mining Apothecary Shop, requesting a salary increase of 800 fl.
1763–1769: Professor of metallurgy and chemistry (salary, 1763: 400 fl) in Idria. He was charged with instructing local mercury miners in these two sciences.
1767: Negotiated with the St. Petersburg Academy concerning its vacant position for chemistry and metallurgy. Despite *A. S. Marggraf's* recommendation, the Academy decided that Scopoli was not qualified.
1769–1777: Professor of chemistry and from 1770 Mining Councillor in the Schemnitz Mining Academy (salary, 1769: 1,500 fl; 1770–1777: 2,000 fl). Among his colleagues was *I. von Born* (1769–1770).
1777–1788: OP of botany and chemistry in Pavia University's Medical Faculty. One inducement for his move was the impending completion of a new laboratory. Among his colleagues was A. Volta (1778–1788) and among his students, L. V. Brugnatelli (1780?–1788). His bitter fights with L. Spallanzani, OP for natural history, may have hastened his death.
**State Honors** None. **Major Scientific Societies** 1774: Berlin NF. **Informal Recognition** 1780: *Weigel*. 1788: *Leonhardi*. 1789: *Gmelin*. 1791: *Göttling*.

**Bibliography**
Antolini, B. (Clergyman, Cavalese). Letter concerning Scopoli's parents (20 December 1969).
Baldinger, pt. 4: 161–170.
DSB, XII: 561–562.
Haubelt, J. "Scopoli a Lomonosov." *Sborník pro dějiny přírodních věd a techniky*, 8 (1963), 181–199.
Lesky, E. *Arbeitsmedizin im 18. Jahrhundert: Werksarzt und Arbeiter im Quecksilberbergwerk Idria*. Vienna: Verlag des Notringes der wissenschaftliche Verbände Österreichs, 1956, pp. 25–31, 54–55.
Luca, II: 123–131.
*Memorie e Documenti per la Storia dell'Università di Pavia e degli Uomini più illustri che v'insegnarono*. Facsimile of the 1877–1878 ed. 3 vols. Bologna: Forni, 1970, I: 423–424; III: 290–294.
Minařik, F. "Die Apotheker- und Botanikerfamilie Freyer in der Quecksilberbergstadt Idrija." Internationale Gesellschaft für Geschichte der Pharmazie: *Veröffentlichungen*, N.F. 24 (1964), 89–102.
Moll, III: 834–838, 932–933.
St. Petersburg Procès-verbaux, II: 625–626, 636, 641.

Wurzbach, XXXIII: 210–215.
Also see Schemnitz Mining Academy (*Gedenkbuch*, pp. 7–8).

### ANDREAE, Johann Gerhard Reinhard
1724, Hanover / 1793, Hanover

**Background** Son of Leopold Andreae (1686, ?–1730, Hanover), owner of the Hirsch-Apotheke in Hanover. Lutheran. **Education** After being tutored at home, he served an apprenticeship in the family apothecary shop. In 1744 he proceeded to Berlin where he took several courses, including chemistry from *J. H. Pott*. Then he traveled by way of the Saxon and Harz mining districts to Frankfurt a. M., where he worked as a journeyman pharmacist until 1746. Upon his return home, Court Physician P. G. Werlhof urged him to study chemistry, mineralogy, and metallurgy with *J. A. Cramer* in Blankenburg. However, when he got to Blankenburg, *Cramer* recommended that he go to Leiden. He studied chemistry there with H. D. Gaubius, went to London for a few months, and then returned home.

**Career**
1747–1793: Administrator, then after his mother's death in 1752, owner of the Hirsch-Apotheke in Hanover. From 1778 to 1781 he employed F. Ehrhart, a pupil of T. Bergman and friend of C. W. Scheele, to organize his extensive collections. Among his correspondents were *J. F. Meyer, J. R. Spielmann, J. C. Wiegleb, W. H. S. Bucholtz, G. A. Scopoli*, and *J. F. Gmelin.*
1753: Visited Frankfurt a. M.
1763: Visited Switzerland.
1765–1769: Analyzed Hanover's soils on government commission.
1785: Analyzed Limmer's spring water on government commission.
1787: Visited Elberfeld on a mineral-collecting trip.

**State Honors** None. **Major Scientific Societies** 1776: Erfurt. (He declined membership in several societies because he regarded such honors as nothing more than "learned charlatanry.") **Informal Recognition** 1775: *Suckow*. 1780: Anon. 1780: *Weigel*. 1784: Scherf. 1786: *Hermbstaedt*. 1787: *Gren*. 1791: *Göttling*.

**Bibliography**
Bergman, 44–56, 76.
DAB, I: 8–9.
*Journal von und für Deutschland*, 3, no. 1 (1786), 373–374.
Leiden, 1011.
NDB, I: 282–283.
Nose, C. W. "Einige mineralogische Nachrichten." *Chemische Annalen*, no. 1 (1786), 373–374.
Schlichtegroll, vol. I for 1793, pp. 164–182.
Also see Klaproth (Dann, pp. 22, 142) and Berlin Medical-Surgical College ("Matrikel," 11: 136).

### VOGEL, Rudolph Augustin
1724, Erfurt / 1774, Göttingen

**Background** Son of Paul Heinrich Vogel (1686, Erfurt–1746, Erfurt), who studied with *G. W. Wedel* at Jena University and *F. Hoffmann* and *G. E. Stahl* at Halle University, then served in Erfurt as a physician, an AOP of medicine (1737–1745), and an OP of medicine (1746). Lutheran. **Education** After studying in Erfurt's Town Council Gymnasium for four years, he entered Erfurt University to study medicine in 1740. While there he probably took chemistry classes from his father, A. E. Büchner, and *H. Ludolf*. In 1744 he proceeded to Leipzig University and in 1747, to Berlin's Medical-Surgical College. His chief interest there was anatomy—he purchased a set of anatomical preparations for 50 Rtl—but he may have also taken *J. H. Pott*'s course on chemistry. In late 1747 he returned to Erfurt University and earned his M.D. with a dissertation on the anatomy of the larynx.

**Career**
1748–1752: PD in Erfurt University. Among his colleagues were *Ludolf* and *C. A. Mangold*, with whom he pursued chemistry.
1753–1774: AOP, then from 1760, OP of chemistry in Göttingen University's Medical Faculty (salary, 1753–1763: 460 Tl; 1764–1774; 560 Tl). Among his students were *J. C. H. Heyer*(?) (ca. 1767) and *C. E. Weigel* (1769–1771).
1763–1774: District Physician in Göttingen.

1767: Nominated to succeed *J. G. Lehmann* in the St. Petersburg Academy by the mathematician-physicist W. Karsten, but nothing came of this.
1774: At his death, his library (3,680 books with a value of ca. 2,000 Rtl) and mineral collection were auctioned.

**State Honors** 1764: Royal British Physician. **Major Scientific Societies** 1754: Erfurt. 1754: Leopoldina. 1758: Stockholm. 1770: Göttingen. **Informal Recognition** 1760: Wallerius. 1767: Baldinger. 1769: Suckow. 1770: *Wiegleb*. 1777: Raspe. 1777: *Scopoli*. 1780: *Gmelin*. 1787: *Gren*. 1791: *Göttling*.

**Bibliography**
Biereye, 115.
Elwert, 588.
Motschmann, J. C., et al. *Erfordia Literate*. Vol. III. Erfurt: Weber, 1748–1753, pt. 1: 131–134; pt. 2: 113.
St. Petersburg Correspondence, 78.
Streich, G. "Die Büchersammlungen Göttinger Professoren im 18. Jahrhundert." *Wolfenbütteler Forschungen*, 2 (1977), 259, 261, 284, 291.
Vogel, R. A., to A. von Haller. Seven letters (1753–1771). Stadtbibliothek, Bern.
Also see Mangold (Baldinger, p. 16), Erfurt University (Hampel, pp. 37–39), Göttingen University (Bärens, pp. 92–93; Ganss, pp. 21–25; Pohlai; Pütter, I: 158–159, 291; II: 45–46), and Leipzig University (*Matrikel*, III: 436).

**JACQUIN, Nicolas Joseph**
1727, Leiden / 1817, Vienna

**Background** Son of Claudius Nicolas Jacquin (1694, Leiden–1743, Leiden), owner of a velvet factory in Leiden. His mother was from the nobility. Catholic. **Education** He attended a private school in Leiden and the Jesuit gymnasium in Antwerp. In early 1745 he entered Louvain University to study philosophy. A few semesters later, he returned to Leiden where he concentrated first on classics, then on medicine and natural science. In 1748, after his mother stopped supporting him because he refused to enter the clergy, he went to Paris. He studied medicine there and in Rouen for the next three years. In 1752, encouraged by Imperial Physician G. van Swieten (an old family friend), he went to Vienna to finish his medical education.

**Career**
1754–1759: Traveled to the West Indies on a botanical expedition which was organized by van Swieten and financed by Emperor Franz at a total cost of 27,000 fl. Upon his return, he was granted a lifelong pension, which, however, was terminated by Emperor Joseph in the mid-1780s.
1759–1763: Curator of the botanical collections in Vienna.
1763: Received calls to Innsbruck for chemistry and botany, St. Petersburg Academy for botany, and the new Schemnitz Mining School for chemistry and metallurgy.
1763–1768: Mining Councillor and professor of chemistry in the Schemnitz Mining School (salary: 2,000 fl, free residence, fodder for four horses, and firewood). Among his students was *I. Born* (ca. 1765).
1769–1796: Mining Councillor (salary: 1,000 fl) and OP of chemistry and botany in Vienna University's Medical Faculty (salary: 2,000 fl plus free residence—first, above the University's chemical laboratory, then once these quarters became inadequate, a 300 fl residential allowance). Among his students were *J. F. Gmelin* (1770–1771), *F. Wurzer* (1787), and his son and successor *J. F. Jacquin* (to 1788). In 1796 he retired with full salary and residential allowance.
1775: His sister married *J. Ingen-Housz*.
1777: Sold his herbarium to J. Banks for 2,000 fl.
1780–1817: Director of the Imperial Gardens at Schönbrunn (free residence and large budget).
1783?: Sold part of his library to Wilna University for 4,000 fl.

**State Honors** 1774: Imperial Ennoblement. 1806: Imperial Baron. **Major Scientific Societies** 1780: St. Petersburg. 1783: Berlin NF. 1783: Stockholm. 1786: Berlin. 1788: London. 1804: Paris (corresponding member for botany). **Informal Recognition** 1787: *Gren*. 1789: *Gmelin*. 1791: *Göttling*. 1802: *Scherer*.

**Bibliography**
DSB, VII: 57–59.

Forster, XII: 108–109, 118–120, 123, 125–126, 131; XIV: 151, 159–160, 190.
Garside, S. "Baron Jacquin and the Schönbrunn Gardens." *Journal of South African Botany*, 8 (1942), 201–224.
Kronfeld, E. M. "Jacquin." *Oesterreichisches Rundschau*, 3 (1905), 237–251.
———. "Jacquin des Jüngeren botanische Studienreise 1788–1790." *Botanisches Centrallblatt: Beihefte*, 38, no. 2 (1921), 132–176.
Luca, I, pt. 1: 208–211, 328–329.
NDB, X: 257–259.
*Ö¢sterreichisches Biographisches Lexikon, 1815–1950*. Vol. III. Graz: Böhlaus, 1965, pp. 52–53.
Sander, H. *Beschreibung seiner Reisen*. Vol. II. Leipzig: Jacobäer, 1784, pp. 532–546.
Stafleu, F. A. "Jacquin and his American Plants." In N. J. Jacquin, *Selectarum Stirpius Americanarum Historia*, facsimile of the 1763 ed., vol. I. New York: Hafner, 1971, pp. F7–F32.
Also see Vienna University (Müller, pp. 14–15, 48–49, 86; Oberhummer, pp. 141–160, 194) and Schemnitz Mining School (*Gedenkbuch*, pp. 5–7).

## INGEN-HOUSZ, Jan
1730, Breda, Netherlands / 1799, Bowood Park near Calne, Wiltshire, England

**Background** Son of Arnoldus Ingen-Housz (1693, Bommel–1764, Breda), a leather merchant and from 1755 apothecary in Breda. Catholic. **Education** After attending the local Latin school, he went to Louvain University around 1747. There he studied medicine, chemistry, and natural philosophy, qualifying for the M.D. in 1753. Then he traveled, evidently to Paris and Edinburgh. In late 1754 he enrolled at Leiden University where he spent a year studying chemistry with H. D. Gaubius, physics, and anatomy.

**Career** 1755–1764: Practiced medicine and established an apothecary shop in Breda.
1765–1768: Practiced medicine in Edinburgh, where he pursued chemistry with W. Cullen, and then London, where he learned D. Sutton's inoculation techniques.
1768–1799: Imperial Physician (salary: 5,000 fl). In 1768 he went to Vienna to inoculate Maria Theresa's children. His success led the Empress to grant him a lifelong pension of 5,000 fl. During the next decade he inoculated other members of the Imperial family and many minor rulers and aristocrats.
1768–1777: Resided in Vienna. During this period he visited Italy, Berlin, Paris, and London. In 1775 he married *N. J. Jacquin*'s sister in Vienna.
1777–1780: Visited France, the Netherlands, and Britain, spending most of his time in Paris and London engaged in electrical and chemical research. In 1780 he invested over £20,000 in a smuggling(?) venture connected with the American Revolution. Ultimately he lost much of this money. He returned to Vienna at the threat of being removed from the Imperial payroll.
1780–1788: Resided in Vienna. Among his collaborators were *Jacquin* and J. A. Scherer. He had a steady, and distracting, stream of aristocratic and scientific visitors who were eager to learn of his researches on the respiration of plants and the combustion of wires.
1788–1789: Resided in Paris.
1789–1799: Resided in and around London enjoying the patronage of Lord Lansdowne.

**State Honors** 1768: Imperial Court Councillor. **Major Scientific Societies** 1769: London. **Informal Recognition** 1787: *Gren*. 1791: *Göttling*. 1798: Frank. 1799: *Scherer*. 1807: Kastner.

**Bibliography**
DSB, VII: 11–16.
Forster, XII: 108–111, 116–123, 126, 131, 135, 137.
Hays, I. M. *Calendar of the Papers of Benjamin Franklin in the Library of the American Philosophical Society*. Vols. I–IV. Philadelphia: American Philosophical Society, 1908.
Ingen-Housz, J. Letter-book (1774–1793). Stedelijk Archief, Breda.
Mugridge, D. H. "Scientific Manuscripts of Benjamin Franklin." Library of Congress: *Quarterly Journal of Current Acquisitions*, 4 (1947), 12–21.
NDB, X: 171–172.
Smit. P. "Jan Ingen-Housz (1730–1799): Some new Evidence about his Life and Work." *Janus*, 67 (1980), 125–139.
Wiesner, J. *Jan Ingen-Housz: Sein Leben und sein Wirken als Naturforscher und Arzt*. Vienna: Konegen, 1905.

## POERNER, Carl Wilhelm
1732, Leipzig / 1796, Meissen

**Background** Son of Johann-David Poerner (?, Annaberg–?, Leipzig), who matriculated at Leipzig University in 1719, Wittenberg University in 1723, and Erfurt University in 1726, took a doctorate in law, and practiced as an attorney in Leipzig. Lutheran. **Education** After instruction from his parents and tutors, he attended the St. Nicolaus School in Leipzig. In 1748 he entered Leipzig University, where he studied the philosophical and medical sciences, including chemistry with A. Ridiger. He earned his Med. Bac. in 1751 and his M.D. in 1754 with a dissertation reporting experiments on egg albumen and blood serum.

**Career**
1754–1768: PD at Leipzig University. He often taught chemistry. Among his students were *J. G. Leonhardi* (1766–1768) and *C. F. Wenzel* (1768).
1768: Received a call from the St. Petersburg Academy which, thanks to an enthusiastic recommendation from *A. S. Marggraf*, wanted him as the successor for *J. G. Lehmann*. He was initially offered a salary of 800 rubles plus 250 rubles for travel money. When he asked for more, he was offered a salary of 1,000 rubles plus free residence and 400 rubles for travel. Nonetheless, he decided to remain in Saxony.
1769–1796: Official at the Saxon Porcelain Works in Meissen. Among his colleagues was *Wenzel* (1781–1793).
    1769–1770: Probationary chemist (salary: 650 Tl).
    1770–1796: Mining Councillor and Commissioner responsible for chemistry (salary, 1770–1776: 1,000 Tl; 1776–1796: 900 Tl [Note: The salary reduction of 1776 paralleled that of most other Saxon officials.]).

**State Honors** None. **Major Scientific Societies** 1768: St. Petersburg (offered him a position, but did not make him a member). **Informal Recognition** 1773: *Cartheuser*. 1774: *Wiegleb*. 1775: *Suckow*. 1777: Raspe. 1780: *Weigel*. 1787: *Gmelin*. 1787: *Gren*. 1788: *Leonhardi*. 1791: *Göttling*.

**Bibliography**
*Leipziger gelehrtes Tagesbuch*. (1796), 111–112.
Poerner, C. W. Autobiography. In S. T. Quellmalz, *Panegyrin Medicam*. Leipzig: Langenheim, 1754, pp. xiv–xvi.
St. Petersburg Procès-verbaux, II: 647, 654, 659.
Staatsarchiv, Dresden. Letter concerning Poerner's employment at Meissen (25 November 1971).
Weiz, F. A. *Das gelehrte Sachsen*. Leipzig: Schneider, 1780, p. 188.
Also see Erfurt University ("Matrikel," 10: 61), Leipzig University (*Matrikel*, III: 309), and Wittenburg University ("Matrikel," 5: 42).

## WIEGLEB, Johann Christian
1732, Langensalza / 1800, Langensalza

**Background** Son of Christian Ludwig Wiegleb (?, Tennstedt–1738, Langensalza), who enrolled at Erfurt University in 1717, took his law examination at Wittenberg University in 1726, then became an attorney in Langensalza. Stepson of J. C. Thilo, also an attorney in Langensalza. Lutheran. **Education** Intended for the clergy, he was tutored at home and then enrolled in the town school. However, in 1748, after three years of schooling, he decided to become a pharmacist like his uncle J. G. Reisig. He served his apprenticeship in Dr. C. F. Sartorius's Marien-Apotheke in Dresden, using his spare time to study chemical, and especially alchemical, books in the shop's library. In 1754 he became a journeyman, serving six months for a salary of 10 Rtl. Then he worked as a journeyman in Quedlinburg's Hof-Apotheke for a year. In 1756 he returned home on account of his uncle's death.

**Career**
1756–1758: Administered Reisig's apothecary shop in Langensalza.
1759–1796: Apothecary in Langensalza. He spent about 1,600 Tl to establish his shop and residence. Among his apprentices was *J. F. A. Göttling* (1767–1775).
1764: Began serious study of chemistry with the encouragement of Prussian Field Physician E. G. Baldinger, a pupil of *C. A. Mangold*. He read the works of *R. A. Vogel*, *C. E. Gellert*, P. Macquer, and *F. A. Cartheuser*.
1770: Toured Germany and the Netherlands, giving special attention to industrial developments.

1779?–1796?: Director of Germany's first successful chemical-pharmaceutical boarding school. He had around fifty students over the years. Among them were *S. F. Hermbstaedt* (ca. 1779–1781) and the Birmingham manufacturer's son M. R. Boulton (1789–1790), to whom he tried selling his method of indigo dyeing for £3,000.
1788–1800: Langensalza Town Treasurer.
**State Honors** None. **Major Scientific Societies** 1776: Leopoldina. 1776: Erfurt. **Informal Recognition** 1778: *Hagen*. 1780: Anon. 1780: *Gmelin*. 1784: Scherf. 1785: *Crell*. 1786: Beseke. 1786: *Hermbstaedt*. 1787: Fuchs. 1787: *Gren*. 1788: *Leonhardi*. 1788: *Westrumb*. 1791: Göttling. 1792: Humboldt. 1792: *Klaproth*. 1793: Scherer. 1793: *Wurzer*. 1794: Baldinger. 1796: S. in M. 1796: Van Mons. 1798: *Bucholz*. 1798: Frank. 1800: *Scherer*. 1802: Salzmann. 1818: *Trommsdorff*. 1821: *Pfaff*.
**Bibliography**
Baldinger, E. G. "Chymische Lehrschule." *Neues Magazin für Aerzte*, 4, no. 1 (1782), 88.
DAB, II: 743–744.
DSB, XIV: 332–333.
Möller, R. "Ein Apotheker und Chemiker der Aufklärung." *Pharmazie*, 20 (1965), 230–239.
Musson, A. E., and E. Robinson. *Science and Technology in the Industrial Revolution*. Toronto: University of Toronto Press, 1969, pp. 213–215.
Stoeller, F. C., and A. N. Scherer. "Johann Christian Wiegleb." *Allgemeines Journal der Chemie*, 4 (1800), 684–720.
Strube, W. "Die Ueberwindung der Alchimie in Deutschland in der zweiten Hälfte des 18. Jahrhunderts." Technologische Hochschule für Chemie, Leuna: *Wissenschaftliche Zeitschrift*, 5 (1963), 217–228.
Wiegleb, J. C., to M. R. Boulton. Two letters (1790–1791). Library, Assay Office, Birmingham, England. [E. Robinson has kindly shared his translations of these letters with me.]
Also see Erfurt University ("Matrikel," 10: 151) and Wittenburg University ("Matrikel," 11: 507).

**BUCHOLTZ, Wilhelm Heinrich Sebastian**
1734, Bernburg / 1798, Weimar
**Background** Son of Georg Ernst Bucholtz (1708, Sondersleben–1747, Bernburg), who, after fleeing his hometown to avoid induction into the Prince of Anhalt-Dessau's mercenary regiment, served freely in the Prince of Anhalt–Bernburg's guards (rising to corporal) and then became the Prince's mill inspector. Lutheran. **Education** He attended the town school until his father's death, then worked as a lawyer's copyist for a year. In 1748 he became a pharmaceutical apprentice in Magdeburg. From 1752 to 1755 he served as a journeyman in Homburg, Giessen, and Hildburghausen, where Court Apothecary Müller aroused his interest in chemistry. From 1755 to 1761 he administered the Hof-Apotheke in Weimar. Encouraged by his master Dr. Jacobi, he studied Latin, French, and medicine. When Jacobi died in 1761, Bucholtz decided to study medicine at Jena University. He lived with J. F. Faselius, an OP of medicine who was teaching chemistry there. He earned his M.D. with a dissertation on mineral soaps in 1763.
**Career**
1763–1767: Practiced medicine privately in Weimar. He had some difficulty qualifying to practice because his examiners were dissatisfied with his knowledge of anatomy and surgery.
1767–1798: Administrator, then from 1773 owner of Weimar's Hof-Apotheke. His wealthy first wife bought the shop in 1765. She gave it to him eight years later, perhaps to console him for his recent failure to obtain the post of District Physician. Soon afterwards, nonetheless, they were divorced. Among his employees were *J. F. A. Göttling* (1775–1785) and *J. B. Trommsdorff* (1783–1787). Upon his death the shop sold for 4,500 Tl.
1777–1798: District Physician for Weimar and Berka. He had a budget for treating those who could not afford a physician. He relinquished this responsibility in 1788 in response to charges that he was improperly increasing his apothecary shop's trade.
1779: Nominated for *F. A. Cartheuser*'s chair at Giessen University, but nothing came of this.
February 1784–May 1785: Supervised the launching of several balloons for Duke Carl August and his court.
**State Honors** 1777: Saxe-Weimar Court Physician. 1782: Saxe-Weimar Mining Councillor. **Major Scientific Societies** 1766: Munich. 1768: Erfurt. 1769: Leopoldina. 1794: St.

Petersburg. **Informal Recognition** 1786: Beseke. 1787: *Gren.* 1791: *Göttling.* 1794: Baldinger. 1798: *Bucholz.* 1799: *Scherer.* 1804: *Gmelin.*

**Bibliography**
Bucholtz, W. H. S. Autobiography (ca. 1797). *Sächsische Provinzialblätter*, 6 (1799), 345–359.
DAB, I: 91–92.
Dann, G. E. "Apotheker der Familien Bucholz und Meissner." *Pharmazeutische Zeitung*, 76 (1931), 45–47.
"Genealogie Bucholz: Erfurter Zweig der anhalt.-thüring. Familie." Stadtarchiv, Erfurt (5/801–B3).
Hessisches Staatsarchiv, Darmstadt (Abt. VI. 1: Giessen; Knov. 28, Fsc. 13).
Möller, R. "Ein Apotheker des klassischen Weimar." *Pharmazie*, 15 (1960), 181–190.
Rust, J. L. A. *Historisch-literarische Nachrichten von den jetzt lebenden Anhaltischen Schriftstellern.* 2 vols. Wittenberg and Zerbst: Zimmermann, 1776–1777, I: 71–77; II: 57.
Scherer, A. N. "Wilhelm Heinrich Sebastian Bucholtz." *Allgemeines Journal der Chemie*, 2 (1799), 591–615.
Trommsdorff, J. B. "Kurze Biographie des verewigten Wilhelm Heinr. Sebastian Bucholz." *Journal der Pharmacie*, 6, no. 2 (1799), 376–384.

### CARTHEUSER, Friedrich August
1734, Halle / 1796, Schierstein near Wiesbaden

**Background** Son of *Johann Friedrich Cartheuser* (1704, Hayn–1777, Frankfurt a. d. Oder), PD at Halle University, then from 1740 OP of medicine at Frankfurt a. d. Oder University. Lutheran. **Education** In 1749, after attending the local lyceum, he entered Frankfurt a. d. Oder University. He pursued the philosophical and medical sciences there for three years, probably taking chemistry from his *father*. Then he went to Berlin where, among other things, he studied with *J. H. Pott* and enjoyed fruitful contacts with *A. S. Marggraf*. In 1753 he returned home and earned his M.D. by defending a materia medica dissertation written by his *father*. Then he spent several months learning mining and smelting in the Saxon mining district. He returned home by way of Bohemia, Franconia, and Hesse in 1754.

**Career**
1754–1766: PD at Frankfurt a. d. Oder University. Among his colleagues was his *father* and among his students was *C. A. Gerhard* (?) (1757–1760).
1760: Nominated by the mathematician L. Euler for the St. Petersburg Academy's empty chair of chemistry, but nothing came of this.
1766–1779: OP of natural philosophy in Giessen University's Philosophical Faculty, of medicine in its Medical Faculty, and from 1777 of physics, botany, and mining in its new Economics Faculty (salary, 1779: 469 fl). In 1779 he retired on account of poor health.
1772–1779: Director of the University's botanical gardens.
1779–1796: Lived on his Idstein estate (1779–1790), then in Schierstein (1790), then in the Landgrave of Hesse-Darmstadt's residence at Birkenbach (1791–1793), and finally, on account of the war, in Schierstein (1793–1796).

**State Honors** 1767: Hesse-Darmstadt Mining Councillor. 1778: Nassau-Usingen Privy Treasury Councillor. 1791: Hesse-Darmstadt Privy Councillor. **Major Scientific Societies** 1755: Erfurt. 1773: Berlin N.F. **Informal Recognition** 1759: Wallerius. 1767: Baldinger. 1777: Raspe. 1781: *Crell.* 1787: Fuchs. 1787: *Gren.* 1791: *Göttling.*

**Bibliography**
Baldinger, pt. 3: 13–15; pt. 4, 136–137.
Elwert, 109–116.
Euler, no. 1: 154.
Hessisches Staatsarchiv, Darmstadt (Abt. VI, 1: Giessen; Konv. 28, Fsc. 13).
Strieder, II: 121–125; III: 537–538; X: 376.
Also see Frankfurt a. d. Oder University ("Matrikel," p. 371) and Heidelberg State Economics School (Stieda, pp. 153, 163, 174–175).

### WEBER, Jacob Andreas
1737, Bietigheim, Württemberg / 1792, Grub near Coburg

**Background** Son of Hans Jörg Weber (?, Nürtingen–1749, Bietigheim), baker and administrator of the Imperial storehouse in Bietigheim. Lutheran. **Education** In 1758 he entered Strasbourg University to study medicine. While there he may have taken chemistry from *J. R. Spielmann*. The following year he proceeded to Tübingen University where he probably attended P. F. Gmelin's chemistry course. In 1760 he qualified for his medical license under Gmelin with a dissertation on the transpiration of the skin.

**Career**
1760–1770: Practiced medicine privately in the countryside, then from 1765 in Tübingen.
1771: Visited Holland where he joined Rotterdam's new scientific society and apparently studied chemical factories.
1772–1791: Founded and, with Dr. W. G. Ploucquet, ran a sal ammoniac and dye works on the outskirts of Tübingen.
1777–1778: Advised saltpeter producers in Magdeburg.
1781–1783: Established sal ammoniac works in Wied and Koblenz at government expense. Soon after arriving in Wied in the summer of 1781, he was also treating victims of a fever epidemic.
1785–1787: Pursued various projects in Austria. In Gmunden he first explored the possibility of establishing a chemical factory that would use residues from the local saltworks. Then he developed and sought recompense for a machine that refined salt, sugar, alum, and vitriol with lower fuel consumption than existing apparatus. He was granted a lifetime award equal to a third of the savings resulting from his invention. In Vienna he and W. F. Kornbeck applied for state support of a new process for making cinnabar. Unsuccessful, he then tried to establish a sal ammoniac works there.
1791–1792: Chemist in A. E. von Sand's Berlin-blue dye works at Grub near Coburg. He tried to establish a sal ammoniac works there. He died as poor as a "day laborer."

**State Honors** 1781?: Court Councillor for the Count of Wied (he probably assumed this title on his own initiative after serving as a consultant to the Count). **Major Scientific Societies** None. **Informal Recognition** 1784: Hahnemann. 1787: *Gren*. 1788: *Leonhardi*. 1791: *Göttling*. 1794: Baldinger. 1804: *Gmelin*.

**Bibliography**
Bauer, F. C. *Die Tübinger Schule und ihre Stellung zur Gegenwart*. Tübingen: Fues, 1859, p. 119.
Cyriaci, E. *Die Coburger Familie von Sand von 1275–1940*. Coburg: Rossteutscher, 1941, pp. 33–34, 62–63.
Fleischhut, G. (Mittelrhein-Museum, Koblenz). Letter bearing on Weber's sal ammoniac factory near Koblenz (30 April 1971).
Kipp, F. *Grub a. F.: Aus vergangenen und gegenwärtigen Tagen einer Pfarrei*. Coburg: Bonsack, 1914, pp. 43–44.
McKay, J. H. (Bataafsch Genootschap der Proefondervindelijke Wijsbegeerte, Rotterdam). Letter concerning Weber's entry into this society in 1771 (3 May 1971).
Merkel, J. (Clergyman, Grub am Forst). Letter with information about Weber's death on 12 January 1792 (10 December 1969).
Mickler, E. (Clergyman, Bietigheim). Letter bearing on Weber's ancestors and birth on 21 July 1737 (20 September 1968).
Müller, V. (Fürstlich Wiedisches Archiv, Neuwied). Letter concerning Weber's title and his possible relations with Count Alexander zu Wied, a patron of alchemy and sal ammoniac works (15 March 1971).
Multhauf, R. "A premature Science Advisor: Jacob A. Weber (1737–1792)." *Isis*, 63 (1972), 356–369.
Sydow, J. (Stadtarchiv, Tübingen). Letter concerning Weber's factory in Tübingen (10 March 1971).
Winkelbauer, W. (Finanz- und Hofkammerarchiv, Vienna). Letter bearing on Weber's relations with the Hapsburg government between 1785 and 1787 (19 March 1971).
Also see Strasbourg University ("Matrikel," 2: 81) and Tübingen University (Beese, p. 84).

**GERHARD, Carl Abraham**
1738, Lerchenborn near Liegnitz / 1821, Berlin

**Background** Son of Wolfgang Abraham Gerhard (d. 1758, Sandewalde, Silesia), pastor in Lerchenborn, then Sandewalde. Lutheran. **Education** He studied medicine and science in Berlin and Frankfurt a. d. Oder from 1755 to 1760. Among his teachers were *A. S. Marggraf*, *J. H. Pott*(?), *J. G. Lehmann*(?), *J. F. Cartheuser*(?), and *F. A. Cartheuser*(?). He earned his

M.D. at Frankfurt a. d. Oder University in 1760 with a mineralogical-chemical dissertation dedicated to his patron *J. T. Eller.*

**Career**
1760–1768: Gave private courses on materia medica, natural philosophy, mechanics, and other topics in Berlin. In 1765 he applied through L. Euler for the chair of "experimental physics" in the St. Petersburg Academy, requesting a salary of 600 rubles plus residence, wood, and candles. However, he lost Euler's support when the Academy insisted on a candidate who could both make new discoveries and establish them mathematically.
March 1768: Passed examinations in mechanics, experimental natural philosophy, and mineralogy, satisfying the Berlin Academy that he was qualified to give instruction in these subjects, to serve as a scientific advisor in the General Directory, and to assume membership in the Academy's Physical Class.
August 1768–1810: Prussian mining official. Among his colleagues was *J. B. Richter* (1801–1807).
   1768–1770: Mining Councillor (salary, 1768: 400 Rtl, paid by the Academy).
   1769: Inspected the Silesian mining district.
   1770: Inspected the Westphalian mining district; visited *C. E. Gellert* and the Mining Academy in Freiberg; and upon his return to Berlin, organized the new Prussian Mining School.
   1770–1779: Senior Mining and from 1771 Finance Councillor.
   1770–1779: Teacher in the Berlin Mining School.
     1770–1774: Taught mineralogy and mining.
     1775–1779: Taught mineralogy, chemistry, and smelting, using an assaying laboratory constructed in his house with government and personal funds.
   1779: Inspected the Silesian mining district again.
   1779–1786: Privy Mining Councillor.
     1779–1786: Teacher of mineralogy and product analysis in the Berlin Mining School. Among his colleagues were *F. C. Achard* (1779–1784) and *M. H. Klaproth* (1784–1786).
     1781: Sold his mineral collection to the Berlin Mining School (pension, 1781–1810: 200 Tl).
   1786–1810: Privy Finance and Mining Councillor (salary, 1806: 3,486 Rtl, including 400 Rtl from the Academy and 200 Rtl from the Law Commission; 1807–1810: 3,086 Rtl because of budget cutbacks following on Prussia's military defeats). From 1795 to 1806 he was a member of the General Directory for Mining and Smelting, of the Salt Department (it was dissolved in 1805), of the Privy Finance Council, and of the Financial Deputation to the Law Commission.
   1810: Forced into retirement (pension, 1810–1821: 2,000 Rtl).
September 1768–1821: Ordinary Member of the Berlin Academy (salary, 1768–1791: none as Academician; 1791–1810: 200 Rtl; 1810–1821: 600 Rtl). Among his colleagues were *Marggraf* (1768–1782), *Achard* (1776–1800), *Klaproth* (1788–1817), *S. F. Hermbstaedt* (1800–1817), and *H. F. Link* (1815–1821).
   1804–1812: Director of the Physical Class as *Achard*'s substitute until 1810, then his replacement.

**State Honors** 1811: Knight of the Prussian Red Noble Order (1811: Third Class; 1818: Second Class with Oak Leaf). **Major Scientific Societies** 1768: Berlin. 1770: Leopoldina. 1783: Berlin NF. 1783: Munich. **Informal Recognition** 1777: Raspe. 1780: *Gmelin.* 1781: *Crell.* 1784: *Suckow.* 1786: Beseke. 1787: *Gren.* 1788: *Westrumb.* 1790: *Leonhardi.* 1791: *Göttling.* 1791: Nicolai. 1800: Anon.

**Bibliography**
AEWK, LX: 476–477.
Euler, no. 3: 234–243.
NDB, VI: 274–275.
Pr. Hdbh., 1795, pp. 55, 112, 164; 1805, pp. 211, 488; 1806, pp. 47, 180, 210.
Wutke, K. *Aus der Vergangenheit des Schlesischen Berg- und Hüttenlebens.* Vol. V in *Festschrift zum XII. Allgemeinen Deutschen Bergmannstage in Breslau 1913: Der Bergbau im Osten des Königreichs Preussen.* Breslau: Nischkowsky, 1913, pp. 9, 11–12, 16, 35, 37, 51–52, 66, 211–214, 432, 437–440.
Also see Berlin Academy (Harnack, I: 382, 469, 491, 545, 560, 645; Kirsten), Berlin Medical-

Surgical College ("Matrikel," 11: 143), Berlin Mining School (Krusch, pp. iv–xxi, xxvii, xxxi), and Frankfurt a. d. Oder University ("Matrikel," p. 396).

**MEYER, Johann Carl Friedrich**
1739, Stettin / 1811, Berlin

**Background** Son of Johann Michael Meyer (d. 1759, Stettin?), who owned the Hof-Apotheke in Stettin and was described by *J. G. Lehmann* in 1752 as "learned and pleasant." Lutheran. **Education** In 1760, after pharmaceutical training in Stettin(?), he enrolled as a pharmacy student in Berlin's Medical-Surgical College. During his stay in Berlin he studied chemistry with *J. H. Pott* and *A. S. Marggraf*. In 1764 he went to Uppsala University, where he took botany from Linnaeus and possibly chemistry from J. G. Wallerius.

**Career**
1766?–1811: Owner of the Hof-Apotheke in Stettin. Among his journeymen was *V. Rose, Jr.* (1783).
1780–1811?: Established and ran works in Stettin for producing French brandy, liqueur, and, from 1783, seltzer water. King Frederick II, desiring to reduce alcohol imports, provided 10,000 Tl (half in 1781, half in 1786) to help meet the capital costs.
1781–1811: Assessor on the Provincial Medical Board for Pomerania.
1784: Visited *M. H. Klaproth* and others in Berlin.
1810–1811: Served as Stettin's deputy in Berlin for discussions of Prussia's new taxation system.

**State Honors** None. **Major Scientific Societies** 1781: Leopoldina. 1782: Stockholm. 1788: Berlin. ?: Berlin NF. ?: St. Petersburg. **Informal Recognition** 1778: *Hagen*. 1784: Scherf. 1785: *Crell*. 1786: Beseke. 1786: Guyton de Morveau. 1786: *Hermbstaedt*. 1787: *Gren*. 1788: *Westrumb*. 1789: *Gmelin*. 1790: *Leonhardi*. 1791: Denina. 1791: *Göttling*. 1817: *Bucholz*. 1818: *Trommsdorff*.

**Bibliography**
*Berlinisches Jahrbuch für die Pharmacie*, 15 (1811), 234–241.
*Bibliothek der neuesten physisch-chemischen, metallurgischen, technologischen und pharmaceutischen Literatur*, 1, no. 3 (1788), 353–354.
DAB, II: 430–431.
Dann, G. E. (Professor, Dransfeld). Letters concerning Meyer's father and career (18, 29 July 1973).
Denina, C. *La Prusse littéraire sous Frédéric II*, Vol. III. Berlin: Rottmann, 1791, pp. 31–32.
Habrich, C. "Johann Carl Friedrich Meyer (1739–1811): Naturwissenschaftler und pharmazeutischer Standespolitiker." In *Pharmazie und Geschichte: Festschrift für Günter Kallinich zum 65. Geburtstag*, edited by W. Dressendörfer et al. Straubing and Munich: Donau, 1978, pp. 85–93.
Joecher, supp. IV: 1619.
"Die Liköre des Hof-Apothekers Meyer in Stettin und Frederich der Grosse." *Pharmazeutische Zeitung*, 80 (1935), 797.
Liphardt, J. C. L. Letter bearing on Meyer's seltzer water works, *Chemische Annalen*, no. 2 (1787), 250–251.
Rudolphi, K. A. *Joanni Fr. Blumenbach*. Berlin: Stark, 1825, pp. 76–77.
*Taschenbuch für Scheidekünstler und Apotheker*, 22 (1801), 39.
Thimon, C. (Universitetsbiblioteket, Uppsala). Letter concerning Meyer's enrollment and correspondents in Uppsala (14 July 1969).
Zywity (Archiv, Kirchenbuchstelle, Evangelische Kirche der Union, Berlin). Letter bearing on the Meyer family based on church records from Stettin (20 August 1975).
Also see Klaproth (Dann, pp. 39, 81, 147, 154–155), Lehmann (Freyberg, p. 135), and Berlin Medical-Surgical College ("Matrikel," 11: 146).

**WINTERL, Jacob (Albert) Joseph**
1739, Eisenerz, Austria / 1809, Budapest

**Background** Son of Johann Jacob Winterl, chief bookkeeper and secretary at the local iron and steel works. His mother was from the nobility. Catholic. **Education** After attending school in Garsten and Kiemsmunster, he went to the Jesuit college in Klosteunerburg to study theology. Later he proceeded to Vienna University, where he studied medicine, probably including chemistry with R. F. Laugier. In 1767 he earned his M.D. there with a dissertation proposing a new theory of inflammation.

## Career

1767–1770: Practiced medicine in various Hungarian mining towns. He probably met *N. J. Jacquin* and *G. A. Scopoli* in Schemnitz.

1770–1809: OP of botany and chemistry in the new Medical Faculty of Tyrnau University, which moved to Budapest in 1777 (salary, 1770: 1,200 fl; 1795: 1,300 fl).

**State Honors** None. **Major Scientific Societies** 1800: Göttingen (corresponding member). 1808: Munich. **Informal Recognition** 1791: *Göttling*. 1804: Kastner. 1809: *Gehlen*. 1817: *Bucholz*. 1818: Hohnbaum.

## Bibliography

Snelders, H. A. M. "The Influence of the Dualistic System of Jakob Joseph Winterl (1732–1809) on the German Romantic Era." *Isis*, 61 (1970), 231–240. [Wrong birth date.]

Stadtamt, Eisenerz. Letter about Winterl's parents and birth on 15 April 1739 (8 October 1968).

Székefalvi-Nagy, Z. "Leben und Werk von J. J. Winterl (1732–1809)." *NTM*; 8, no. 1 (1971), 37–45. [Wrong birth date.]

Wurzbach, LVII: 89–91.

Also see Budapest University (Endre, p. 365; Györy).

## BORN, Ignaz

1742, Kapnik, Transylvania / 1791, Vienna

**Background** Son of Ludwig Born (?, Saxony–1748, Karlsburg, Transylvania), a military officer, mining entrepreneur, and estate holder. His mother was from the nobility. Catholic. **Education** He first attended school in Hermannstadt, then in 1753 went to the Jesuit gymnasium in Vienna. In 1760 he became a Jesuit novitiate. Two years later, after withdrawing from the order, he proceeded to Prague University. He studied law for a year, then took a grand tour of Germany, Holland, France, and perhaps Spain. Upon his return he studied the mining sciences including chemistry with J. T. Peithner at Prague University and, around 1766, with *N. J. Jacquin* at the Schemnitz Mining School. In the mid-1760s he also visited the mining districts of Idria (where he met *G. A. Scopoli*), Hungary, Bohemia, and Saxony (where he probably met *C. E. Gellert*).

## Career

1768–1769: Managed his estates at Altzedlitz, Inchau, and Lukawetz in Bohemia.

1769–1770: Mining Councillor in Schemnitz. His appointment was motivated by his broad knowledge of "theoretical and practical metallurgy." Among his colleagues was *Scopoli*. In 1770 he was nearly killed in an accident while inspecting a mine in Transylvania.

1770–1772: Government and Mining Councillor in Prague (salary: 1,700 fl). In 1772 he resigned in protest against an edict forbidding mining officials to publish about mining matters.

1772–1776: Pursued the "metallurgical and mineralogical" sciences in Prague and at his Altzedlitz estate, which he enlarged in 1772 by purchasing an adjacent estate for 69,000 fl.

1776–1779: Catalogued parts of the Imperial natural history collections in Vienna (salary: 2,000 fl).

1777–1791: Mining, then from 1779 Court Councillor in the Imperial Mining Service in Vienna (salary, 1777: 2,000 fl; 1779: 3,000 fl; 1788–1791: 4,000 fl). Among his colleagues was *Jacquin*.

December 1777: Visited London, meeting G. Forster and others.

1778–1784: Vice-Director of the Theresian "Ritterakademie" in Vienna.

1782?–1791: Developed and tried to reap the profits from a new amalgamation process for removing gold and silver from their ores.

 1782?–1784: Developed the process in private experiments at an alleged cost of 20,000 fl.

 1785: Emperor Joseph II awarded him one-third of any savings that might be realized from the process over the next ten years and charged him with introducing it internationally so that the sales of Austrian mercury would increase.

 1785–1786: Supervised the establishment of a full-scale model plant at Glashütte near Schemnitz.

 1786: Presided over an international meeting of mining specialists at Glashütte. These men found that his process was better than smelting and subsequently sought, generally with but modest success, to introduce the process in Latin America, Scandinavia, Russia, and Germany. He received 12,000 fl from the Spanish Crown and a diamond from the Russian Empress for his efforts.

 1787–1791: Struggled with the Hapsburg mining bureaucracy, first to introduce his process, then to obtain the resulting income. As of 1790 he had only received 18,000 fl to counterbalance his expenses of roughly 60,000 fl.

1786–1791: Developed a new chlorine bleaching process, then in 1790 with the aid of Count G. Festetics de Tolna began to set up bleaching works near Vienna.

1791: Died with debts of over 270,000 fl.

**State Honors** 1774: Imperial Baron. **Major Scientific Societies** 1771: Stockholm. 1774: Berlin NF. 1774: Leopoldina. 1774: London. 1774: Munich. 1776: Göttingen. 1776: St. Petersburg. 1786: Berlin. **Informal Recognition** 1786: Beseke. 1787: *Gren.* 1791: *Göttling.* 1807: Kastner.

**Bibliography**

Born, I. "Dopisy Ignáce Borna D. G. a J. Ch. D. Schreberüm," edited by J. Beran. *Fontes scientiarum in Bohemia florentium historiam illustrantes.* 1 (1971).

Born, I., to J. J. Baier. Letter accepting membership in the Leopoldina (4 December 1774). Archiv, Deutsche Akademie der Naturforscher (Leopoldina), Halle.

Deutsch, A. "Ignaz von Born, 1742–1791." *Das Freimauer-Museum,* 6 (1968), 158–168.

DSB, II: 315–316.

Forster, XII: 104–105, 112–166; XIII: 300–301, 957; XIV: 159, 162–166, 172–177, 182.

Gicklhorn, R. "Die Bergexpedition des Freiherrn von Nordenflycht und die deutschen Bergleute in Peru." *Freiberger Forschungshefte: Kultur und Technik,* D40 (1963), 30–37, 41, 43, 46–47, 126–128.

Haubelt, J. "Studie o Ignáci Bornovi." Prague University: *Acta Universitatis Carolinae: Philosophica et Historica Monographia,* 39 (1971).

Hofer, P. "Ignaz von Born: Leben-Leistung-Wertung." Diss., Vienna University, 1955.

Luca, I: 40–46.

NDB, II: 466–467.

Schlichtegroll, vol. II for 1791, pp. 219–249.

*Staats-Anzeigen,* 13 (1789), 349–352; 14 (1790), 170–171.

Teich, M. "Born's Amalgamation Process and the International Metallurgic Gathering at Skleno in 1786," *Annals of Science,* 32 (1975), 305–340.

Vávra, J. "Osvícenská éra v česko-ruských vědeckých stycích: Ignác Born, Česká společnost nauk a Petrohradská akademie věd v letech 1774–1791." Čzechoslovenska akademie věd: *Studie,* 10 (1975).

Whitaker, A. P. "The Elhuyar Mining Missions and the Enlightenment." *Hispanic American Historical Review,* 31 (1951), 557–585.

Also see Vienna Theresianum (*Album,* p. 4).

**KLAPROTH, Martin Heinrich**

1743, Wernigerode / 1817, Berlin

**Background** Son of Johann Julius Klaproth (1712, Wernigerode–1767, Wernigerode), tailor and, from time to time, town councilman in Wernigerode. Lutheran (probably Pietist). **Education** Intended for the clergy, he entered Wernigerode's Latin school in 1755. Three years later, however, he withdrew from the school as the result of some unpleasant incident. Most likely guided by his former science teacher, J. C. Meier, he became an apprentice in Quedlinburg's Rats-Apotheke in the spring of 1759. His master, F. V. Bollman, was hard, giving him little, if any, theoretical training and less spare time. In 1766, two years after becoming a journeyman, he moved on to the Hof-Apotheke in Hanover. Here, probably encouraged by the Hirsch-Apotheke's owner, *J. G. R. Andreae*, he read the chemistry texts of *J. F. Cartheuser* and *J. R. Spielmann* and began making his own experiments. Among the apprentices under him was *J. F. Westrumb*. In 1768 he proceeded to Berlin's Apotheke zum goldenen Engel and in 1770, as a dispenser to Danzig's Rats-Apotheke.

**Career**

1771–1780: Administered Berlin's Apotheke zum weissen Schwan at the special request of his deceased friend, the minor chemist V. Rose, Sr. During this period he instructed *V. Rose, Jr.* and apparently enjoyed fruitful contacts with *A. S. Marggraf*, whose wealthy niece he married in 1780.

1780–1800: Owner of *Marggraf*'s father's old shop, the Apotheke zum Bären in Berlin (1780: purchased it with 9,500 Tl from his wife; 1800: sold it for 28,500 Tl).

1780: Petitioned unsuccessfully for permission to give private lectures on chemistry under the auspices of Berlin's Medical-Surgical College.

1782–1817: Prussian medical official. Among his colleagues in Berlin were *S. F. Hermbstaedt* (1794–1817), *Rose* (1797–1807), and *H. F. Link* (1815–1817).
1782–1797: Assessor on the Higher Medical Board.
1797–1799: Councillor on the Higher Medical Board, and Court Apothecary Shop Commissioner (salary: 160 Tl).
1799–1817: Senior Councillor on the Higher Medical and Sanitation Board, which in 1809 became the Scientific Deputation for Medical Affairs (salary: 150 Tl), and Court Apothecary Shop Commissioner (salary: 160 Tl).
1782–1810: Private lecturer of chemistry under the auspices of Berlin's Medical-Surgical College. Among his students were *A. F. Gehlen* (ca. 1800) and *J. N. Fuchs* (ca. 1803).
January 1784: Launched a hydrogen balloon on King Frederick II's birthday.
1784–1817: Chemistry teacher at Berlin's Mining School (salary: 200 Tl). Among his colleagues were *C. A. Gerhard* (1784–1786) and *Hermbstaedt* (1797–1804). After sickness prevented him from lecturing in 1816, *Hermbstaedt* substituted for him.
1787–1812?: Professor of chemistry in G. F. von Tempelhoff's artillery school, which became the Royal Artillery Academy in 1791 and was absorbed into the new General War School in 1810 (salary, 1787–1791: none; 1791–1809: 400 Tl plus 200 Tl for experiments; 1810: combined with his salary from Berlin University). The General War School closed in 1812 because of the war. When it reopened in 1815, he was probably not on the faculty.
1788–1817: Extraordinary member, then from 1800 ordinary member for chemistry of the Berlin Academy (salary, 1788–1800: none; 1800–1817: 200 Rtl plus, from 1803, a free residence). Among his colleagues were *Gerhard* (1788–1815), *F. C. Achard* (1788–1800), *Hermbstaedt* (1800–1817), and *Link* (1815–1817).
1788–1798: Traveled extensively in Germany, visiting, among others, *C. E. Gellert* in Freiberg (1788), *J. C. F. Meyer* in Stettin (1795), and *W. A. Lampadius* in Freiberg (1797).
1792: Investigated the Royal Porcelain Works with A. von Humboldt.
1799: Sat on a Royal Commission which reviewed *Achard*'s proposals for producing sugar from beets, reaching a favorable conclusion.
1810–1817: OP of chemistry in Berlin University's Philosophical Faculty (salary: 1,200 Tl). Among his colleagues were *Hermbstaedt* and *Link* (1815–1817). Among his students was H. Rose (1816). In 1816 he stopped lecturing on account of poor health.

**State Honors** 1811: Knight of Prussian Red Noble Order (Third Class). **Major Scientific Societies and Honorary Degrees** 1786: Erfurt. 1788: Berlin. 1791: Berlin NF. 1795: London. 1804: Paris (one of six foreign associates). 1805: St. Petersburg. 1806: Honorary Ph.D. from Erlangen University. 1808: Munich. 1810: Stockholm. 1815: Göttingen. **Informal Recognition** 1784: Scherf. 1786: Beseke. 1786: *Crell*. 1786: Guyton de Morveau. 1786: *Hermbstaedt*. 1787: *Gren*. 1788: *Westrumb*. 1789: *Gmelin*. 1790: *Leonhardi*. 1791: Denina. 1791: *Göttling*. 1791: Nicolai. 1793: *Wurzer*. 1794: Baldinger. 1796: Fourcroy. 1796: S. in M. 1797: Mann. 1798: *Bucholz*. 1800: Anon. 1803: Ed. Board. 1804: Kastner. 1806: *Trommsdorff*. 1811: Sertürner. 1817: Schweigger. 1819: Brandes. 1821: *Pfaff*.

**Bibliography**
DAB, I: 322–324.
Dann G. E. *Martin Heinrich Klaproth (1743–1817): Ein deutscher Apotheker und Chemiker: Sein Weg und seine Leistung.* Berlin: Akademie Verlag, 1958.
DSB, VII: 394–395.
Fischer, E. G. "Denkschrift auf Klaproth," Akademie der Wissenschaften, Berlin: *Abhandlungen* (1819–1820), 11–26.
Hein, W.-H "Alexander von Humboldt und Martin Heinrich Klaproth." *Beiträge zur Geschichte der Pharmazie*, 29, no. 2 (1977), 9–16.
Lemper, E.-H. *Adolf Traugott von Gersdorf (1744–1807): Naturforschung und soziale Reformen im Dienste der Humanität.* Berlin: VEB, 1974, pp. 280, 326, 355.
NDB, XI: 707–709.
Also see Lampadius (Seifert, pp. 28–30), Marggraf (Crell, p. 192), and Berlin Artillery Academy (Scharfenort, pp. 9–11, 377).

**CRELL, (Florens) Lorenz (Friedrich)**
1745, Helmstedt / 1816, Göttingen

**Background** Son of Johann Friedrich Crell (1707, Leipzig–1747, Helmstedt), who, after earning an M.D. at Leipzig University in 1732, was substitute OP of medicine at Wittenberg University (1736–1741), then OP of anatomy, physiology, and pharmacy at Helmstedt University (1741–1747). Lutheran. **Education** His initial education was in the home of his maternal grandfather, L. Heister (d. 1758), OP of medicine at Helmstedt University. In 1759 he began attending courses at the University. After six years of the philosophical sciences, he took up medicine. During his medical studies, he was introduced to chemistry by his favorite teacher, the polymath G. C. Beireis. In 1768 he took his M.D. In 1768 he took his M.D., then traveled to Strasbourg where he apparently met *J. R. Spielmann*, to Paris where he spent a few months studying, to Edinburgh where he took chemistry from J. Black in the winter semester of 1769/70, and to London where he attended Royal Society meetings in November 1770. The following month he returned to Brunswick.

**Career**
1771–1774: OP of metallurgy in Brunswick's Collegium Carolinum (salary, 1773: 200 Rtl).
1774–1810: OP of theoretical medicine and materia medica in Helmstedt University's Medical Faculty. Among his students were *F. A. C. Gren* (1782–1783) and *F. Wurzer* (1792). After he took his Ph.D. there in 1783, he also became an OP in the Philosophical Faculty.
1778–1804: Edited several periodicals for chemistry, the most notable of which were the *Chemisches Journal* . . . (1778–1781), *Die neuesten Entdeckungen* . . . (1781–1784), and the *Chemische Annalen* . . . (1784–1804).
1779: Nominee for *F. A. Cartheuser*'s chair in Giessen. Nothing came of this because he was "not known" and because he was thought to have a high salary.
1785: Visited Berlin where he met *F. C. Achard* and others.
1787: Visited Göttingen where he stayed with *J. F. Gmelin*.
1790: Declined a call to Vienna (salary: 3,000 fl plus free residence). Nevertheless, Emperor Leopold made him an absentee prebendary of the Hamburg Cathedral.
1791: Visited Hameln where he witnessed many of *J. F. Westrumb*'s experiments.
1802: Met *A. N. Scherer* in Leipzig to discuss the possibility of merging their journals. His competitor was so difficult that nothing came of the talks.
1803: Visited Berlin where, after negotiations with *A. F. Gehlen* and other chemists, he agreed to merge his journal with the *Neues allgemeines Journal der Chemie*.
1810–1816: OP of chemistry in Göttingen University's Medical Faculty (salary, 1816: 1,400 Rtl). He was not appointed co-director of the University's chemical institute because F. Stromeyer, the director, opposed any dilution of the director's authority.

**State Honors** 1780: Brunswick Mining Councillor. 1791: Imperial Ennoblement. 1814: Hanoverian Court Councillor. **Major Scientific Societies** 1778: Leopoldina. 1778: Göttingen. 1779: Berlin NF. 1779: Erfurt. 1784: Berlin. 1784: Stockholm. 1786: Paris (corresponding member). 1786: St. Petersburg. 1788: London, 1807: Munich. **Informal Recognition** 1780: *Gmelin*. 1784: *Suckow*. 1786: *Beseke*. 1786: *Wiegleb*. 1787: *Fuchs*. 1787: *Gren*. 1791: *Göttling*. 1793: *Wurzer*. 1794: *Baldinger*. 1796: *Fourcroy*. 1796: *Girtanner*. 1797: *Mann*. 1798: *Bucholz*. 1798: *Frank*. 1800: *Anon*. 1804: *Gehlen*. 1806: *Trommsdorff*. 1811: *Sertürner*.

**Bibliography**
ALZ: IB (21 April 1792), 401.
Blumenbach, J. F. *Memoria Laurentii de Crell*. Göttingen: Dieterich, 1822.
Crell, L., to A. von Haller. One letter (1777). Stadtbibliothek, Bern.
——— to J. Black. Seven letters (1779–1782). Library, Edinburgh University.
DSB, III: 464–466.
Forster, G., to S. T. Sömmering. Letter mentioning Crell's recent visit to Göttingen (17 Oct. 1787). In *Georg Forster: Werke*, edited by G. Steiner, vol. IV. Frankfurt a. M.: Insel 1970, p. 481.
Göttingen Universitäts-Archiv ([4]IV b, 43).
Hessisches Staatsarchiv, Darmstadt (Abt. VI, 1: Giessen; Konv. 28, Fsc. 13).
Kaye, I. (Royal Society, London). Letter concerning Crell's attendance at Royal Society meetings (23 February 1968).
König, J. (Niedersächsisches Staatsarchiv, Wolfenbüttel). Letter with information about Crell's salary in 1773 (8 November 1968).
Niedersächsisches Staatsarchiv, Wolfenbüttel (2 Alt 14703, 169–187; 37 Alt 2396).
Rotermund, I: 404.
Scherer, A. N. "Vorläufige Erinnerungen über den Fortgang und einige Erweiterungen des allgemeinen Journals der Chemie." *Allgemeines Journal der Chemie*, 7 (1801 [1803]), vii.
Schrader, W. "Professor Johann Friedrich Crell und Lorenz v. Crell: Zwei Helmstedter Mediziner." *Alt-Helmstedt: Blätter der Heimatkunde für die Stadt und den Landkreis Helmstedt*, 19, no. 2 (March 1955), 1–2.

Westrumb, J. F. "Einige Bemerkungen, verschiedene Gegenstände der neuen Chemie betreffend." *Chemische Annalen*, no. 2 (1792), 3.
Also see Achard (Stieda, p. 172), and Göttingen University (Ganss, p. 36; Pütter, III: 80–85).

## HEYER, Justus Christian Heinrich
1746, Halberstadt / 1821, Brunswick

**Background** Son of Johann Friedrich Heyer (1695,?–1774, Halberstadt) pastor of Halberstadt's St. Spiritus Hospital Church and co-rector, then rector of the Martin-School. Lutheran. **Education** After serving an apprenticeship in Halberstadt's Rats-Apotheke, he worked as a journeyman pharmacist in Göttingen, Hamburg, and Halberstadt. In Göttingen he may have heard *R. A. Vogel*'s lectures on chemistry.

**Career**
1776–1817: Apothecary in Brunswick. Upon the death of his father-in-law in 1791, he became the owner of the Apotheke am Hagenmarkt (value: 11,000 Rtl). He retired in 1817.
January–February 1784: Collaborated with Prof. E. A. W. Zimmermann in launching two hydrogen balloons commissioned by Duke Carl Wilhelm Ferdinand.
1787: Member of a ducal commission that evaluated J. J. W. Dedekind's process for extracting sugar from beets, reaching an unfavorable conclusion.
ca. 1810: Developed a successful method for extracting sugar from prunes, winning a prize of 4,000 francs.
1821: Upon his death, his mineral collection was sold for 2,000 Tl.
**State Honors** None. **Major Scientific Societies** 1786: Erfurt. 1790: Leopoldina. 1796: Berlin NF. **Informal Recognition** 1784: Scherf. 1786: *Hermbstaedt*. 1788: *Westrumb*. 1789: *Gmelin*. 1791: *Göttling*. 1794: Baldinger.

**Bibliography**
Bohlmann, R. "Geschichte der Apotheken des Landes Braunschweig: Die Apotheke am Hagenmarkt." Braunschweig Technical University: Pharmaziegeschichtliches Seminar: *Veröffentlichungen*, 2 (1959), 28–31.
DAB, I: 273–274.
Schneider, W. "Apotheker Heyer und der braunschweiger Ballonaufstieg 1784." *Der niedersächsische Apotheker*, 4 (1953) 9–11.
Schreiner (Clergyman, Halberstadt). Letter about Heyer's father based on Halberstadt's church records (20 May 1976).

## LEONHARDI, Johann Gottfried
1746, Leipzig / 1823, Dresden

**Background** Son of David Leonhardi (?, Reichenbach, Vogtland–1764/71, Leipzig), who studied at Königsberg University, at a Dutch university, and at Leipzig University (M.D., 1728), then practiced medicine in Leipzig. Lutheran (probably Pietist). **Education** After being instructed at home by a tutor and his elder brother, he studied at Leipzig's Thomas School from 1759 to 1764. Then he enrolled at Leipzig University where he pursued both the philosophical and medical sciences. His favorite teacher was *C. W. Poerner*, with whom he made medical rounds and carried out chemical investigations. He earned his bachelor of medicine in 1767, his medical license in 1769, master of arts in 1770, and M.D. in 1771 with a dissertation on the effects of atmospheric cold on the human body.

**Career**
1770–1782: PD, then in 1781 AOP in Leipzig University's Medical Faculty. When he was made AOP, he was evidently promised the next vacant medical chair at Wittenberg University.
1782–1791: OP of botany and anatomy for a few months, then of pathology and surgery in Wittenberg University's Medical Faculty. From 1791 until the University was closed in 1816, he was on leave in Dresden with the right to appoint substitutes to his chair.
1791–1823: Saxon Crown Physician in Dresden (salary: 1,200 Tl). He promised his former pupil C. F. S. Hahnemann that he would not neglect chemistry in his new position, a promise that he was unable to keep. In Dresden he was a member of the Sanitation Board and from 1797 an inspector of the Court Apothecary Shop. From 1813 to 1815, he resided with the royal family in Prague. In 1821 he visited Leipzig.

**State Honors** 1791: Saxon Court Councillor. 1815: Knight's Cross of the Saxon Civil Service Order. **Scientific Societies** 1789: Leopoldina. **Informal Recognition** 1786: Beseke. 1786: *Wiegleb.* 1787: *Gren.* 1791: *Göttling.* 1796: *Link.* 1796: Van Mons. 1798: Frank. 1800: Anon. 1804: *Gmelin.*
**Bibliography**
DSB, VIII: 246–248.
Flemming, L. F. F. *Socero Christiano Gotthelf Pienitzio . . . Agitur de Vita et Meritis beati Joh. Gottf. Leonhardi.* Dresden, 1823, pp. 5–24.
Kläbe, J. G. A. *Neuestes gelehrtes Dresden.* Leipzig: Voss, 1796, pp. 86–89.
Leonhardi, C. D. Autobiography. In J. C. Pohl, *Panegyrin medicam.* Leipzig: Langenheim, 1764, pp. ix–xii.
Leonhardi, D. Autobiography. In P. G. Schacher, *Panegyrin Medicam.* Leipzig: Roth, 1729, p. [8].
Leonhardi, J. G. Autobiography. In A. G. Plaz, *Panegyrin medicam.* Leipzig: Langenheim, 1771, pp. xii–xvi.
*Neuer Nekrolog der Deutschen*, 1, no. 2 (1823), 770–774.
Staatsarchiv, Dresden. Letter giving Leonhardi's salary as Saxon Crown Physician (8 May 1972).
Zaunick, R. "Ein übersehener Brief Samuel Hahnemanns aus seiner Leipziger chemischen Arbeitsperiode." In *Studien zur Geschichte der Chemie: Festgabe Edmund O. v. Lippmann*, edited by J. Ruska. Berlin: Springer, 1927, pp. 120–121.
Also see Leipzig University (*Matrikel*, III: 235) and Wittenberg University (Friedensburg, p. 882; "Matrikel," 5: 273).

## WENZEL, Carl Friedrich
1747, Dresden / 1793, Freiberg

**Background** Son of Friedrich Gottlob Wentzel, who became a master bookbinder in Dresden in 1739 and was the bookbinder in Electoral Saxony's General Excise Office. Lutheran (probably Pietist). **Education** Around 1762 he ran away from his bookbinding apprenticeship with 14 fl and plans to visit relatives at the Cape of Good Hope. Earning his keep by engraving seals, he made his way to Hamburg, then Amsterdam. There, unable to arrange passage to Africa, he apprenticed himself to an apothecary-surgeon. He accompanied his master on two voyages to Greenland, becoming ship's surgeon upon his master's death. At the end of the second voyage, he was forced to serve a stint as surgeon in the Dutch army. He returned to Saxony around 1766 and matriculated at Leipzig University in 1768. He studied mathematics, physics, and chemistry (probably with *C. W. Poerner*) there for three years.
**Career**
1771?–1779: Pursued chemistry privately in Dresden. In 1776 he won a gold medal from the Royal Danish Academy of Sciences for a paper on the disintegration of metals by reverberation.
1779–1793: Auditor, then from 1786 Senior Smelting Officer in the Freiberg Smelting Bureau (salary, 1779: 300 Tl; 1793: 600 Tl). Among his colleagues was *C. E. Gellert*, for whose chair in the Mining Academy he was considered a possible successor. He began offering private courses on chemistry in Freiberg in the mid-1780s.
1781–1793: Consulting chemist for the Saxon Porcelain Works in Meissen. Among his colleagues there was *Poerner.*
**State Honors** None. **Major Scientific Societies** 1779: Berlin NF. **Informal Recognition** 1780: *Gmelin.* 1781: *Wiegleb.* 1784: *Suckow.* 1787: *Gren.* 1791: *Göttling.* 1807: Kastner. 1821: *Pfaff.* 1822: *Lampadius.*
**Bibliography**
ALZ: IB (31 August 1793), 706–707.
Forster, XII: 77.
Schlichtegroll, vol. II for 1793, pp. 291–294.
Staatsarchiv, Dresden. Letter bearing on Wenzel's employment in the Saxon Mining Service (25 November 1971).
Wensch, K. (Genealogist, Dresden). Letter based on church records in Dresden and Freiberg giving information on Wenzel's father, Wenzel's exact birthdate—20 August 1747, not 1740—and Wenzel's titles at death (28 August 1970).
Also see Lampadius (Seifert, pp. 24, 28) and Leipzig University (*Matrikel*, III: 453).

## GMELIN, Johann Friedrich
1748, Tübingen / 1804, Göttingen

**Background** Son of Philipp Friedrich Gmelin (1721, Tübingen–1768, Tübingen), apothecary, physician, and, from 1755, OP of chemistry and botany at Tübingen University. Lutheran (probably Pietist). **Education** After attending the local school, he entered Tübingen University in 1762. There he studied medicine, taking chemistry from his father who used P. Macquer's text. In 1768 he wrote a dissertation on the irritability of plants. The following year he completed the requirements for his M.D., then embarked on an international study tour. Passing through Paris where he apparently met Macquer, he enrolled at Leiden University. While there he took chemistry from H. D. Gaubius. In 1770 he proceeded to London where he spent a few months studying botany and therapy and attended one Royal Society meeting. Then he journeyed by way of Göttingen, where he met *R. A. Vogel*, to Vienna. After a semester there with *N. J. Jacquin* and others, he returned to Tübingen.

**Career**
1771–1775: PD, then from 1772 AOP in Tübingen University's Medical Faculty.
1775–1804: OP of chemistry in Göttingen University (salary, 1775–1782: 300 Rtl). His initial appointment was as OP in the Philosophical Faculty and AOP in the Medical Faculty. However, in 1778 he shifted completely into the Medical Faculty, probably so that once he had sufficient seniority, he could share in the fees from M.D. candidates. In 1783 he was provided with a large laboratory, lecture hall, free residence, and regular budget for experiments. Among his students were *C. Girtanner* (1780–1782), *G. F. Hildebrandt* (1780–1783, colleague: 1785), J. T. Lowitz (1780–1783), *J. F. A. Göttling* (1785–1787), *F. Wurzer* (1786), *H. F. Link* (1786–1789, colleague: 1790–1791), *W. A. Lampadius* (1791–1792), *C. H. Pfaff* (1793–1794), his successor, F. Stromeyer (1793–1799), and his son, L. Gmelin (1803–1804).

**State Honors** 1788: Hanoverian Court Councillor. **Major Scientific Societies** 1774: Leopoldina. 1778: Göttingen. 1781: Erfurt. 1794: St. Petersburg. 1796: Berlin NF. **Informal Recognition** 1786: Beseke. 1786: *Wiegleb*. 1787: *Gren*. 1791: *Göttling*. 1793: *Wurzer*. 1794: Baldinger. 1796: *Girtanner*. 1796: Van Mons. 1798: *Bucholz*. 1798: Frank. 1800: Anon. 1806: *Trommsdorff*. 1821: *Pfaff*.

**Bibliography**
"Briefwechsel zwischen Albrecht von Haller und Eberhard Friedrich von Gemmingen," ed. H. Fischer. Litterarischer Verein in Stuttgart: *Bibliothek*, 219 (1899), 78–84.
DSB, V: 429.
*Die Familie Gmelin*. Edited by Familienverband Gmelin. Neustadt a. d. Aisch: Degener, 1973, pp. 46–47, 256.
Gmelin, J. F., to A. von Haller. Thirty-four letters (1768–1777). Stadtbibliothek, Bern.
——— to P. Macquer(?), One letter (28 February 1770). Staatsbibliothek Preussischer Kulturbesitz, West Berlin (Darmst. G. 1 1800).
Heyne, C. G. "Memoria Io. Friderici Gmelin." Gesellschaft der Wissenschaften, Göttingen: *Commentationes*, 16 (1808), i–viii.
Leiden, 1097.
Lichtenberg, I: 359; II: 53, 249, 259–260, 353; III: 260.
Moll, I: 220–233.
Pietsch. E. H. E. *Die Familie Gmelin und die Naturwissenschaften*. Frankfurt a. M.: Gmelin Institut, 1964, pp. 13–17.
Royal Society Minute Book (5 July 1770). Royal Society, London.
Also see Pfaff (Pfaff, pp. 69–70), Göttingen University (Ganss, pp. 26–32; Hamann; Pütter, II: 146–148; III: 75–78), and Tübingen University (Beese, pp. 97–98; Bök, pp. 197–198, 250–251, 290; *Matrikel*, III: 198, 241, 253).

## WEIGEL, Christian Ehrenfried
1748, Stralsund / 1831, Greifswald

**Background** Son of Bernhard Nicolaus Weigel (1721, Woigdehagen near Stralsund–1800, Stralsund), who earned his M.D. at Greifswald University in 1745, became a Town Physician in Stralsund soon afterwards, and devoted much spare time to chemical experiments. (He invented Weigel's drops.) Lutheran. **Education** Carefully educated by tutors, his father, and his maternal uncle Charisius (also a physician), he entered Greifswald University in 1764 to study medicine. He concentrated on botany, chemistry, and anatomy, depending on his father and uncle to make up for the faculty's deficiencies. In particular, he learned both the chemical literature and chemical operations from his father. In 1769 he proceeded to Göttingen University where he continued his study of the natural sciences, taking chemistry from *R. A. Vogel* and *J. C. P. Erxleben*, and began serious work in the medical disciplines. During vacations he made excursions to Kassel, Pyrmont, and the Harz mining district where he botanized, collected minerals, and observed assaying and smelting. In 1771 he earned his M.D. at Göttingen University with a chemical-mineralogical dissertation, then journeyed by way of the Harz and Hanover, where he visited *J. G. R. Andreae*, back home to Stralsund. For the next year, while awaiting an opportunity to enter the academic world, he accompanied his father on medical rounds and sharpened his experimental skills in his father's laboratory.

**Career**
1772–1831: PD, then from 1773 Adjunct for botany, and finally from 1774 OP of chemistry and pharmacy in Greifswald University's Medical Faculty (salary, 1773: 200 Rtl; 1774: 400 Rtl; 1810: 926 Rtl). Among his students was A. G. Ekeberg (ca. 1788).
1773–1781: Supervisor of the University's botanical garden and natural history collection.
1796–1831: Director of the University's chemical institute.
1780–1806: Member, then from 1794 Director of the Medical Board for Pomerania and Rügen.
1821: Accompanied the Swedish Crown Prince on a trip to Beckaskog.
1830: Attended the meeting of the Society of German Scientists and Physicians in Hamburg.

**State Honors** 1795: Swedish Royal Physician. 1806: Imperial Ennoblement. 1811: Honorary Member of the Royal Swedish Medical Board. 1814: Knight of the Swedish North Star Order. 1821: Knight of the Prussian Red Noble Order (Third Class). 1821: Commander of the Swedish Wasa Order. **Major Scientific Societies** 1777: Berlin NF. 1778: Erfurt. 1790: Leopoldina. 1792: Stockholm. **Informal Recognition** 1780: *Gmelin*. 1781: *Wiegleb*. 1784: *Suckow*. 1786: Beseke. 1787: Fuchs. 1787: *Gren*. 1791: *Göttling*. 1794: Baldinger. 1800: Anon. 1821: *Pfaff*.

**Bibliography**
Callisen, XX: 493–497; XXXIII: 249–250.
DSB, XIV: 224–225.
Ewe (Stadtarchiv, Stralsund). Letter concerning Weigel's family (8 January 1970).
Fester, G. A. "Zur Geschichte des Gegenstromkühlers." *Sudhoffs Archiv*, 45 (1961), 341–350.
Ludwigs, F. (Riksarkivet, Stockholm). Letter bearing on Weigel's career (21 January 1972).
*Neuer Nekrolog der Deutschen*, 9 (1831), 699–705.
Weigel, C. E. Autobiography (1771). In R. A. Vogel, *Praemissae sunt observationes binae de Asthmate*. Göttingen: Rosenbusch, 1773, pp. xv–xvii.
———Autobiography (1774). In A. Westphal, *Einladungs-Schrift zu der öffentlichen Rede in welcher . . . Weigel . . . am 14. May . . . den Antritt Seiner akademischen Aemter feierlich machen wird*. Greifswald: Röse, 1774, pp. 7–15.
Also see Greifswald University (Anselmino, pp. 177–120, 123–124, 129–130; Boriss, p. 521; Seth, pp. 157, 238, 295; Valentin, pp. 472–475).

**HAGEN, Karl Gottfried**
1749, Königsberg / 1829, Königsberg

**Background** Son of Heinrich Hagen (1709, Schippenbeil–1772, Königsberg), a student of *C. Neumann*, *J. H. Pott*, and *J. T. Eller* in the late 1720s, apothecary in Königsberg from 1746, owner of the Hof-Apotheke there from 1757, and correspondent of *A. S. Marggraf*, *J. G. Model*, *J. C. F. Meyer*, and *J. C. Wiegleb*. Lutheran (probably Pietist). **Education** Initially he was tutored by an uncle who, as a parson, imbued him with the desire to become a preacher. However, upon his father's insistence he served an apprenticeship in the family apothecary shop until 1769 when his father, fearful that he would be conscripted, allowed him to enroll in Königsberg University as a medical student. Soon after his father's death, he went to Berlin to take the examination required of Court Apothecaries. While there he met *M. H. Klaproth*, with whom he subsequently exchanged journeymen.

**Career**
1773–1816: Owner of the Hof-Apotheke in Königsberg. Among those serving in his shop were *V. Rose, Jr.* (1784–1785) and *A. F. Gehlen* (1790s).
1775–1829: PD, AOP, Adjunct, then OP at Königsberg University. His teaching career began in 1775 when, in response to the Medical Faculty's expression of interest, he qualified for private lecturing by taking the M.D. with a dissertation on tin. Among his students were

*J. B. Richter* (1783–1789) and *Gehlen* (1790s).
1775–1788: PD, then from 1779 AOP, then from 1783 Adjunct in the Medical Faculty (salary: none until 1787, then 263 Tl).
1788–1807: OP of chemistry in the Medical Faculty (salary: 290 Tl).
1807–1829: OP of chemistry, physics, and natural history in the Philosophical Faculty (salary, 1807–1808: 450 Tl; 1808–1829?: 950 Tl). From 1808 to 1809, he lectured the Prussian princes Frederick William, William, and Frederick on physics, chemistry, and botany. He also gave a course on chemistry at this time to the artillery officers.
1815?–1829: Director of the University's coin, mineral, and instrument collections.
1786/1790–1829: Member of the East Prussian Provincial Sanitation Board, then from 1799 Medical Councillor on the Provincial Medical Board (salary, 1799–1829: 100 Tl).
1804: Traveled through Germany, visiting *J. F. A. Göttling* in Jena and *Klaproth*, *Rose*, and *Gehlen* in Berlin.

**State Honors** 1819: Knight of the Prussian Red Noble Order (1819: Third Class; 1825: Second Class with Oak Leaf). **Major Scientific Societies** 1776: Leopoldina. 1782: Berlin NF. **Informal Recognition** 1780: *Weigel*. 1786: Beseke. 1786: *Hermbstaedt*. 1787: *Gren*. 1789: *Gmelin*. 1791: *Göttling*. 1794: Baldinger. 1800: Anon. 1811: Sertürner. 1818: *Trommsdorff*.

**Bibliography**
DAB, I: 240–241.
Hagen, A. "K. G. Hagen's Leben und Wirken," *Neue preussische Provinzial-Blätter*, 9 (1850), 46–86, 116–138.
Hagen, H. *Abhandlungen Chemischen und Physikalischen Inhalts*. Edited by K. G. Hagen. Königsberg: Hartung, 1778, pp. iii–xviii.
NDB, VII: 470–474.
Pr. Hdbh. 1796, p. 147; 1803, p. 121; 1828, pp. 77, 79, 190.
Schneider, W. "Carl Gottfried Hagen." *Die pharmazeutische Industrie*, 16 (1954), 111–114.
Also see Königsberg University (Matthes, pp. 1041–1045).

### SUCKOW, George Adolph
1751, Jena / 1813, Heidelberg

**Background** Son of Lorenz Johann Daniel Suckow (1722, Schwerin–1801, Jena), PD at Jena University (1746–1755), then a teacher in Hamburg's gymnasium (1755–1756), and finally OP of natural philosophy at Jena University (1756–1801). Lutheran. **Education** Having received a good primary education at home, he entered Jena's gymnasium in 1761. Five years later he matriculated at Jena University, where he studied the philosophical sciences, then medicine. He took chemistry from his father, who regarded this subject as a branch of natural philosophy, and E. A. Nicolai. In 1772 he earned his M.D. with a dissertation on the chemical composition of Jena's waters.

**Career**
1774–1784: OP for physics, mathematics, natural history, and chemistry in the new Cameral School in Kaiserslautern (salary, 1774–1775: 400 fl; 1775–1784?: 600 fl; income from lecture fees, 1780?: ca. 300 fl). He was also responsible for state economy during his first year and for civil engineering, forestry, and mining the next three years. He served on the "manufacturing commission" that oversaw the new linen factory established by the school's backers.
1779: Nominated by *F. A. Cartheuser* for his chair at Geissen University, but nothing came of this.
1783: Declined a call to Freiburg im Breisgau University (salary: 1,000 fl) after the government agreed to move the Cameral High School to Heidelberg.
1784–1813: OP of physics, chemistry, and natural history in the State Economics High School in Heidelberg, which in 1803 became the State Economics Section of Heidelberg University (salary, 1798: 1,000 fl plus free residence; 1804: 2,000 fl). During this period he turned down calls to St. Petersburg, Würzburg, Stuttgart, Freiburg i. Br., Giessen, Greifswald (ca. 1799), Jena, and elsewhere.

**State Honors** 1784: Palatine-Zweibrück Court Councillor. 1805: Palatine Privy Court Councillor. **Major Scientific Societies** 1775: Berlin NF. 1798: Leopoldina. 1807: Munich. **Informal Recognition** 1786: Beseke. 1786: *Wiegleb*. 1787: *Gren*. 1788: *Westrumb*. 1790: *Leonhardi*. 1791: *Göttling*. 1800: Anon. 1804: *Gmelin*.

**Bibliography**
ADB, XXXVII: 105–106.
Baldinger, pt. 2: 125–136.
Hessisches Staatsarchiv, Darmstadt (Abt. VI, 1: Giessen; Konv. 28, Fsc. 13).
Moll, III: 898–900.
Suckow, G. A. Autobiography. In E. G. Baldinger, *Praemittuntur Observationes de Morbis ex Metastasi Lactis in Puerperis*. Jena: Mauk, 1772, pp. ix–xi.
Also see Heidelberg State Economics School (Stieda), Heidelberg University (Winkelmann, II: 281, 316–317; Weisert), and Kaiserslautern Cameral School (Freitag, pp. 23–28, 30; Webler).

### WESTRUMB, Johann Friedrich
1751, Nörten-Hardenberg, Hanover / 1819, Hameln

**Background** Son of Gustav Westrumb, regimental surgeon with the Hanoverian dragoons until 1778. Lutheran. **Education** On account of his mother's death, he was sent as a boy to live with her brother, Pastor Hantelmann in Dannenberg. At the age of fourteen he went to Hanover, his mother's hometown, to become an apprentice in the Hof-Apotheke. During his service there, he came under the influence of the aspiring journeyman, *M. H. Klaproth*. He completed his apprenticeship in 1770, then served as a journeyman pharmacist in Frankfurt a. d. Oder and Brandenburg.

**Career**
1773–1779: Administrator of Hanover's Hof-Apotheke. Among his friends in Hanover were *J. G. R. Andreae* and the botanist-chemist F. Ehrhart, who was corresponding with T. Bergman and C. W. Scheele.
1779: Proceeded to Hameln, his father's hometown, to direct the Rats-Apotheke. When he married the former lessee's widow, he became the lessee of the shop.
1779–1819: Lessee of the Rats-Apotheke in Hameln. Among those who came to study under him was *F. Wurzer* (1792).
1787: Investigated Pyrmont's salt springs on a commission from the Prince's Rent-Board in Arolsen.
1789–1790: Attempted to establish commercial bleaching with chlorine in Hameln.
1790–1818: Town Senator in Hameln. In 1806 he was responsible for arranging military billeting in Hameln and for managing the town's forests.
1791–1819: Hanoverian Mining Commissioner and, at least until 1800, member of the Hanoverian Commerce Board. As Mining Commissioner, he carried out chemical analyses for Hanover and other states (e.g., Magdeburg in 1794).

**State Honors** None. **Major Scientific Societies and Honorary Degrees** 1786: Erfurt. 1788: Berlin NF. 1788: Göttingen. 1793: Leopoldina. 1811: Honorary M.D. from Marburg University. **Informal Recognition** 1784: Scherf. 1786: Beseke. 1786: *Hermbstaedt*. 1787: *Gren*. 1788: *Leonhardi*. 1789: *Gmelin*. 1791: *Crell*. 1791: *Göttling*. 1791: Hahnemann. 1792: Humboldt. 1792: *Klaproth*. 1793: Scherer. 1793: *Wurzer*. 1794: Baldinger. 1796: Fourcroy. 1796: *Girtanner*. 1796: *Link*. 1796: Van Mons. 1796: S. in M. 1798: *Bucholz*. 1800: Anon. 1806: *Trommsdorff*. 1811: Sertürner. 1820: Schweigger. 1821: *Pfaff*.

**Bibliography**
Börsche (Stadtarchiv, Hameln). Letter bearing on Westrumb's role in Hameln's town government (12 January 1971).
DAB, II: 741–742.
Du Ménil, A. P. J. "Westrumb als Mensch und Gelehrter." *Journal für die Chemie und Physik*, 28 (1820), 1–8.
Hamann (Niedersächsisches Staatsarchiv, Hanover). Letter clarifying Westrumb's status as Mining Commissioner (27 November 1970).
Kerstein, G. "Ein vergessener Hamelner Gelehrter." Heimatsmuseum, Hameln: *Jahrbuch* (1968), 26–29.
Kirchenbuchamt, Hanover. Letter revealing Westrumb's father's name on the basis of the church record of his marriage on 2 October 1749 (11 December 1974).
Oersted, II: 586–588.
Reu, E. (Clergyman, Nörten-Hardenberg). Letter transmitting Westrumb's baptismal entry which reveals his parents' hometowns (25 November 1974).

Walter (Niedersächsisches Staatsarchiv, Hanover). Letter tracing his father's career from 1754 to 1778 on the basis of the Hanoverian *Staatskalender* (4 October 1974).

Westrumb, A. "Johann Friedrich Westrumb." *Neues vaterländisches Archiv des Königreichs Hannover*, 7 (1825), 23–42.

## ACHARD, Franz Carl
1753, Berlin / 1821, Kunern, Silesia

**Background** Son of Guillaume Achard (1716, Geneva–1755, Berlin), pastor in Berlin's Werder Church from 1743. Stepson of Charles Vigne from 1759. Calvinist. **Education** Aided by relatives in the Berlin Academy, he secured the instruction of various academicians, especially *A. S. Marggraf*.

**Career**

January 1776: Having impressed King Frederick II with an account of his research, he obtained the Berlin Academy's recommendation for membership as *Marggraf*'s collaborator.

June 1776–1800: Extraordinary member, then from 1779 ordinary member of the Berlin Academy (salary, 1776–1779: none; 1779–1782: 400 Rtl; 1782–1787: 1,250 Rtl; 1787–1800: 1,450 Rtl). Among his colleagues were *Marggraf* (1776–1782), *C. A. Gerhard* (1776–1800), and *M. H. Klaproth* (1788–1800). In 1782 he succeeded *Marggraf* as director of the Academy's laboratory and as director of the Academy's Physical Class.

1779–1784: Teacher of chemistry and physics in Berlin's Mining School (salary: 100 Rtl). Among his colleagues was *Gerhard*.

1779–1782?: Demonstrated recent chemical and physical discoveries to Frederick II on a fairly regular basis.

1780–1782?: Experimented with tobacco on an estate near Lichtenberg, in hopes of winning a lifelong annuity of 1,000 Rtl for successful domestication of tobacco. He was awarded half the prize (pension, 1782?–1806: 500 Rtl.)

December 1783–February 1784: Launched several balloons for Berlin's public. Despite charges to the contrary, his costs (962 Rtl) exceeded his income (538 Rtl at 1 Rtl per ticket).

1785: Established a system of regular meteorological observations (pension, 1785?–1810: 200 Rtl).

1786: Lectured Berlin's dyers on the chemistry of dyeing (honorarium: 200 Tl from the Academy; 200 Tl from the General Factory Department).

1786–1788?: Attempted to develop a commercially viable process for producing sugar from beets on his estate at Caulsdorff near Berlin. A fire brought his efforts to a halt.

1788: Indebted to numerous creditors who, by threatening arrest, managed to attach his salary from the Academy.

1795: Demonstrated his optical telegraph to Frederick William II, sending several messages, including "The King is loved by his subjects just as much as he is feared by his enemies." Frederick William rewarded him with a gift of 500 Rtl.

1795?: Visited the sugar refinery in Königsaal, Bohemia, in preparation for another round of experiments on sugar beets.

1796–1800: Developed a process for producing sugar from beets on his estate at Bucholz near Berlin. In early 1799, having secured a testimonial from *Klaproth*, he appealed for royal assistance. Frederick William immediately promised him a monopoly, a royal salary, and an estate in southern Prussia worth 100,000 Rtl for further work. But skeptical advisors soon arranged for a royal commission to investigate the profitability of Achard's process. The commission, which included *Klaproth* and *Gerhard* carried out further tests with beets from Halberstadt and considered the value of various by-products. The results were sufficiently impressive to sustain the Prussian crown's support. In late 1799 Frederick William granted Achard a salary increase of 1,500 Tl to develop large-scale production. Soon afterwards arrangements were made for him to take a leave from the Academy and yet retain 500 Rtl of his Academy salary. The monarch also agreed to help with the purchase of an estate where the project could be undertaken.

1801–1815: Struggled, ultimately unsuccessfully, to establish profitable production of sugar from beets. In the spring of 1801, he purchased an estate at Kunern, using a grant of 12,000 Tl plus an interest-free loan of 50,000 Tl to meet the cost of 46,000 Tl and to clear old debts of 16,000 Tl. In the spring of 1802, he moved from Berlin to Kunern in order to supervise his factory which began operation that year. However, the operating costs were so high that he soon exhausted supplemental grants of 9,000 Tl and an inheritance of 30,000 Tl. With Prussia's defeat in 1806, his salaries and pensions were apparently reduced from 2,700 Rtl to 700 Rtl. Around the same time, his factory at Kunern was destroyed by fire. For a while, he got along by helping Baron Koppy establish a works at Krayn near Strehlen. But in 1808/09 his situation became so desperate that he explored the possibility of migrating to Russia. In 1810, in response to Russian expressions of interest, the Prussian government reduced his debt of 50,000 Tl to 20,000 and converted his remaining salaries and pensions into a lump-sum payment. In return, he promised to establish a factory-school for those wishing to learn his techniques. He opened the school in 1811, charging tuition of 100 Rtl. He had 38 students during the next three years, then closed the school for want of students. Around 1815 he was forced to close his factory because of poor health.

1816–1821: In retirement at Kunern.

**State Honors** None. **Major Scientific Societies** 1774: Berlin NF. 1776: Berlin. 1778: Erfurt. 1778: Leopoldina. 1778: Munich. 1782: Stockholm. **Informal Recognition** 1780: *Gmelin*. 1784: *Suckow*. 1785: *Crell*. 1786: Beseke. 1787: *Gren*. 1790: *Leonhardi*. 1791: Denina. 1791: *Göttling*. 1791: Nicolai. 1800: Anon. 1800: *Scherer*. 1821: *Pfaff*.

**Bibliography**
ALZ: IB (25 April 1795), 371–372.
Baxa, J. "Franz Carl Achard." *Zucker*, 6 (1953), 109–114.
DSB, I: 44–45.
Eggebrecht, H. "Ueber Achards Reise nach Halberstadt im Januar 1799." *Centralblatt für die Zuckerindustrie*, 36 (1928), 475, 505–507.
Grottkass, R. E. *Franz Carl Achards Beziehungen zum Auslande: Seine Anhänger und Gegner*. Magdeburg: Baensch, 1930.
König, pt. 2: 269–270.
Lippmann, E. O. von. *Beiträge zur Geschichte der Naturwissenschaften und der Technik*. Vol. I. Berlin: Springer, 1923, pp. 266–275; Vol. II. Weinheim: Verlag Chemie, 1952, pp. 298–309.
———. *Geschichte der Rübe (Beta) als Kulturpflanze von den ältesten Zeiten an bis zum Erscheinen von Achard's Hauptwerk (1809)*. Berlin: Springer, 1925, pp. 110–111, 129, 134.
NDB, I: 27–28.
Rümpler, A. "Archivalische Studien über die Anfänge der Rübenzuckerindustrie in Schlesien." *Deutsche Zuckerindustrie*, 27 (1902), 1638–1642, 1711–1714, 1755–1757, 1803–1807, 1880–1883; 28 (1903), 63–67, 113–114, 121–123, 323–326, 413–416, 519–523.
———. "Die Rübenzuckerindustrie in Schlesien vor hundert Jahren." *Deutsche Zuckerindustrie*, 26 (1901), 1693–1697, 1735–1739, 1773–1777, 1805–1809, 1846–1849, 1888–1893, 1917–1921, 1959–1962.
St. Petersburg Correspondence, 465–466.
Speter, M. "Achard." *Centralblatt für die Zuckerindustrie*, 46 (1938), 169–172, 198–200, 218–220, 493–502.
———. "Bibliographie von Zeitschriften-, Zeitungs-, Bücher-, Broschüren- u. dgl. Veröffentlichungen Franz Carl Achards." *Deutsche Zuckerindustrie*, 63 (1938), 69–74, 152–154, 315–318, 407–409, 592–593.
Stieda, W. "Franz Karl Achard und die Frühzeit der deutschen Zuckerindustrie." Sächsische Akademie der Wissenschaften: Philologisch-Historische Klasse: *Abhandlungen*, 39, no. 3 (1928).
Ulrich, K. "F. C. Achard als Aerostatiker." *Zeitschrift für die Zuckerindustrie*, 80 (1955), 387–389.
———. "Zur Geschichte der Zuckerfabrik Cunern." *Centralblatt für die Zuckerindustrie*, 43 (1935), 463–467.
Vobly, K. "Neue Mitteilungen über die Begründer der Rübenzuckerindustrie in UdSSR." Verein der Deutschen Zuckerindustrie Berlin: *Zeitschrift*, 79 (1929), 696–709.
Volta, A. *Epistolario*. Vol II. Bologna: Zanichelli, 1951, pp. 306–307, 315, 490.
Welsch, F. "Die Anfänge der Rübenzuckerindustrie in Deutschland." *NTM*, 2, no. 6 (1965), 66–81.
Also see Klaproth (Dann, pp. 41–42, 44), Marggraf (Crell, p. 192; Grottkass), Berlin Academy (Harnack, I: 382, 386–387, 545, 645; Kirsten; Müller, pp. 64–65, 130–131), and Berlin Mining School (Krusch, pp. xvii–xix).

**GÖTTLING, Johann Friedrich August**
1753, Derenburg near Halberstadt / 1809, Jena

**Background** Son of Johann Friedrich Göttling (d. 1758, Derenburg), parson in Derenburg from 1746. Stepson of Wiegand, parson in Grüningen. Lutheran. **Education** After instruction from his stepfather, he became an apprentice in *J. C. Wiegleb*'s apothecary shop in Langensalza around 1767. He served there until 1775, benefiting from *Wiegleb*'s instruction in the day and secretly copying his master's manuscripts at night. Then he went as a journeyman to Prätorius's shop in Neustadt an der Orla. However, the situation there was so bad that he soon left, returning by way of Weimar, where he met *W. H. S. Bucholtz*, to Langensalza.

**Career**
1775: Overseer of the laboratory in Reisig's apothecary shop in Langensalza.
1775–1785: Administrator of *Bucholtz*'s Hof-Apotheke in Weimar. Among the apprentices who served under him was *J. B. Trommsdorff* (1784–1785).
1780–1803: Published the *Almanach, oder Taschenbuch für Scheidekünstler und Apotheker*, an annual journal which, though not intended for the "chemist by profession," carried many original articles on topics related to chemistry.
1785–1789: Ducal trainee for a new chair of chemistry at Jena University. In 1785 *Bucholtz* and *J. W. Goethe* persuaded Duke Carl August of Saxe-Weimar to improve chemical instruction at Jena University by appointing Göttling, after appropriate formal education, to teach the subject there. Assisted by an annual stipend of 250 Rtl, Göttling studied at Göttingen University for two years, enjoying close contacts with *G. C. Lichtenberg* and *J. F. Gmelin*. Then he journeyed at ducal expense to England, where he met *J. Priestley*. In 1788 he returned to Weimar by way of Holland where he met *A. P. Troostwyck* and other Dutch chemists. The following year he was awarded his Ph.D. at Jena University in recognition of his prior publications.
1789–1809: AOP, then in 1799 Honorary Professor, and finally in 1809 OP of chemistry and technology in Jena University's Philosophical Faculty (salary, 1789: 300 Rtl; 1809: 350 Rtl, 10 bushels of wheat, 16 bushels of barley). Among his students were *A. N. Scherer* (1790–1794; colleague: 1794–1797) and *J. W. Ritter* (1796–1798; colleague: 1803–1804). His salary was so small that he had to engage in various side activities. He began selling "chemical assaying kits" in 1789 and an improved white lead in 1796. He ran a private chemical-pharmaceutical institute from 1794 to 1809; but it was never well attended. Starting in 1799 he tried, with the encouragement and support of Goethe, to develop a profitable process for extracting sugar from beets.
1792: Visited *Trommsdorff* in Erfurt.

**State Honors** None. **Major Scientific Societies** 1787: Erfurt. 1789: Leopoldina. **Informal Recognition** 1784: Scherf. 1786: Beseke. 1787: *Gren*. 1788: *Westrumb*. 1789: *Gmelin*. 1793: *Wurzer*. 1794: Baldinger. 1796: Fourcroy. 1796: *Girtanner*. 1796: *Link*. 1796: S. in M. 1798: *Bucholz*. 1800: Anon. 1811: Sertürner. 1818: *Trommsdorff*.

**Bibliography**
ALZ: IB (25 April 1789), 481–483; (2 October 1790), 1019–1021.
DAB, I: 216–218.
Gädicke, J. C. *Fabriken- und Manufacturen- Address-Lexicon von Teutschland*. 2 vols. Weimar: Industrie-Comptoir, 1798–1799, I; 29; II: 245.
Goethe & Carl August, 42, 55–56, 161, 163, 213–214, 373.
Goethe & Voigt, 53: 75, 104, 184, 232, 523; 54: 86, 148, 397–398.
Göttling, J. F. A., to W. H. S. Bucholtz. Letter relating his impressions of Holland (March 1788). *Teutsche Merkur*, no. 2 (1788), 473–484.
Möller, R. "Chemiker und Pharmazeut der Goethezeit." *Pharmazie*, 17 (1962), 624–634.
NDB, VI: 580–581.
Pohl, 34–37.
Schiff, J. "Johann Friedrich August Göttlings Briefe an Goethe." Goethe-Gesellschaft, Weimar: *Jahrbuch*, 14 (1928), 130–146.
Schneider, W. "Der Apotheker Göttling." *Die pharmazeutische Industrie*, 17 (1955), 28–31.
Stein, R. "Goethes Weinflaschen-Ausblühung und Göttlings Probierkabinett." *Archiv für die Geschichte der Naturwissenschaften und der Technik*, 8 (1928), 187–205.
Also see Erfurt University ("Matrikel," 9: 97) and Jena University (Chemnitius, pp. 24–27, 80–91; Gutbier, pp. 6, 9–10, 33–37).

**GIRTANNER, Christoph**
1760, Sankt Gallen / 1800, Göttingen

**Background** Son of Hironymus Girtanner (1730, Sankt Gallen–1773, Sankt Gallen), a wealthy merchant-banker in Sankt Gallen. Calvinist. **Education** After attending U. von. Salis's Philanthropic Institute in Marschlin, he went to Lausanne University where he studied botany and chemistry. In 1779 he enrolled at Strasbourg University for medicine. While there, he took chemistry from *J. R. Spielmann*. In 1780 he proceeded to Göttingen University where he pursued medicine and the natural sciences. In 1782 he earned an M.D. with a chemical-mineralogical dissertation dedicated to *J. F. Gmelin* and G. C. Lichtenberg.

**Career**
1782–1785: Practiced medicine and pursued natural history at Sankt Gallen.
1785–1787: Toured Europe. He started with the Alps and southern France, meeting J. A. Chaptal in Montpellier. He proceeded to Paris where he attended medical and scientific lectures and became acquainted with B. G. Sage and J. C. de la Métherie. He then went on to London and Edinburgh, where he evidently became F. X. Schwediauer's financial partner in a firm producing salt and other chemicals. He returned to Göttingen by way of London and Holland in the summer of 1787.
1787–1788: Free-lance author in Göttingen.
1788–1789: Traveled through Holland, Britain, and France, returning to Göttingen shortly after the French Revolution began.
1789–1790: Free-lance author in Göttingen.
Spring–Summer 1790: Traveled to Scotland on business. While there, he presented his new theory of irritability to the Royal Society of Edinburgh. He then visited London, where he and A. von Humboldt discussed chemistry and physiology. He returned to Germany by way of Paris where he observed political developments and began translating Lavoisier's chemical terminology into German with J. F. Jacquin.
1790–1800: Free-lance author in Göttingen, writing chemical, medical, scientific, and anti-revolutionary books. In 1799 he visited Sankt Gallen.

**State Honors** 1793: Saxe-Coburg Court Councillor. **Major Scientific Societies** 1786: Göttingen. **Informal Recognition** 1791: *Crell*. 1791: *Göttling*. 1793: *Richter*. 1794: Baldinger. 1796: Fourcroy. 1798: Frank. 1800: *Scherer*.

**Bibliography**
DSB, V: 411.
Forster, XVI: 151–152, 185.
*Historisch-Biographisches Lexikon der Schweiz*. Vol. III. Neuenburg: Attinger, 1926, pp. 529–530.
Humboldt, 75, 96, 167–168, 236–237.
Lichtenberg, I: 377; II: 231–233, 264–266; III: 29–30, 42, 47–49, 94.
NDB, VI: 411–412.
Scherer, A. N. "Christoph Girtanner." *Allgemeines Journal der Chemie*, 4 (1800), 535; 6 (1800), 79–98, 352.
Schlichtegroll, vol. I for 1800, pp. 116–131.
Wegelin, C. "Dr. Med. Christoph Girtanner (1760–1800)." *Gesnerus*, 14 (1957), 141–168.
Also see Strasbourg University (*Matrikel*, I: 140; II: 112).

**GREN, Friedrich Albrecht Carl**
1760, Bernburg / 1798, Halle

**Background** Son of Johann Magnus Gren (1715?, Jönköping, Sweden–1773, Bernburg), a traveling salesman who settled in Bernburg and became a hatter. Lutheran. **Education** He attended Bernburg's Latin school until his father's death diverted him from the clergy to pharmacy. He served his apprenticeship in the local Grüne Apotheke, which was owned by *R. Vogel*'s student Dr. J. G. F. Schulze. In his spare time he read C. Linnaeus's botany text and J. C. P. Erxleben's chemistry text. In 1779 he went to C. G. Wunderlich's shop in Offenbach as a journeyman. However, finding this shop's large-scale production of nitric acid bad for his health, he soon proceeded to Erfurt. There he served in the Schwan-Apotheke, which was owned by W. B. Trommsdorff, OP of chemistry and botany at Erfurt University. He soon won the confidence of his new master, who instructed him in chemistry, gave him free access to the shop's laboratory, and entrusted him with introducing young *J. B. Trommsdorff* to botany. Upon the senior Trommsdorff's death in 1782, Gren returned to Bernburg in hopes of setting up a chemical factory. When this scheme foundered, he readily accepted a scholarship to Helmstedt University that had been arranged by *L. Crell*. He studied medicine there for a year, taking chemistry from G. Bereis and *Crell* and editing for *Crell*. Then he went to Halle University where, besides continuing his medical studies, he tutored the physicist W. Karsten in chemistry. In spring 1784 he visited Weimar in order to witness W. H. S. Bucholtz's launching of a large hot-air balloon. A year later he was supporting himself by giving private chemistry courses (with the Medical Faculty's permission!) and manufacturing sal ammoniac and other salts. Finally in 1786 he earned his M.D. with a dissertation on the formation of fixed and phlogisticated air.

**Career**
1786–1798: PD, AOP, then OP at Halle University.
   1786: PD in the Medical Faculty.
   1787: AOP in the Medical Faculty and, after taking his Ph.D., the Philosophical Faculty. In the latter faculty, he was responsible for physics.
   1788–1798: OP of physics in the Philosophical Faculty for a few months, then of physics and chemistry in the Medical Faculty (salary, 1788: 150 Rtl; 1798: 300 Rtl) (budget, 1791: 403 Tl, including 200 Tl for a free course on physics). His reason for transferring to the Medical Faculty may have been a desire to share in fees from M.D. candidates. He was curator of the University's natural history collection until 1793 (stipend, 1791: 11 Tl).
1789: Declined a call to Rostock to be OP of medicine (salary: 500 Rtl) and Town Physician.
1790–1798: Edited the *Journal der Physik*, which covered the "mathematical and chemical parts of natural philosophy."
1791: Visited Mecklenburg.
1798: Declined a call to Freiberg where the Saxon authorities wanted him to establish his new process for removing silver from copper ores after successful tests at Mansfeld. On the basis of this offer, he requested a salary increase of 200 Rtl. The government was so slow to respond that he wrote M. van Marum in Holland inquiring about a position. His salary was eventually increased—about the same day as he was being interred in Halle.

**State Honors** None.   **Major Scientific Societies** 1791: Erfurt. 1792: Berlin. 1796: Berlin NF.   **Informal Recognition** 1788: *Leonhardi.* 1789: *Gmelin.* 1791: *Crell.* 1791: *Göttling.* 1792: *Klaproth.* 1793: *Richter.* 1793: Scherer. 1793: *Wurzer.* 1794: Baldinger. 1796: Fourcroy. 1796: *Girtanner.* 1796: *Link.* 1796: S. in M. 1797: Freund. 1798: *Bucholz.* 1798: Frank. 1798: *Scherer.* 1802: Salzmann. 1806: *Trommsdorff.* 1807: Kastner. 1811: Sertürner.

**Bibliography**
ALZ: IB (28 February 1789), 225; (25 February 1792), 202; (31 October 1798), 1268–1269.
DAB, I: 223.
DSB, V: 531–533.
Elwert, 171–185.
Gren, F. A. C., to M. van Marum. Six letters (1791–1798). Van Marum Collection, Part I: The Letters (Hollandische Maatschappij der Wettenschappen), microfiche ed. Micro-Methods Limited, East Ardsley, Wakefield, England. Library, American Philosophical Society, Philadelphia.
Kaiser, W. "Friedrich Albert Karl Gren (1760 bis 1798)." *Zahn-, Mund- und Kieferheilkunde*, 64 (1976), 601–610.
Karsten, D. L. G. "Kurze Nachrichten von dem Leben des Professors Gren zu Halle." Gesellschaft naturforschender Freunde zu Berlin: *Neue Schriften*, 2 (1799), 404–413.
NDB, VII: 45–46.
Partington, III: 575–577, 620–625, 632–636.
Scherer, A. N. "Friedrich Albrecht Carl Gren." *Allgemeines Journal der Chemie*, 2 (1799), 357–416, 615–618.
Trommsdorff, J. B. "Kurze Biographie des verewigten Friedrich Albrecht Carl Gren." *Journal der Pharmacie*, 6 (1799), 367–375.
Also see Halle University (Conrad, p. 51; Kaiser and Krosch, 12: 160–176; Kaiser and Krosch, "Statuten," p. 78).

### HERMBSTAEDT, Sigismund Friedrich
1760, Erfurt / 1833, Berlin

**Background** Son of Hieronymus Friedrich Hermbstaedt (1728, Erfurt–1788, Erfurt) who enrolled at Erfurt University in 1745 and later became an actuary and magistrate in Erfurt. Lutheran.   **Education** After instruction from private tutors and at St. Michael's School in Erfurt, he entered the Town Council Gymnasium in 1773. Two years later he began attending philosophical and medical lectures at Erfurt University. Attracted to chemistry by W. B. Trommsdorff, he became an assistant in J. C. Wiegleb's new chemical boarding school in Langensalza around 1779. After a year or so, he proceeded to Hamburg where he continued his training in the Rats-Apotheke under J. A. H. Reimarus. In late 1783 he went to Berlin where, soon after arriving, he was selected by *M. H. Klaproth* to administer the Rose family's Apotheke zum Weissen Schwan. In his spare time he attended lectures at the Medical-Surgical College and established useful connections with, among others, Royal Physicians C. A. Cothenius and C. G. Selle. Perhaps encouraged by these men, he gave up his pharmaceutical career in 1786 and traveled through Saxony, Thuringia, and Hanover, inspecting mines and factories and visiting *C. E. Gellert* in Freiberg, *W. H. S. Bucholtz* in Weimar, *Wiegleb* in Langensalza, and *J. F. Gmelin* in Göttingen.

**Career**

1786–1791?: Technical chemist for J. G. Wegely, an entrepreneur who owned several textile and chemical factories in Berlin.

1787–1791: Private lecturer in Berlin. He gave courses on physics, chemistry, technology, and pharmacy. In 1788 he sought appointment as OP of chemistry and pharmacy in Berlin's Medical-Surgical College, hoping to replace the chair's occupant, C. H. Pein, who was widely regarded as incompetent. However, the commission that examined his qualifications found his knowledge of materia medica inadequate.

1787–1795: Edited the *Bibliothek der neuesten physisch-chemischen, metallurgischen, technologischen und pharmaceutischen Literatur*, a reviewing journal for chemistry and related fields.

1789–1797: Established and directed a "chemical boarding school" in Berlin.

1790–1797: Court Apothecary in Berlin.

1792–1810: OP of chemistry and pharmacy in the Berlin Medical-Surgical College (salary: 300 Rtl).

1794–1833: Prussian medical official. Among his colleagues in Berlin were *Klaproth* (1794–1817), his brother-in-law *V. Rose, Jr.* (1797–1807), and *H. F. Link* (1815–1833).

1794–1809: Senior Councillor on the Higher Sanitation Board (salary, 1798?: 150 Tl), which in 1799 was merged into the Higher Medical and Sanitation Board, which in 1809 was reconstituted as the the Scientific Deputation for Medical Affairs.

1797–1807?: Court Apothecary Shop Commissioner.

1810–1833: Privy Senior Medical Councillor in the Scientific Deputation for Medical Affairs.

1795–1833: Professor of physics, chemistry, and pharmacy in the new Surgical Pépinière, and from 1804 of chemistry and pharmacy in its successor the Medical-Surgical Pépinière, which in 1811 became the Medical-Surgical Academy for the Military (salary, 1795–1797: none; 1797: 150 Tl; 1811: 400 Tl). He had access to the laboratory and supplies of the Court Apothecary Shop for instructional purposes. Among his colleagues was *Link* (1816–1833).

1796–1833: Member of the Factory and Salt Department of the Manufacturing and Commerce Board, then from 1808 of the Technical Deputation in the Ministry of Industry and Trade (salary, 1797: 1,200 Rtl; 1831: 1,650 Rtl plus free residence). His primary responsibility was evaluating innovations in chemical technology. From 1801 he also taught dyeing and other branches of technical chemistry to artisans and aspiring entrepreneurs. In 1802 a laboratory and residence were constructed for him at a cost of 20,000 Rtl.

1797–1804: Teacher of physics in Berlin's Mining School (salary: 100 Rtl). Among his colleagues was *Klaproth*.

1800–1833: Extraordinary, then from 1808 Ordinary Member of the Berlin Academy (salary, 1800–1811: none; 1812–1831: 200 Rtl). Among his colleagues were *C. A. Gerhard* (1800–1821), *Klaproth* (1800–1817), *Link* (1815–1833), E. Mitscherlich (1822–1833), and H. Rose (1832–1833).

1806–1833: Owned a large summer residence at Pankow (cost: 6,500 Rtl). In 1812 he enlarged it by adding a 20-acre botanical garden.

1810–1833: AOP of technical chemistry, then from 1811 OP of chemistry and technology in Berlin University's Philosophical Faculty (salary, 1810: 600 Tl; 1833: 700 Tl). Among his colleagues were *Klaproth* (1810–1817), *Link* (1815–1833), Mitscherlich (1821–1833), H. Rose (1822–1833), and G. Magnus (1831–1833).

1817–1833: Teacher of chemistry and from 1831 physics in Berlin's Mining School (salary, 1817: 200 Tl plus 21 Tl for fuel).

1820–1833: Teacher of chemistry in Berlin's General War School (salary, 1833: 500 Rtl?).

1827: Attended the meeting of the Society of German Scientists and Physicians in Munich.

**State Honors** 1804: Prussian Privy Councillor. 1814: Knight of the Prussian Red Noble Order (1814: Third Class; 1833: Third Class with Bow). 1817: Knight of the Belgian Lion Order. **Major Scientific Societies** 1786: Erfurt. 1786: Leopoldina. 1799: Berlin NF. 1800:

Berlin. 1811: St. Petersburg. 1812: Stockholm. 1832: Munich. **Informal Recognition** 1787: *Gren.* 1788: *Westrumb.* 1789: *Gmelin.* 1791: *Crell.* 1791: *Göttling.* 1793: *Richter.* 1793: *Wurzer.* 1794: *Baldinger.* 1796: *Fourcroy.* 1796: *Girtanner.* 1796: S. in M. 1797: *Freund.* 1797: *Mann.* 1798: *Bucholz.* 1798: *Frank.* 1800: Anon. 1803: Ed. Board. 1806: *Trommsdorff.* 1811: *Sertürner.*

**Bibliography**
Anft, B., "Friedlieb Ferdinand Runge, sein Leben und sein Werk." *Abhandlungen zur Geschichte der Medizin und der Naturwissenschaften*, 23 (1937), 15, 156, 160–164.
*Berlinisches Jahrbuch für die Pharmacie*, 3 (1797), 233; 4 (1798), 293–299.
Berzelius, 15, 212–213, 220.
Biereye, 43.
Borchardt, A. "Die Entwicklung der Pflanzenanalyse zur Zeit Hermbstaedts." Braunschweig Technical University: Pharmaziegeschichtliches Seminar: *Veröffentlichungen*, 13 (1974).
Callisen, XXVIII: 498–502.
DAB, I: 266–267.
DSB, XV: 205–207.
Hitzig, J. E. *Verzeichnis im Jahre 1825 in Berlin lebender Schriftsteller.* Berlin: Dümmler, 1826, pp. 313–322.
Mieck, I. "Sigismund Friedrich Hermbstaedt (1760 bis 1833): Chemiker und Technologe in Berlin." *Technik-Geschichte*, 32 (1965), 325–382.
Müller-Jahncke, W.-D. "Die Bildnisse auf Sigismund Friedrich Hermbstaedt." *Deutsche Apotheker Zeitung*, 120 (1980), 367–369.
NDB, VIII: 666–667.
*Neuer Nekrolog der Deutschen*, 11 (1833), 704–707.
Pohl, 30–33.
Stürzbecher, M. "Die Vorlesungsankündigen von Sigismund Friedrich Hermbstaedt am Collegium medico-chirurgicum und der Universität Berlin." In *Pharmazie und Geschichte: Festschrift für Günter Kallinich zum 65. Geburtstag*, edited by W. Dressendörfer et al. Straubing and Munich: Donau, 1978, pp. 189–202.
Also see Klaproth (Dann, pp. 39–40, 145), Berlin Academy (Kirsten), Berlin Medical-Surgical College (Lehmann, pp. 61–62, 64, 72–75), Berlin Mining School (Krusch, pp. xxiv, xxvi, xxviii–xxix, xxxiii, xxxviii, xl), Berlin Surgical Pépinière (Preuss, pp. 164–165; Schickert, pp. 41, 43, 83, 88, 91), and Erfurt University ("Matrikel," 9: 120).

**RICHTER, Jeremias Benjamin**
1762, Hirschberg, Silesia / 1807, Berlin

**Background** Son of Carl Friedrich Richter, a merchant in Hirschberg. Lutheran. **Education** As a boy of twelve, he went to Breslau to live with his paternal uncle, the town's construction master. Three years later he became a trainee in the Prussian Engineering Corps. He evidently found the military regimen too limiting and, after fifteen requests, finally secured his discharge in 1783. He then enrolled in Königsberg University. While attending mathematical and philosophical lectures, he taught himself chemistry by studying P. Macquer's dictionary and *K. G. Hagen*'s text. In 1789 he earned his Ph.D. and qualified for lecturing by writing and defending a dissertation on the uses of mathematics in chemistry.

**Career**
1789–1790?: PD at Königsberg University.
1790?–1794?: Surveyor and technical chemist at von Lestwitz's estate near Glogau, Silesia. He pursued pure chemistry in his spare time. In 1793, after it became clear that he was not in line for a chair at the Potsdam Engineering Academy, he petitioned the authorities in Berlin for a teaching post. In 1794, abandoning his search for an academic position, he approached F. W. von Reden for a job in the Silesian mining service. Although he was affronted by the terms offered—one-year probation at a salary of 300 Tl—, he accepted the position when granted the right to wear the service's uniform and free living quarters.
1795–1797: Mining Secretary in the Silesian Higher Mining Bureau in Breslau (salary: 1795–1796: 300 Tl plus free residence; 1797: 400 Tl). He was responsible for routine assaying, special investigations in the laboratory, and mathematical calculations.
1798–1807: Second Arcanist in the Royal Porcelain Works and from 1801 Assessor with vote on the Prussian Mining and Smelting Board in Berlin (salary: 1798–1801: 750 Rtl; 1801–1807: 1,000 Tl). Among his colleagues was *C. A. Gerhard*. In 1800, in order to qualify for

active participation in the Mining and Smelting Board, he visited the Prussian iron, bronze, copper, and lime works in Kurmark, Neumark, and Pomerania and prepared a report on the mechanical apparatus at these works. As Assessor he was responsible for reporting his formulae for porcelain dyes and for instructing young employees in his techniques.

1803–1807: Responsible for the copper works at Saigerhall (salary: 100 Rtl plus 20 Rtl in perquisites, 11 Rtl 6 Gr for lighting, and 6 Pf. for every hundred-weight of copper produced.

**State Honors** 1797: Prussian Assessor (honorary until 1801). **Major Scientific Societies** 1796: Munich. 1796: Göttingen. 1800: St. Petersburg (corresponding member). **Informal Recognition** 1796: Fourcroy. 1796: *Link*. 1796: Van Mons. 1797: Freund. 1798: *Bucholz*. 1798: Frank. 1800: Anon. 1803: Ed. Board. 1804: *Gmelin*. 1806: *Trommsdorff*. 1807: *Gehlen*. 1821: *Pfaff*. 1822: *Lampadius*.

**Bibliography**
DSB, XI, 434–438.
"Dienstcurriculum des Bergassessors Dr. Benjamin Jeremias Richter" (ca. 1806). Staatsbibliothek Preussischer Kulturbesitz, West Berlin (Darmst. G 1* 1792).
Löwig, K. J. *Jeremias Benjamin Richter, der Entdecker der chemischen Proportionen*. Breslau: Morgenstern, 1875.
Also see Königsberg University (*Matrikel*, II, 579).

**ROSE, Valentin Jr.**
1762, Berlin / 1807, Berlin

**Background** Son of Valentin Rose (1736, Neuruppin–1771, Berlin), who served as a journeyman pharmacist with H. Hagen in Königsberg, studied with *J. H. Pott* and *A. S. Marggraf* in Berlin, visited Holland, took over Berlin's Apotheke zum weissen Schwan in 1761, and pursued chemistry avidly for the rest of his short life. Lutheran. **Education** After instruction from his father and then his guardian, *M. H. Klaproth*, he became an apprentice in P. Salzwedel's Schwanen-Apotheke in Frankfurt a. M. in 1778 (his apprentice premium was 100 Rtl, half paid at the beginning, half paid two years later). Upon completing his apprenticeship in 1782, he served as a journeyman in the family shop in Berlin. He also took several courses at the Medical-Surgical College. From 1783 to 1785 he served as a journeyman in *J. C. F. Meyer*'s Hof-Apotheke in Stettin and *K. G. Hagen*'s in Königsberg. Then he returned home to take control of the family shop which was being administered by *S. F. Hermbstaedt*, who soon became his brother-in-law.

**Career**
1785–1807: Administrator, then from 1790 full owner of Berlin's Apotheke zum weissen Schwan (in 1791 the shop was valued at 12,000 Rtl). In 1792 he was licensed as a first-class apothecary after successfully analyzing three drugs for the state board, which included *Klaproth*. Among his journeymen were *A. F. Gehlen* (1800?–1805).
1797–1807: Assessor on the Higher Medical Board, which in 1799 became the Higher Medical and Sanitation Board. He also belonged to the Court Apothecary Shop Commission. Among his colleagues were *Klaproth* and *Hermbstaedt*.
1807: Visited *J. B. Trommsdorff* and *C. F. Bucholz* in Erfurt.

**State Honors** None. **Major Scientific Societies** 1804: Berlin NF. **Informal Recognition** 1804: *Gmelin*. 1804: Kastner. 1808: *Gehlen*. 1809: *Bucholz*. 1811: Sertürner. 1818: *Trommsdorff*. 1821: *Pfaff*.

**Bibliography**
*Berlinisches Jahrbuch für die Pharmacie*, 3 (1797), 231; 16 (1815), 334–337.
Bucholz, C. F. Biography. *Taschenbuch für Scheidekünstler und Apotheker*, 30 (1809), 249–256.
DAB, II: 540–542, 553.
Dann, G. E. "Deutsche Apothekerfamilien: Die Familie Rose." *Pharmazeutische Zeitung*, 71 (1926), 629–632.
*Journal für die Chemie, Physik und Mineralogie*, 5 (1808), 291–302.
Justi, 566.
Trommsdorff, J. B. Death notice. *Journal der Pharmacie*, 16, no. 2 (1808), 468.
Also see Hermbstaedt (Borchardt, pp. 10–11), Klaproth (Dann, pp. 57, 144, 149), and Berlin Medical-Surgical College ("Matrikel," 11: 141; 12: 112).

**HILDEBRANDT, (Georg) Friedrich**
1764, Hanover / 1816, Erlangen
**Background** Son of Johann Georg Hildebrandt (?, Hanover–1768, Hanover), who enrolled in Berlin's Medical-Surgical College in 1748, became Hanover's Town Surgeon in 1759, and ended his career as Electoral Court and Crown Surgeon. Lutheran. **Education** His early instruction was from tutors and family friends, including Dr. Mensching who taught him English and natural history. At the age of twelve he entered Hanover's gymnasium where he was an outstanding student. In 1780, though he would have preferred to enter the military, he proceeded to Göttingen University to study medicine. Intending to become a professor rather than a doctor, he focused on anatomy, physiology, and chemistry, which he studied with *J. F. Gmelin*. In 1783 he earned his doctorate in medicine and surgery with an anatomical dissertation. He then embarked on a study tour, visiting mines, factories, and hospitals, and, in Paris and Berlin, studying anatomy. He returned to Göttingen in 1785.

**Career**
1785: PD at Göttingen University. Among his colleagues was *Gmelin*.
1785–1793: OP of anatomy, physiology, and from 1792 chemistry in Brunswick's Anatomical-Surgical College and Assessor on the Duchy of Brunswick's Higher Sanitation Board.
1789–1793: PD for chemistry in Brunswick's Collegium Carolinum. His many requests to be made an OP of chemistry in this school were all denied.
1793–1816: OP at Erlangen University. His appointment was arranged by C. A. von Hardenberg, the new Prussian Minister in Bayreuth. In 1794 he received an offer to return to Brunswick as OP of chemistry, Court Councillor, and Ducal Physician. Hardenberg, who refused to accept his resignation, secured both a salary increase and the title of Court Councillor for him. Among his students was the chemist C. G. C. Bischof (1810–1814; colleague: 1814–1816).
1793–1796: OP of medicine and chemistry in the Medical Faculty.
1796–1816: OP of medicine in the Medical Faculty and of chemistry and from 1799 physics in the Philosophical Faculty. He was shifted to the Philosophical Faculty for chemistry so that he could teach "chemical technology." From 1799 he had a free residence adjacent to the physics laboratory.
1803?: Declined a call to Charkow University in Russia.
1803–1804: Negotiated with Heidelberg University. In response to a call to the chair of physiology, he indicated he would only be interested if he could have chemistry and physics as well. Though this condition was met, he was persuaded to stay in Erlangen with a salary increase of 800 fl and the title of Privy Court Councillor.
1812: Declined a call to the chair of chemistry at Heidelberg University.
1814: Wrote C. E. Moll of the Munich Academy, complaining that his three lectures a day and pressing medical practice had prevented him from proving his worthiness of membership.
1815: Declined to be considered as a candidate to succeed *A. F. Gehlen* in the Munich Academy.

**State Honors** 1794: Prussian Court Councillor. 1804: Prussian Privy Court Councillor. **Major Scientific Societies** 1793: Leopoldina. 1793: Göttingen. 1796: Erfurt. 1809: Berlin NF. 1812: Berlin (corresponding member). **Informal Recognition** 1796: Fourcroy. 1796: *Link*. 1796: S. in M. 1798: *Bucholz*. 1804: *Gmelin*. 1806: *Trommsdorff*. 1808: *Gehlen*. 1819: Schweigger.

**Bibliography**
ALZ: IB (20 September 1794), 848; (4 October 1794), 907; (31 August 1796), 1004.
Bischof, C. G. C. "Kurzer Bericht über Hildebrandts Leben." *Journal für Chemie and Physik*, 25 (1819), 1–16.
DSB, VI: 395.
Hohnbaum, C. "Friedrich Hildebrandt." Deutsche Akademie der Naturforscher, Leopoldina: *Verhandlungen*, 1 (1818), 17–54.
Moll, II: 328–329; III: 747–748.
Mundhenke (Stadtarchiv, Hanover). Letter concerning Hildebrandt's father (25 March 1970).
Niedersächsisches Staatsarchiv, Wolfenbüttel (2 Alt 14703, pp. 287–335).
Rotermund, II, supp. ciii–cix.
Also see Berlin Medical-Surgical College ("Matrikel," 11: 137), Brunswick Anatomical-Surgical College (Döhnel; lecture catalogues for 1785–1793), Erlangen University (Gastauer, pp. 116–127; Kaulbars-Sauer, pp. 1–10; Schleebach, pp. 13–15, 82–130), and Heidelberg University (Stübler, pp. 185–186).

**WURZER, (Francis) Ferdinand (Joseph)**
1765, Brühl near Cologne / 1844, Marburg
**Background** Son of Mathias Nicolaus Wurzer (d. 1795, Bonn), who was chief artillery-officer in the regiment stationed in Brühl until the late 1760s, then acting captain, and finally captain in the Elector of Cologne's army in Bonn. Catholic. **Education** After studying in Bonn at the gymnasium and "Maxische Academie" from 1776 to 1783, he enrolled at Heidelberg University to study medicine. During his year there, he heard D. W. Nebel lecture on chemistry. In the fall of 1784 he toured the Palatinate and Alsace on foot, then proceeded to Würzburg University. He studied medicine there for two years, becoming close friends with the chemistry professor G. Pickel. Next he went to Göttingen University where he attended *J. F. Gmelin*'s courses. In 1787 he proceeded to Vienna University in order to round out his medical education with work in the hospitals. While there, he heard *N. J. Jacquin*. Finally in 1788 he returned home and earned his M.D. at Bonn University with a dissertation on brain fever.
**Career**
1789–1791: Practiced medicine in Bonn. Late in 1789, while on an excursion near his family vineyard, he discovered the mineral springs at Godesburg. His published description of the springs was dedicated to Elector Max Franz, who subsequently developed Godesburg into a famous spa.
1791–1793: Designate OP of chemistry in Bonn University's Medical Faculty. He prepared himself for this post at his own cost by studying chemistry with *L. Crell* in Helmstedt (eight months), *J. F. Westrumb* in Hameln (a few weeks), and his friend Pickel who had set up a chemical factory in Würzburg (two months). He also toured the Harz and Saxon mining districts.
1793–1797: OP of chemistry and materia medica in Bonn University's Medical Faculty (salary, 1793: 400 tl; 1794–1797: none). Because of the French occupation, he did not lecture from late 1794 to late 1795. Instead, he was forced to serve as a physician in hospitals established in the former Electoral palaces in Bonn and Poppelsdorf. In 1797 he was among the vast majority of professors who refused to sign an oath of loyalty to the French.
1798–1799: Teacher of experimental physics, chemistry, natural history, and botany in the new Central School in Bonn (salary: 1,200–1,500 Fr—he actually received only 27 Fr). In the spring of 1799, he resigned after learning that the government would not accept his offer to give two-thirds of his salary to additional faculty.
1800–1804: Professor of physics and chemistry in the reorganized Central School in Bonn (salary: 2,500 Fr—he actually received little, if any, salary). In 1801 he wrote a friend in Brussels (probably J. B. van Mons), indicating his desire to have a chair in the new medical school there. Though his first choice was chemistry, he was also willing to take on physiology, materia medica, pathology, or therapy in that order.
1801?–1805: Inspector of the Godesburg mineral springs and member of the Medical Jury for the Rhine and Mosel Departments. In this second capacity, he carried out an intensive inoculation program.
1805–1844: OP of chemistry and pharmacy in Marburg University's Medical Faculty and from 1806 of chemistry in the State Economics Institute (salary, 1800: 400 Rtl; 1806: 600 Rtl; 1813: 900 Rtl; 1820: 1,200 Rtl). Among his students was F. Wöhler (1820–1821) and among his colleagues, R. W. Bunsen (1839–1844).
1809–1839: Director of the University's chemical institute (budget, 1813: 308 Rtl). He relinquished control of the institute to Bunsen in 1839.
1814–1844: Director of Upper Hesse's Medical Deputation. From 1821 he was the chief Medical Advisor for Upper Hesse and from 1839 Privy Senior Medical Councillor for Upper Hesse.
1816: Declined a call to go to Holland as OP of chemistry at Louvain, Ghent, or Lüttich.
1817–1819: Declined a call to Bonn University as OP of chemistry after protracted negotiations in which he finally yielded to pressure from the Hessian Elector.
**State Honors** 1805: Hesse-Cassel Court Councillor. 1819: Knight of the Belgian Lion Order (1832: Commander Second Class). 1825: Hessian Privy Court Councillor. 1840: Knight of the Prussian Red Noble Order (Third Class). **Major Scientific Societies and Honorary Degrees** 1792: Leopoldina. 1801: Erfurt. 1801: Göttingen (corresponding member). 1818: Berlin NF. 1838: Honorary Ph.D. from Marburg University. **Informal Recognition** 1816: *Bucholz*. 1817: Döbereiner. 1817: Schweigger. 1821: *Pfaff*.
**Bibliography**
Callisen, XXXIII: 352.

Diergart, P. "Begründung der Bonner Chemie durch Ferdinand Wurzer." *Sudhoffs Archiv*, 29 (1936), 116–120.

Elwert, 676–679.

Gottbert, H. E. von. "Stammlisten der kurkölnischen Armee im 18. Jahrhundert." Westdeutsche Gesellschaft für Familienkunde: *Mitteilungen*, 1 (1913–1917), 91, 143.

Horn (Personenstandsarchiv, Brühl). Letter containing a copy of Wurzer's baptismal entry (7 January 1970).

Justi, 816–820.

*Kurfürstliches Hessisches Hof- und Staats-Handbuch auf das Jahr 1844*. Kassel: Waisenhaus, 1844, pp. 19, 202, 290, 307, 310, 352.

Riemer (Stadtarchiv und Wissenschaftliche Stadtbibliothek, Bonn). Letter bearing on Wurzer's father (12 May 1970).

Strieder, XVII: 311–321.

Wurzer, F., to ? in Brussels. Letter (1801). Staatsbibliothek Preussischer Kulturbesitz, West Berlin (Darmst. Gl 1789).

"Wurzer's Jubiläum." Apotheker-Verein im nördlichen Teutschland: *Pharmazeutische Zeitung*, 12 (1838), 305–308.

Also see Bonn University (Braubach; "Quellen") and Marburg University (Gundlach, p. 462; Hessisches Staatsarchiv; Meinel, pp. 3–24, 424–426).

**LINK, Heinrich Friedrich**
1767, Hildesheim / 1851, Berlin

**Background** Son of August Heinrich Linck (1738, Wendeburg near Brunswick–1783, Hildesheim), who matriculated at Helmstedt University in 1759, served as a tutor to a wealthy family in Brunswick in the 1760s, resided in Hildesheim from around 1767, and became a pastor there around 1773. Lutheran. **Education** As a youth in Hildesheim, he not only attended the Andreas Gymnasium but also studied the natural sciences with his father, who owned a natural history cabinet, Canon F. von Beroldingen, an amateur chemist and mineralogist, and Dr. J. E. D. Schnecker. At the age of ten, he accompanied his father and Schnecker on a botanical excursion into the Harz. He took two subsequent trips into the Harz with Schnecker, probably meeting Goethe atop the Brocken in September 1784. Two years later he went to Göttingen University to study medicine and the natural sciences. Though his chief mentor there was the naturalist, J. F. Blumenbach, he most likely took chemistry from *J. F. Gmelin*. Among his student friends was A. von Humboldt, with whom he and others formed a private "physical society" in the fall 1789. That same fall he earned his M.D. with a botanical dissertation to which he appended several theses defending the antiphlogistic theory.

**Career**

1790–1792: PD at Göttingen University. Among his colleagues was *Gmelin*. In 1790 he geologized near Goslar. In 1792 he almost became a practicing physician in a southern German town. However, just as he was deciding to leave academia, he received a call to Rostock University.

1792–1811: OP of physics, chemistry, and botany in Rostock University's Philosophical Faculty (budget, 1811: 50 Rtl). Shortly after arriving, he was granted the Ph.D. so that he would be formally qualified for his chair.

1797–1799: Traveling with Count J. C. von Hoffsmannegg through Germany (meeting *A. N. Scherer* in Brunswick), France (meeting A. F. Fourcroy in Paris), and Spain to Portugal, where they spent a year or so botanizing plants.

1799: Declined a call from Halle University to succeed *F. A. C. Gren* as OP of chemistry and physics (salary: 700 Rtl) when his salary at Rostock was increased by 250 Rtl.

1802?: Declined a call to Dorpat University.

1811: Received calls from Breslau University for botany and chemistry, Halle University for chemistry and physics, and the Munich Academy for physics. Rostock University countered these calls by transferring him into the Medical Faculty as OP of natural history and chemistry so that he could share in the fees from M.D. candidates. Nevertheless, he decided to go to Breslau, because his primary responsibility there would be botany.

1811–1815: OP of botany and chemistry in Breslau University's Medical Faculty. He was provided with a laboratory and from 1812 a free residence. Eager to focus on botany, he probably supported the unsuccessful campaign to have an OP appointed solely for chemistry.

1815–1851: OP of botany in Berlin University's Medical Faculty. Among his students was E. Mitscherlich (1818–1819), who worked in his private laboratory. Among his colleagues were

*M. H. Klaproth* (1815–1817), *S. F. Hermbstaedt* (1815–1833), Mitscherlich (1821–1851), H. Rose (1822–1851), and G. Magnus (1831–1851).

1815–1851: Ordinary member of the Berlin Academy (salary: 950 Rtl). Among his colleagues were *C. A. Gerhard* (1815–1821), *Klaproth* (1815–1817), *Hermbstaedt* (1815–1833), Mitscherlich (1822–1851), Rose (1832–1851), and Magnus (1840–1851).

1815–1851: Director of the Royal Botanical Gardens and curator of the Royal Herbarium in Berlin.

1815–1851: Member, then from 1823 Privy Medical Councillor in the Scientific Deputation for Medical Affairs. Among his colleagues were *Klaproth* (1815–1817) and *Hermbstaedt* (1815–1833).

1815–1851: OP of botany in the Medical-Surgical Academy for the Military. Among his colleagues was *Hermbstaedt* (1816–1833).

1823: Journeyed with Mitscherlich to Sweden where they visited J.J. Berzelius.

**State Honors** 1825: Knight of the Prussian Red Noble Order (1825: Third Class; 1833: Third Class with Bow; 1843?: Second Class with Oak Leaf). 1846?: Prussian Order of Merit (Civil Class). **Major Scientific Societies** 1792: Göttingen. 1800: Leopoldina. 1807: Berlin NF. 1808: Munich (corresponding member; 1829: foreign member). 1812: Berlin (corresponding member; 1815: ordinary member). 1828: Paris (corresponding member for botany). 1840: Stockholm (for zoology and botany). **Informal Recognition** 1791: *Crell*. 1804: *Gmelin*. 1815: Schweigger. 1821: *Pfaff*.

**Bibliography**
ADB, XVIII: 714–720.
*Berlinisches Jahrbuch für die Pharmacie*, 7 (1801), 145, 148.
Berzelius, 34.
DSB, VIII: 373–374.
Ebeling, F. (Kirchenbuchamt, Hildesheim). Letter containing information about Link's family and his probable meeting with Goethe in 1784 (26 March 1970).
Humboldt, 73–74.
Lenz, I: 547–549; IV: 115.
Martius, K. F. P. "Denkrede auf Heinrich Friedrich Link." *Münchener gelehrte Anzeigen*, 32 (1851), 474–536, 553–555.
Pr. Hdbh., 1821, pp. 97–98, 121; 1831, p. 71; 1843, p. 132; 1846, p. 144; 1848, pp. 79, 92–93, 102, 119, 120, 146, 149, 150, 160, 228.
Schiff, J. "Das erste chemische Institut der Universität Breslau." *Archiv für die Geschichte der Naturwissenschaften und der Technik*, 9 (1922), 29–38.
*Zwei Gedächtnissreden gehalten bei der Trauerfeier für den Hchw. Gr.-M. Br. Heinrich Friedrich Link am 2. Februar 1851 in der Grossen Loge von Preussen genannt Royal-York zur Freundschaft*. Berlin: Sittenfeld, 1851.
Also see Scherer (Scherer to Fries, 26 June 1797), Berlin Academy (Kirsten), Berlin Surgical Pépinière (Preuss, p. 166), and Rostock University (*Matrikel*, V: 19, 22, 42, 73–74; Ruser, pp. 33–35; Schott, p. 987).

## BUCHOL(T)Z, Christian Friedrich
1770, Eisleben / 1818, Erfurt

**Background** Son of Christian Friedrich Bucholtz (1741, Bernburg–1775, Erfurt), a younger brother of *W. H. S. Bucholtz* and owner of the Mohren-Apotheke in Eisleben, then the Römer-Apotheke in Erfurt. Stepson from 1777 of C. F. Voigt (d. 1818?, Erfurt), a minor chemist who administered the family shop until Christian could take it over. Lutheran. **Education** As a boy, besides going to school in Erfurt, he enjoyed instructive contacts with his uncle *Bucholtz* and *J. F. A. Göttling* in Weimar. In 1784 he became an apprentice in C. W. Fiedler's shop in Kassel. After five years there, he served as a journeyman for two years in Ochsenfurt and for three years in Klauer's shop in Mühlhausen. Finally in 1794 he returned to Erfurt to take charge of the family shop.

**Career**
1794–1818: Owner of Erfurt's Römer-Apotheke.
1809–1816: PD, then from 1810 AOP, and finally from 1813 OP of chemistry in Erfurt University's Philosophical Faculty. Among his colleagues was *J. B. Trommsdorff*.
1810–1813?: Member of Erfurt's Medical and Sanitation Board.
1813: Imprisoned by the French, along with 30 other Erfurters, for refusing to pay his war taxes. He went blind as a result of this ordeal.

1815: On the list of candidates to succeed *A. F. Gehlen* in the Munich Academy, but nothing came of this.

**State Honors** 1816: Schwarzburg-Sondershausen Court Councillor.   **Major Scientific Societies and Honorary Degrees** 1800: Erfurt. 1808: Berlin NF. 1808: Munich (corresponding member). 1808: Honorary Pharm D. from Rinteln University. 1809: Honorary Ph.D. from Erfurt University. 1812: Berlin (corresponding member).   **Informal Recognition** 1804: *Gehlen*. 1804: *Gmelin*. 1804: Kastner. 1811: Hausmann. 1811: Sertürner. 1818: Schweigger. 1821: *Pfaff*.

**Bibliography**
AEWK, XIII: 303–305.
Brandes, R. "Bucholz Leben." *Archiv der Pharmazie und Berichte der deutschen pharmazeutischen Gesellschaft*, 2, no. 1 (1819), 1–22.
DAB, I: 90–91.
DSB, II: 564–565.
Erfurt Stadtarchiv (5/955–1; 5/801–B3).
Moll, I: 80; III: 747–748.
Trommsdorff, J. B. "Christian Friedrich Bucholz." *Neues Journal der Pharmacie*, 2 (1818), 526–538.
Also see W. H. S. Bucholtz (Dann, p. 46).

### TROMMSDORFF, Johann Bartholomäus
1770, Erfurt / 1837, Erfurt

**Background**   Son of Wilhelm Bernhard Trommsdorff (1738, Erfurt–1782, Erfurt), who studied the medical and natural sciences (including chemistry at Erfurt University with *C. A. Mangold* and at Göttingen University with *R. A. Vogel*), earned his M.D. at Erfurt University in 1765, bought Erfurt's Schwan-Apotheke in 1768 (among his employees was *F. A. C. Gren*), became OP of chemistry and botany at Erfurt University in 1772 (among his students was *S. F. Hermbstaedt*), and served along with *W. H. S. Bucholtz* as judge for the Erfurt Academy's annual chemistry prize from 1776. Stepson of Johann Jacob Planer (1743, Erfurt–1789, Erfurt), OP of chemistry and botany at Erfurt University. Lutheran.   **Education**   In 1783, after schooling and private instruction in botany from *Gren*, he entered Erfurt's Town Council Gymnasium. However, his widowed mother's financial position soon became so bad that he had to take up an apprenticeship in the family apothecary shop. In 1784 he proceeded to Weimar where he served under *Bucholtz* and *J. F. A. Göttling* in the Hof-Apotheke. Upon completing his apprenticeship in 1787, he returned to Erfurt and studied mathematics and languages with his new stepfather, who apparently wanted him to become a merchant. The following year he went to Stettin where, besides serving as a journeyman in C. F. G. Zittelmann's apothecary shop, he studied trade and enjoyed fruitful contacts with *J. C. F. Meyer*. In 1789 he went on to Stargard where he served in *S. F. Fischer*'s apothecary shop and continued his study of trade. On account of his stepfather's death, however, he soon had to return home to manage the family's affairs.

**Career**
1790–1837: Administrator, then from 1797 lessee, and finally from 1826 owner of the family apothecary shop in Erfurt (net income, 1801: 704 Rtl). At his death he also owned the apothecary shop in nearby Rudolstadt.

1793–1834: Editor of the *Journal der Pharmacie* and its successor the *Neues Journal der Pharmacie*. His earnings from this and other publications were high (e.g., 1801: 596 Rtl).

1794: Petitioned the government for appointment as professor of chemistry at Erfurt University. The faculty agreed to his appointment as AOP of chemistry, providing that he first earned a degree in medicine or philosophy. He qualified for the Ph.D. that September.

1795–1816: AOP, then from 1811 OP of chemistry and pharmacy in Erfurt University's Medical Faculty (salary, 1795–1799: none; 1800–1816: 60 Tl). Among his colleagues was *C. F. Bucholz* (1809–1816).

1795?–1831: Produced and sold chemical kits.

1795–1828: Director of a "chemical-physical-pharmaceutical boarding school" in Erfurt (net income, 1800: 517 Tl). At least 13 of his more than 300 students subsequently established chemical factories. In 1798 and again in 1805 he took his students on field trips to the Harz mining district. On the second trip he visited the chemist F. Stromeyer in Göttingen. In 1814 he closed the school for a year so that his students could fight against that "son of hell, the cowardly tyrant Bonaparte." In 1823 the Prussian government announced that attendance at

Trommsdorff's school was equivalent to university attendance. It provided the school with 95 Tl for constructing a laboratory and 100 Tl per year for operating costs.
1799: Applied for *A. N. Scherer*'s vacated chemical lectureship in Weimar, requesting a salary of 500 Rtl plus free residence. Nothing came of this, probably because Duke Carl August was still angry about *Scherer*'s unauthorized departure.
1803: Declined a call to St. Petersburg.
1803–1810: Member of Erfurt's Medical and Sanitation Board.
1804–1811: Erfurt's Commissioner of Brewing (salary: 300 Tl). He resigned when the French government cut off his salary.
1809: Declined a call to succeed *Göttling* at Jena University.
1811–1820: Co-owner of a drug and chemical factory at Teuditz near Lützen. Despite his investments of 10,000 Rtl and other investments of 8,000 Rtl, the factory had a deficit of over 4,000 Rtl in 1815. He dropped out of the venture in 1820 when the factory was moved to Kösen near Naumburg.
1811: Declined calls to the chairs of chemistry and technology at Hildburghausen and of chemistry and physics at Warsaw.
1814–1837?: Member of Erfurt's Sanitation Board.
1815: Considered as a possible successor to *A. F. Gehlen* in the Munich Academy.
1816: Declined a call to St. Petersburg.
1817: Declined a call to succeed *M. H. Klaproth* in Berlin.
1827: Visited England.
1828, 1830, 1835, 1836: Attended meetings of the Society of German Scientists and Physicians in Berlin, Hamburg, Bonn, and Jena. At the meeting in Bonn, he was elected president of the section on physics, chemistry, and pharmacy after J. J. Berzelius refused the post in his favor.

**State Honors** 1811: Schwarzburg-Rudolstadt Court Councillor. 1820: Knight of the Prussian Red Noble Order (1820: Third Class; 1833: Third Class with Bow). 1823: Imperial Honor Medal. 1834: Prussian Privy Court Councillor. **Major Scientific Societies and Honorary Degrees** 1792: Erfurt. 1795: Leopoldina. 1802: Berlin NF. 1805: Honorary M.D. from Erfurt University. 1812: Berlin (corresponding member). **Informal Recognition** 1794: Baldinger. 1796: *Girtanner*. 1796: S. in M. 1796: Van Mons. 1798: *Bucholz*. 1800: Anon. 1803: Ed. Board. 1804: *Gmelin*. 1804: Kastner. 1808: *Göttling*. 1811: Sertürner. 1817: Döbereiner. 1821: *Pfaff*.

**Bibliography**
Abe, H. R. "Johann Bartholomäus Trommsdorff (1770–1837): Aus meinem Leben." *Leopoldina* (1976), 178–200.
———, H. Schmidt, A. Säubert, K. Heinig, W. Strube, and F. Wiegand. "Johann Bartholomäus Trommsdorff und die Begründung der modernen Pharmazie." *Beiträge zur Geschichte der Universität Erfurt*, 16 (1971/72).
ALZ: IB (19 November 1794), 1047–1048.
Caumnitz, C. "Johann Bartholomäus Trommsdorff (1770–1837): Ein Begründer der wissenschaftlichen deutschen Pharmazie." Archiv, Deutsche Akademie der Naturforscher (Leopoldina), Halle.
DAB, II: 692–695.
DSB, XIII: 465–466.
Gittner, H. "Die Rheinreise des Johann Bartholomai Trommsdorff zur 13. Naturforscherversammlung in Bonn anno 1835." *Deutsche Apotheker-Zeitung*, 99 (1959), 31–36.
Götz, W. "Die Beziehungen J. B. Trommsdorffs zur französischen Chemie und Pharmazie." *Beiträge zur Geschichte der Pharmazie*, 31, no. 3 (1979), 1–5.
———. "Zu Leben und Werk von Johann Bartholomäus Trommsdorff (1770–1837): Darstellung anhand bisher unveröffentlichten Archivmaterials." *Quellen und Studien zur Geschichte der Pharmazie*, 16 (1977).
*Die Harzreisen des Johann Bartholomä Trommsdorff 1798 und 1805*. Edited by H. Gittner. Oberhausen: Storck, 1957.
Pohl, 38–69, 208–209.
Rosenhainer, O., and H. Trommsdorff. *Johann Bartholomäus Trommsdorff 1770–1837*. Jena: Vopelius, 1913.
Teschke, W. "Trommsdorff in Pommern." *Pharmazeutische Zeitung* 91–100 (1955), 1340–1341.

Trommsdorff, H. "Johann Bartholomäus Trommsdorff und seine Zeitgenossen." Akademie gemeinnütziger Wissenschaften zu Erfurt: *Jahrbücher*, N.F. 53 (1937), 5–55; 55 (1941), 131–234.

## SCHERER, Alexander Nicolaus
1771, St. Petersburg / 1824, St. Petersburg

**Background** Son of Johann Benedict Scherer (1740, Strasbourg–1828, Barr, Alsace), who studied philosophy and law at Strasbourg, Jena, and Leipzig, journeyed by way of Cracow to Russia in 1765, resided in St. Petersburg as legal councillor in the Russian Ministry for Baltic Affairs from 1766 to 1769 and then as member of the French Embassy until 1773, deserted his family and went by way of Uppsala to Paris where he served in the Foreign Ministry from 1775 to 1787, and spent his remaining years as an official in Strasbourg (1787–1791), an émigré in several German towns (1792–1807), a French teacher in Tübingen (1808–1824), and a pensioner in Barr (1824–1828). Lutheran. **Education** From 1783 he lived with his maternal uncle in Riga, attending the local cathedral school. In 1791 he went to Jena University to study theology. Once there he soon switched to the natural sciences even though this decision meant an end to his family's support. His chief enthusiasm was chemistry, which he studied with *J. F. A. Göttling*. In 1793 he was among the founding members of Jena's new Scientific Society and its most ardent advocate of Lavoisier's theory. Early the next year, even though he could pay␣but half the customary fee, he was awarded the Ph.D.

**Career**

1794–1797: PD of chemistry at Jena University. Among his colleagues was *Göttling* and among his students, J. W. Ritter (1796–1797).

1797–1799: Saxe-Weimar Mining Councillor. His appointment was arranged by A. von Humboldt and J. W. von Goethe.

June 1797–May 1798: Traveled by way of Brunswick (where he met *H. F. Link*, who was on his way to Portugal), through France, to Britain. There he investigated ovens for brick making, machines for stripping bark from standing trees, clays used for ceramics, and techniques for removing sulfur from coal. He also made the acquaintance of many scientists. He returned to Weimar in the spring of 1798.

May 1798–October 1799: Resided in Weimar (salary: 300 Tl, plus residence, wood, and fodder for one horse). In early 1799 he was charged with giving free public lectures on chemistry (budget: 50 Tl).

1798–1803: Editor of the *Allgemeines Journal der Chemie*. Among his assistant editors were Ritter (1798–1799) and C. J. B. Karsten (1801–1802; salary: 250 Rtl). In 1802 he met with *L. Crell* in Leipzig to explore the possibility of merging their journals, but nothing came of their talks. The following year he relinquished control of his journal to an editorial board consisting of *S. F. Hermbstaedt*, *M. H. Klaproth*, and *J. B. Richter*, who delegated the actual editing to the rising young chemist *A. F. Gehlen*.

1798: Nominated by the physician P. Best for a chair of chemistry in the proposed new medical school in Cologne. For want of funds, nothing came of this proposal.

1798–1799: Aspirant for a chair in Halle University. In September 1798 he secretly applied for a position in Halle, evidently hoping to succeed the ailing J. R. Forster, OP of natural history. Three months later, after *F. A. C. Gren*'s unexpected death, his candidacy was strengthened by *Hermbstaedt*'s recommendation that he be appointed to a chair for "chemistry, pharmacy, and natural history." In January 1799, however, he was passed over in favor of *Link* because he did not have an M.D. Eight months later, after *Link*'s decision to stay in Rostock, he was offered a chair in Halle's Philosophical Faculty. He eagerly accepted. However, his resignation was denied by his resentful patron, Duke Carl August, so he had to leave Weimar by night.

1799–1800: OP of chemistry and physics in Halle University's Philosophical Faculty (salary: 500 Tl). Apparently short of funds, he announced in the spring of 1800 that he would soon be opening a "chemical institute." However, before anything could come of this scheme, he resigned his chair. He recommended that K. W. Juch or Ritter be appointed to succeed him.

1800–1803: Resided in Berlin. For a while he was associated with Baron Eckartstein's faience factory in Potsdam (salary: 1,500 Rtl). But he was soon traveling about Germany in the hope of finding a better position.

1803–1804: OP of chemistry and pharmacy in Dorpat University's Medical Faculty (salary: 2,000 rubles).

1804–1823: OP of chemistry and pharmacy at St. Petersburg's Medical-Surgical Academy.

1804–1817: Teacher of chemistry at St. Petersburg's Mining Cadet School.
1805–1824: Salaried member for chemistry at the St. Petersburg Academy of Sciences.
1805–1807: Adjunct.
1807–1815: Extraordinary member.
1815–1824: Ordinary member.
1806–1815: Teacher of chemistry(?) at St. Petersburg's Teacher's School.

**State Honors** 1803: Russian Court Councillor. 1809: Knight of the St. Vladimir Order (Fourth Class). 1814: Russian State Councillor. 1819: Knight of the St. Anna Order (Second Class). **Major Scientific Societies** 1793: Erfurt. 1795: Berlin NF. 1797: St. Petersburg (corresponding member; 1807: extraordinary member; 1815: ordinary member). **Informal Recognition** 1796: Fourcroy. 1796: *Girtanner.* 1796: S. in M. 1798: *Bucholz.* 1800: Anon. 1803: Ed. Board. 1806: *Trommsdorff.* 1807: Kastner.

**Bibliography**
ADB, XXXI: 99–104.
ALZ: IB (23 December 1797), 1384; (8 August 1798), 945–948; (7 May 1800), 504.
Crell, L. "Vorbericht." *Chemische Annalen,* no. 2 (1803), vii.
Friedel, H. "Aus dem Leben des Tübinger Professors Johann Benedikt von Scherer." *Heimatkundliche Blätter für den Kreis Tübingen,* N.F. no. 60 (May 1974).
Goethe & Carl August, 214–223, 234, 261, 268, 433–435, 445.
Goethe & Voigt, 53: 341, 343, 356, 360, 362; 54: 66, 75, 77, 79, 188, 440.
"Goethe's Briefwechsel mit den Gebrüdern von Humboldt (1795–1832)," edited by F. T. Bratranek. In *Neue Mittheilungen aus Johann Wolfgang v. Goethes handschriftlichem Nachlasse,* vol. III. Leipzig: Brockhaus, 1876, pp. 313–314.
Humboldt, 572–573, 577, 580, 584–585.
Karsten, G. "Umrisse zu Carl Johann Bernhard Karsten's Leben und Wirken." *Archiv für Mineralogie, Geognosie, Bergbau und Hüttenkunde,* 26; no. 3 (1854), 224–226.
Koch, H. "Beiträge zur Biographie A. N. Scherers: Ein Nachtrag." *NTM,* 5; no. 12 (1968), 113–115.
Levshin, B. V. (Arkhiv, Akademiia nauk SSSR, Moscow). Letter illuminating Scherer's career in Russia (12 March 1971).
Möller, R. "Beiträge zur Biographie A. N. Scherers." *NTM,* 2; no. 6 (1965), 37–55.
*Neuer Nekrolog der Deutschen,* 2 (1824), 1208–1211.
Pohl, 73.
Scherer, A. N., to J. F. Fries. Six letters (1797–1816). Bibliothek, Carl-Sudhoff-Institut, Leipzig University.
Scherer, to M. van Marum. Ten letters (1798–1803). Van Marum Collection, Part I: The Letters (Hollandische Maatschappij der Wettenschappen), microfiche ed. Micro-Methods Limited, East Ardsley, Wakefield, England. Library, American Philosophical Society, Philadelphia.
Steffens, H. *Lebenserrinerungen aus dem Kreis der Romantik.* Edited by F. Gundelfinger. Jena: Diederich, 1908, p. 109.
Also see Bonn University ("Quellen," p. 883) and Rostock University (*Matrikel,* V: 42).

**LAMPADIUS, Wilhelm August (Eberhard)**
1772, Hehlen near Holzminden / 1842, Freiberg

**Background** Son of Wilhelm Christoph Lampadius (?, Brunswick–1777, Atlantic Ocean), a lieutenant who went down with a shipload of troops bound for America. Lutheran. **Education** Upon his father's death, he was sent to live with his maternal uncle, pastor J. C. F. Prössel, at Bofzen near Fürstenberg an der Weser. He attended the local school until the age of thirteen. Then, because his mother could not afford to see him through an education, he became an apprentice in Göttingen's Rats-Apotheke under Stössel. During his apprenticeship he first studied languages and natural sciences on his own, then began attending lectures at Göttingen University. When he finished his pharmaceutical training in 1791, he threw himself into scientific studies. He worked closely with G. C. Lichtenberg and *J. F. Gmelin,* serving as an experimental assistant in their courses. In the spring of 1792, he matriculated as a chemistry student. That summer Lichtenberg introduced him to his future employer, the minor chemist Count J. von Sternberg.

**Career**
1793: Personal secretary to Count Sternberg, who engaged him as a traveling companion for an expedition through Russia to China. On his way to Russia, he met *M. H. Klaproth* and

*S. F. Hermbstaedt* in Berlin. Upon arriving in St. Petersburg, he learned that the Russian government had denied permission for the expedition. He stayed for a time, meeting the chemist T. Lowitz, then accompanied Sternberg back to his Bohemian estates. He established a new ironworks for Sternberg at Radnitz near Pilsen. During this time, A. G. Werner, who was seeking a possible successor to *C. E. Gellert* in the Freiberg Mining Academy, interviewed him on *Klaproth*'s recommendation. Favorably impressed, Werner recommended his appointment. Testimonials were obtained from Sternberg, *Klaproth*, and *Gmelin*. Then he was offered the job.

1794–1842: Professor and mining official in Freiberg.
    1794–1795: AOP of chemistry at the Freiberg Mining Academy (salary: 400 Tl). During this period he was responsible for learning from and assisting *Gellert*. His plan to augment his income by selling collections of "chemical products" at 30 and 40 Rtl failed for want of sufficient orders.
    1796–1842: OP of chemistry and smelting and member of the Freiberg Higher Smelting Office (salary, 1796–1800: 600 Tl; 1801–1802: 1,080 Tl; 1803–1815: 1,380 Tl; 1816–1836: 1,545 Tl; 1837–1842: 1,685 Tl). Among his students was J. N. Fuchs (1803), who later became a chemist in Bavaria.
1796: Visited the Harz mining district.
1799–1803: Investigated the feasibility of the commercial production of sugar from beets. He carried out successful small-scale tests in early 1799. Then he visited the sugar refinery at Königsall in Bohemia to observe large-scale refining methods. That summer he established pilot plants at Waltersdorff and Bottendorf. Even when the by-products were sold, however, these plants could not be made to yield steady profits.
1803, 1810: Visited the Harz mining district, then the Rhine and Weser mining districts.
1812: Instructed over a hundred people (mostly women) on the production of sugar from potatoes and of coffee substitutes from beets and chestnuts. In May he demonstrated his processes to Napoleon.
1816: Established gas lighting at the Halsbrücke amalgamation works, introducing this means of lighting to Germany.
1826, 1827, 1828: Attended meetings of the Society of German Scientists and Physicians in Dresden, Munich, and Berlin.

**State Honors** 1816: Saxon Mining Commission Councillor. 1832: Knight of the Saxon Civil Service Order. **Major Scientific Societies** 1809: Berlin NF. **Informal Recognition** 1802: *Göttling*. 1804: *Gmelin*. 1804: Kastner. 1814: Schweigger. 1816: *Bucholz*. 1818: *Trommsdorff*. 1821: *Pfaff*.

**Bibliography**
ALZ: IB (20 December 1794), 1179–1180; (13 April 1796), 399–400.
Callisen, XXIX: 428–429.
DAB, I: 358–359.
Heisterbergk, H. "Briefe von Wilh. August Lampadius." Freiberger Altertumsverein: *Mittheilungen*, 62 (1932), 40–50.
Lampadius, G. "Aus den Lebenserinnerungen des weil. Diakonus zu St. Nicolai in Leipzig Wilhelm Adolph Lampadius." Ibid., 65 (1935), 67–76.
Lichtenberg, III: 65–67, 149–150.
———. *Briefe an Johann Friedrich Blumenbach*. Edited by A. Leitzmann. Leipzig: Dieterich, 1921, pp. 55, 116.
*Neuer Nekrolog der Deutschen*, 20 (1842), 303–314.
Richter, J. "Wilhelm August Lampadius–ein Lehrer und Forscher an der Bergakademie Freiberg," *Freiberger Forschungshefte: Kultur und Technik*, D90 (1975), 15–27.
Schiffner, C. "Wilhelm August Lampadius." *Beiträge zur Geschichte der Technik und Industrie*, 12 (1922), 40–50.
Schwarz, H.-D. "Wilhelm August Lampadius zum 200. Geburtstag am 18. August 1972." *Pharmazeutische Zeitung*, 117 (1972), 1223–1225.
Seifert, A. *Wilhelm August Lampadius: Ein Vorgänger Liebigs*. Berlin: Verlag Chemie, 1933.
Staatsarchiv, Dresden. Letter with information on Lampadius's salary as professor and mining official (8 May 1972).
Ulrich, K. "Die Anfänge der Rübenzuckerindustrie im Freistaat Sachsen." *Die Deutsche Zuckerindustrie*, 54 (1929), 578–585.
Zeller, H. "Lampadius' 'Reise zu den sieben Schwestern': Ein Beitrag zur Familiengeschichte." *Archiv für Stamm- und Wappenkunde*, 4 (1903/04), 35–36.

Also see Göttingen University. ("Matrikel," 1: 336) and Prague University (Wraný, pp. 246–247, 348).

**PFAFF, Christoph Heinrich**
1773, Stuttgart / 1852, Kiel

**Background** Son of Burckhard Pfaff (1738, Usingen–1817, Stuttgart), who, though he only attended Latin school, enjoyed a good career, rising from a tax secretary, to mayor of Balingen in 1761, to Treasury Councillor in Stuttgart in 1767, and eventually to General Treasurer for Württemberg. Lutheran. **Education** After being tutored at home, he entered Duke Carl's Academy in Stuttgart in 1782. During his first four years there, he prepared for work at the university level. Then from 1786 to 1788, he studied in the Academy's philosophical section and from 1788 to 1793, in its medical section. From the mid-1780s, natural history was his chief enthusiasm. Still, he received an introduction to chemistry, which was to become his "favorite science," from the tedious lectures of C. G. Reuss, from the exciting text of F. A. C. *Gren*, from the letters of G. Cuvier (a former companion who had embraced A. L. Lavoisier's theory), and from the stimulating course of C. F. Kielmeyer. In 1793 he earned his M.D. at Stuttgart with a dissertation on animal electricity. That fall he journeyed to Göttingen University for training in obstetrics, his chosen profession. While there, he also worked closely with the physicist G. C. Lichtenberg, carried out experiments in *J. F. Gmelin*'s laboratory, and became acquainted with *C. Girtanner* and his brother's colleague in Helmstedt, *L. Crell*. In the fall of 1794 he proceeded to Copenhagen where he studied practical obstetrics for a year.

**Career**
1795–1797: Personal physician to Count F. Reventlow, who later was his patron as Curator of Kiel University from 1800 to 1808. He accompanied the Count and his family to Italy where they spent two winters in Rome. In Florence he met the chemist F. Fontana.
1797–1798: Practiced medicine at Heidenheim a. d. Brenz until late 1797. Upon receiving a call to Kiel University, he visited home and then went by way of Jena, where he met *J. F. A. Göttling*, to Kiel.
1798–1852: AOP, OP, then Emeritus OP at Kiel University.
  1798–1800: AOP in the Medical Faculty. Shortly after arriving, he used a call to Stuttgart's mining office to obtain responsibility for physics (salary: 300 Rtl).
  1801–1802: OP of physics in the Philosophical Faculty. In 1801 he went to Paris at government expense to prepare for the chair of chemistry. He enjoyed fruitful contacts there with L. B. Guyton de Morveau, A. F. Fourcroy, C. L. Berthollet, E. J. B. Bouillon la Grange, L. N. Vauquelin, L. J. Thénard, and A. Volta. On his return trip he experimented in M. van Marum's laboratory in Haarlem.
  1802–1846: OP of chemistry and physics in the Medical Faculty.
  1846–1852: Emeritus OP.
1804–1852: Secretary, then from 1828 Director of the Schleswig-Holstein Sanitation Board. As the member responsible for pharmacy, he drew up the new pharmacopoeia in 1831.
1804: Declined a call to Halle University for theoretical medicine.
1804, 1809: Visited his family in Stuttgart.
1811: Declined a call to Halle University for chemistry and physics. He was also on the candidate list for the chair of chemistry at Breslau University.
1814: Visited his family in Stuttgart and *A. F. Gehlen* in Munich.
1815: On the list of candidates for *Gehlen*'s successor in the Munich Academy.
1817: Declined a call to Tübingen University for chemistry.
1818?: Declined a call to Bonn University for materia medica.
1819: Visited E. Mitscherlich in Berlin and *W. A. Lampadius* and J. J. Berzelius in Frieberg.
1824: Visited C. G. C. Bischof in Bonn and E. E. Brunner in Bern.
1829: Visited Thénard, L. J. Gay-Lussac, J. B. Dumas, M. E. Chevreul, and A. C. Becquerel in Paris and M. Faraday in London.
1830: Attended the meeting of the Society of German Scientists and Physicians in Hamburg where he was elected chairman of the section on chemistry and physics after Berzelius refused the honor.
1838: Visited *H. F. Link* and G. Magnus in Berlin.
1841: Went to Vienna for an unsuccessful glaucoma operation.
1847: Visited C. G. Gmelin in Tübingen and C. J. Löwig in Zürich.

**State Honors** 1795: Württemberg Court Physician. 1815: Knight of the Order of Denmark (1815: Silver Cross; 1840: Commander's Cross). 1843: Danish Conference Councillor. **Major**

**Scientific Societies and Honorary Degrees** 1807: St. Petersburg (corresponding member). 1808: Munich (corresponding member; 1820: foreign member). 1812: Berlin (corresponding member). 1843: Honorary M.D. from Tübingen University. 1843: Honorary Ph.D. from Kiel University. **Informal Recognition** 1804: *Gmelin*. 1808: *Gehlen*. 1819: Du Mênil.
**Bibliography**
Callisen, XXXI: 210–211.
*Georg Cuviers Briefe an C. H. Pfaff aus den Jahren 1788 bis 1792*. Edited by W. F. G. Behn. Kiel: Schwers, 1845.
Moll, III: 748.
Oersted, I: 349; II: 69, 466–476.
Pfaff, C. H. *Lebenserinnerungen*. Kiel: Schwers, 1854.
*Sammlung von Briefen gewechselt zwischen J. F. Pfaff und Herzog Carl von Würtemberg . . . und Anderen*. Edited by C. Pfaff. Leipzig: Hinrich, 1853, pp. 52. 75–166, 267.
Also see Kiel University (Schipperges, pp. 57, 120–123, 130, 156).

### GEHLEN, Adolph Ferdinand
1775, Bütow, Pomerania / 1815, Munich

**Background** Son of Jacob Ferdinand Gehlen, a wealthy apothecary with agricultural holdings near Bütow. Lutheran. **Education** Around 1790, after schooling in his hometown, he became an apprentice in *K. G. Hagen*'s Hof-Apotheke in Königsberg. During his apprenticeship, in late 1793, he matriculated at Königsberg University. Though he was hard of hearing, he mastered several foreign languages while studying the natural sciences and medicine. Around 1797 he proceeded to Berlin where he served as a journeyman in *V. Rose*'s Apotheke zum weissen Schwan. He evidently made such a good impression that *M. H. Klaproth* gave him private instruction.

**Career**
1800: Gave a private course on pharmacy in Berlin.
1803–1810: Editor of Germany's leading chemistry journal, the *Neues allgemeines Journal der Chemie* (1803–1806), which became the *Journal für Chemie und Physik* (1806–1807), then the *Journal für Chemie, Physik, und Mineralogie* (1808–1810).
1806: Earned the Ph.D. at Halle University so he could lecture there.
1806–1807: PD at Halle University and chemist in J. C. Reil's clinical institute.
1807–1815: Academician for chemistry and secretary of the section for physics and chemistry in the Munich Academy of Sciences (salary, 1807: 1,800 fl). Among his colleagues was his closest friend, J. W. Ritter (1807–1810), who probably arranged for his appointment. He visited *C. F. Bucholz* in Erfurt en route to Munich.
1812: Declined a call to Breslau University for chemistry.
1813: Visited Vienna where, in the presence of the leading local chemists, he carried out a series of experiments in the Imperial Glass Factory.
1815: Died in Munich of arsenic poisoning.

**State Honors** 1809?: Bavarian Court Councillor. **Scientific Societies** 1805: Berlin NF. 1807: Munich. **Informal Recognition** 1803: Ed. Board. 1804: Kastner. 1808: *Göttling*. 1811: Sertürner. 1815: Schweigger. 1816: *Bucholz*. 1818: *Trommsdorff*. 1819: Brandes.
**Bibliography**
Akademie der Wissenschaften, Munich: *Denkschriften*, 5 (1814–1815), xxix–xxxv.
Buchner, J. A. "Hauptzüge aus dem Leben . . . Gehlen." *Repertorium für die Pharmazie*, 1 (1815), 435–446.
DAB, I: 194.
DSB, XV: 171–173.
Karsten, D. L. G. "Vorrede." In F. A. C. Gren, *Grundriss der Chemie*, 2d ed., vol. II. Halle: Waisenhaus, 1800, p. iv.
Lenz, IV: 42.
Moll, I: 193–213; II: 359, 726–727.
NDB, VI: 132–133.
Oersted, II: 96–98, 106, 108, 187, 208, 211–212, 219, 221, 256, 260, 355, 370.
Schubert, G. H. von. *Der Erwerb aus einem vergangenen und Erwartungen von einem zukünftigen Leben*. Vol. II. Erlangen: Palm and Enke, 1855, pp. 391–396.
Also see Königsberg University (*Matrikel*, II: 625) and Munich Academy (Prandtl, pp. 9–15).

# Appendix II

# INSTITUTIONAL HISTORIES

This appendix focuses on chemistry's place in schools and academies within the German-speaking parts of Central Europe from the sixteenth to the early nineteenth centuries. It includes every learned institution that employed a professor or academician for chemistry between 1700 and 1800. It also includes Basel and Wittenberg Universities because, despite their failures to obtain salaried chemical positions before the nineteenth century, both played some role in disseminating information about chemistry in earlier times.

For easy reference, the institutions are discussed alphabetically by town and, for those few towns that had more than one institution with a chemical position, chronologically by founding date. Table A2, which summarizes the findings, should facilitate special investigations.

Each institutional history presents the following information:

*IDENTIFICATION:* The history begins by identifying the school or academy under consideration, giving important dates in the institution's development and indicating the ruler(s) with jurisdiction over the institution during the eighteenth century.

*FIRST CHEMICAL INSTRUCTION:* The history next gives available information about the earliest chemical instruction offered at the institution. (This heading is omitted for those schools founded with a salaried chemical position and for the two academies.)

*CHEMICAL POSITION:* The history then briefly recounts the establishment and subsequent development of the institution's chemical position. This narrative is followed by a table listing all the institution's chemical representatives, giving their dates of service, their salaries, and their publishing activity in chemistry. (The names of chemists with biographical profiles in Appendix I are italicized.)

*BIBLIOGRAPHY:* Every history closes with a bibliography to the sources of information regarding chemistry at the institution. In addition to these specific works, the following works, listed in chronological order, have proved helpful in the initial stages of investigating most of the universities and specialized schools:

*Acta Academica Praesentem Academiarum Societatum Litterariarum Gymnasiorum et Scholarum.* 1–6 (1733–1738).

Hagelgans, J. G. *Orbis Literatus Academicus Germanico-Europaeus.* Frankfurt a. M.: Hocker and Hutter, 1737.

Börner, F. "Ein Verzeichniss aller gegenwärtig lebenden Professorum Medi-

TABLE A2: Salaried Chemical Positions and Laboratories in German Schools and Academies, 1700–1800

|  | 1700 | 1720 | 1740 | 1760 | 1780 | 1800 |
|---|---|---|---|---|---|---|
| **Medical Faculties and Schools** | | | | | | |
| Altdorf University | 1,L | 1,L | 1,L | 1,L | 1,L | 1,L |
| Bamberg University | x | x | x | x | 0 | 1 |
| Berlin Medical-Surgical College | x | x | 1 | 1 | 1 | 1 |
| Berlin Veterinary Medicine School | x | x | x | x | x | 1 |
| Berlin Surgical Pépinière | x | x | x | x | x | 1 |
| Budapest University | x | x | x | x | 1 | 2,L? |
| Cologne University | 0 | 0 | 0 | 0 | 1,L | x |
| Duisburg University | 0 | 0 | 1? | 1? | 1? | 1 |
| Erfurt University | 0 | 0 | 0 | 2 | 1 | 1 |
| Erlangen University | x | x | x | 1 | 1 | 0 |
| Frankfurt a. d. Oder University | 0 | 0 | 1 | 1 | 1 | 1 |
| Freiburg i. Br. University | 0 | 0 | 0 | 1 | 1,L | 1,L |
| Fulda University | x | x | 0 | 0 | 1 | 1 |
| Giessen University | 0 | 0,L | 0 | 0 | 1 | 1,L? |
| Göttingen University | x | x | 0 | 1 | 1 | 1,L |
| Graz Lyceum | x | x | x | x | x | 1 |
| Greifswald University | 0 | 0 | 0 | 0 | 1 | 1 |
| Halle University | 1 | 0 | 0 | 0 | 0 | 0 |
| Heidelberg University | x | 0 | 1 | 1 | 1 | 1 |
| Helmstedt University | 1 | 0 | 1 | 1 | 1 | 1 |
| Herborn High School | 0 | 0 | 0 | 0 | 0 | 1 |
| Ingolstadt University | 0 | 0 | 0,L? | 1 | 1,L | 1,L |
| Innsbruck University | 0 | 0 | 1 | 1 | 1,L | 1 |
| Jena University | 1 | 1 | 1 | 1 | 1 | 1 |
| Kassel College | x | x | 0 | 0 | 1,L | x |
| Kiel University | 0 | 0 | 0 | 0 | 1 | 1 |
| Königsberg University | 1? | 1 | 1 | 0 | 1 | 1 |
| Leipzig University | 0 | 0 | 1 | 1 | 1 | 1 |
| Marburg University | 0,L | 0,L | 0 | 0 | 0 | 1,L |
| Münster University | x | x | x | x | x | 1 |
| Olmütz Lyceum | x | x | x | x | x | 1 |
| Prague University | 0 | 0 | 0 | 1 | 1 | 1,L |
| Rostock University | 1 | 1 | 1 | 0 | 0 | 0 |
| Strasbourg University | 1 | 1 | 1 | 1 | 1 | x |
| Strasbourg Medical School | x | x | x | x | x | 1 |
| Stuttgart University | x | x | x | x | 1 | x |
| Tübingen University | 0 | 1 | 1? | 1,L | 1,L | 1,L |
| Vienna University | 0 | 0 | 0 | 1,L | 1,L | 1,L |
| Vienna Medical-Surgical Academy | x | x | x | x | x | 1,L? |
| Würzburg University | 0 | 0 | 1 | 1 | 1 | 1,L |
| **Administrative Faculties and Schools** | | | | | | |
| Berlin Artillery Academy | x | x | x | x | x | 1 |
| Berlin Mining School | x | x | x | x | 1 | 1 |
| Freiberg Mining Academy | x | x | x | x | 1 | 1,L |
| Giessen University Economics Faculty | x | x | x | x | 1 | x |
| Heidelberg State Economic School | x | x | x | x | x | 1,L |
| Kaiserslautern Cameral School | x | x | x | x | 1,L | x |
| Schemnitz Mining Academy | x | x | 0 | 0 | 1,L | 1,L |
| Vienna Theresianum | x | x | x | 0 | 1 | 1 |
| **Philosophical Faculties and General Schools** | | | | | | |
| Bonn Central School | x | x | x | x | x | 1 |
| Colmar Central School | x | x | x | x | x | 1 |
| Cologne Central School | x | x | x | x | x | 1 |

## Institutional Histories / 227

| Philosophical Faculties and General Schools (cont.) | 1700 | 1720 | 1740 | 1760 | 1780 | 1800 |
|---|---|---|---|---|---|---|
| Erlangen University Philosophical Faculty | x | x | x | 0 | 0 | 1,L |
| Halle University Philosophical Faculty | 0 | 0 | 0 | 0 | 0 | 1 |
| Jena University Philosophical Faculty | 0 | 0 | 0 | 0 | 0 | 1 |
| Mainz Central School | x | x | x | x | x | 2,L |
| Rostock University Philosophical Faculty | 0 | 0 | 0 | 0 | 0 | 1 |
| Strasbourg Central School | x | x | x | x | x | 1 |
| Academies | | | | | | |
| Berlin | 0 | 0 | 0 | 1,L | 2,L | 2,L |
| Munich | x | x | x | 0 | 0 | 1 |

Note: This table is limited to those institutions that had salaried chemical positions on one of the dates indicated. Hence it does not include Basel University, Brunswick Collegium Carolinum, Brunswick Anatomical-Surgical College, Mainz University, and Wittenberg University, even though they are included in the appendix itself. x = the institution was not in existence on the given date. 0, 1, 2 = the number of salaried chemical positions on the given date, not including laboratory assistants. L = the institution had a laboratory on the given date.

cinae auf den sämtlichen Universitäten Deutschlands und einer angränzenden Lande enthält." In his *Nachrichten von den vornehmsten Lebensumständen und Schriften jeztlebender berühmter Aerzte und Naturforscher in und um Deutschland*, vol. II. Wolfenbüttel: Meissner, 1751, pp. 564–578.

*Akademischer Addresscalender auf das Jahr 1755*. Erlangen: Kammerer, 1755.

*Akademischer Address-Kalender auf das Jahr 1759*. Erlangen: Kammerer, 1758.

*Akademischer Address-Kalender auf das Jahr 1761. u. 62*. Erlangen: Kammerer, 1761.

*Akademischer Address-Kalender auf das Jahr 1767 u. 1768*. Erlangen: Kammerer, 1767.

*Akademischer Addresskalender auf das Jahr 1769 und 70*. Erlangen: Kammerer, 1769.

Ekkard, F. *Litterarisches Handbuch der bekannten höhern Lehranstalten in und ausser Teutschland*. 2 vols. Erlangen: Schleich, then Palm, 1780–1782.

"'Der Universitäts-Bereiser' Friedrich Gedike und sein Bericht an Friedrich Wilhelm II," edited by R. Fester. *Archiv für Kulturgeschichte*, 1st supp. (1905).

Heun, C. *Allgemeine Uebersicht sämmtlicher Universitäten Deutschlands*. Leipzig: Heinsius, 1792.

Colland, F. *Kurzer Inbegriff von dem Ursprunge der Wissenschaften, Schulen, Akademien und Universitäten in ganz Europa, besonders aber der Akademien und hohen Schule zu Wien*. Vienna: Trattner, 1796.

Justi, K. W., and F. S. Mursinna. *Annalen der deutschen Universitäten*. Marburg: Akademische Buchhandlung, 1798.

[List of Germany's professors of chemistry and botany], *Berlinisches Jahrbuch für die Pharmacie*, 17 (1816), 310–313.

Erman, W., and E. Horn. *Bibliographie der deutschen Universitäten*. 3 vols. Leipzig and Berlin: Teubner, 1904–1905.

Lockemann, G. "Der chemische Unterricht an den deutschen Universitäten im ersten Viertel des neunzehnten Jahrhunderts." In *Studien zur Geschichte der Chemie: Festgabe Edmund O. v. Lippmann*, edited by J. Ruska. Berlin: Springer, 1927, pp. 148–158.

Schröder, K. *Vorläufiges Verzeichnis der in Bibliotheken und Archiven vorhandenen Vorlesungsverzeichnisse deutschsprachiger Universitäten aus der Zeit vor 1945.* Saarbrücken: Anglisches Institut der Universität des Saarlandes, 1964.

Goldmann, K. *Verzeichnis der Hochschulen.* Neustadt an der Aisch: Degener, 1967.

Schmitz, R. *Die deutschen pharmazeutisch-chemischen Hochschulinstitute: Ihre Entstehung und Entwicklung in Vergangenheit und Gegenwart.* Ingelheim am Rhein: Boehringer, 1969.

Eulner, H.-H. *Die Entwicklung der medizinischen Spezialfächer an den Universitäten des deutschen Sprachgebietes.* Stuttgart: 1970.

# HISTORIES

**ALTDORF UNIVERSITY**

Founded as an academy in 1580; raised to university status in 1623. Under the jurisdiction of Nuremberg's Town Council. Closed by Bavaria in 1809.

*First Chemical Instruction*

In 1665 M. Hoffmann, OP of anatomy, surgery, and botany, offered to teach chemical manipulations in the University Apothecary Shop.

*Chemical Position*

Medical Faculty: In 1677 *J. M. Hoffmann*, son of M. Hoffmann, was appointed AOP of chemistry and anatomy. Four years later he was promoted to a new ordinary chair of anatomy. In 1682, once the authorities had agreed to construct a laboratory, he resumed responsibility for chemistry, giving an inaugural lecture on the "necessity and utility of chemistry." At the May 1683 dedication of the laboratory, he spoke on "the best method of advancing chemistry." From then on, one of the medical OPs was responsible for teaching chemistry and maintaining the laboratory. However, a strong interest in chemistry was evidently not a precondition for receiving this assignment.

| Chemical representative | Years | Institutional salary | Chemical publications | |
|---|---|---|---|---|
| | | | Before | During |
| J. M. Hoffmann | 1677–1681, 1682–1713 | Yes, from 1682 | No | Yes |
| J. J. Baier | 1713–1735 | Yes | Yes | Yes |
| J. J. Kirsten | 1737–1765 | Yes | Yes | Yes |
| J. N. Weiss | 1765–1768 | Yes | No | No |
| A. Nietzki | 1768 | Yes | Yes | No |
| J. N. Weiss | 1769–1783 | Yes | No | No |
| P. L. Wittwer | 1784–1785 | Yes | No | No |
| J. C. Ackermann | 1786–1801 | Yes | No | No |
| K. W. Juch | 1801–1805 | Yes | Yes | Yes |
| J. Feiler | 1805–1809 | Yes | No | No |

## Bibliography

Baier, J. J. *Ausführliche Nachricht von der Nürnbergischen Universität-Stadt Altdorff.* 2d ed. Nuremberg: Tauber, 1717, pp. 74, 98–100.

Flessa, D. *Die Professoren der Medizin zu Altdorf von 1580–1809.* Diss., Erlangen University; Erlangen: Hogl, 1969.

Henrich, F. "Über das chemische Laboratorium der ehemaligen Nürnbergischen Universität in Altdorf." *Zeitschrift für angewandte Chemie,* 39 (1926), 92–98.

Kämmerer, H. "Chemie." In *Nürnberg: Festschrift dargeboten den Mitgliedern und Teilnehmern der 65. Versammlung der Gesellschaft deutscher Naturforscher und Aerzte.* Nuremberg: Schrag, 1892, pp. 37–57.

*Die Matrikel der Universität Altdorf.* Edited by E. von Steinmeyer. 2 vols. Würzburg: Stürtz, 1912. (Cross-referenced as *Matrikel.*)

Sauer-Haeberlein, K. *Personalbibliographien der Professoren der Medizin zu Altdorf von 1580–1809.* Diss., Erlangen University; Erlangen: Hogl, 1969.

Will, G. A. *Geschichte und Beschreibung der Nürnbergischen Universität Altdorf.* Altdorf: Kussler, 1795, pp. 90, 94, 118, 207–210.

## BAMBERG UNIVERSITY

Founded as an academy in 1648; provided with a medical faculty in 1735; raised to university status in 1773. Under the jurisdiction of the Prince-Bishop of Bamberg until 1802. Closed by Bavaria in 1803.

### First Chemical Instruction

At the instigation of J. I. J. Döllinger, OP of medicine, the local apothecary was ordered in 1769 to announce all his chemical operations so that medical students could watch them.

### Chemical Position

Medical Faculty: In 1775 Döllinger was assigned therapy, chemistry, and materia medica. He seems soon to have passed chemistry on to J. R. M. Joachim, the unsalaried OP of medical institutions. Joachim taught the subject almost every year from 1775 and gave his title as OP of chemistry and medical institutions in 1789. Upon the redistribution of teaching responsibilities in 1790, however, neither he nor any other OP received chemistry, which was reserved for a future appointee. In 1794 B. Sippel, after studying chemistry in Würzburg at state expense, was named OP of chemistry with a modest salary. Eight years later he resigned to take over his brother's apothecary shop in Brückenau. Meanwhile, the apothecary E. F. Rumpf had been named AOP of chemistry and pharmacy. After the University's closing, he taught chemistry in the series of medical schools that succeeded the Medical Faculty.

| Chemical representative | Years | Institutional salary | Chemical publications | |
|---|---|---|---|---|
| | | | Before | During |
| J. I. J. Döllinger | 1775 | 80 fl | No | No |
| J. R. M. Joachim | 1775?–1790 | None | No | No |
| B. Sippel | 1794–1802 | 200 fl | No | No |
| E. F. Rumpf | 1800–1803 | None | No | No |

## Bibliography

Böhmer, P. *Die medizinischen Schulen Bambergs in der ersten Hälfte des 19. Jahrhunderts.* Diss., Erlangen University; Erlangen: Hogl, 1971, pp. 19, 26, 38, 59, 66–67, 118.

Lehmann, C. *Über die Medizin an der Academia Ottoniana und Universitas Ottoniano-Fridericiana Bambergensis 1735–1803.* Diss., Erlangen University; Erlangen: Hogl, 1967, pp. 50–51, 58–59, 101, 114.

Weber, H. "Geschichte der gelehrten Schulen im Hochstift Bamberg von 1007–1803." Historischer Verein zu Bamberg: *Bericht,* 42–44 (1879–1881), 285–298, 604.

## BASEL UNIVERSITY

Founded in 1459. Under the jurisdiction of the canton of Basel.

### First Chemical Instruction

In 1639 the Medical Faculty gave permission to a foreign "Chymiatros" to give a private course on the preparation of chemical drugs.

### Chemical Position

Medical Faculty: Although the OPs frequently offered chemistry courses during the seventeenth and eighteenth centuries, they never assigned the subject to one of the chairs. Hence from 1732 to 1758, and again from 1762 to 1777, it was not taught.

### Bibliography

Burckhardt, A. *Geschichte der medizinischen Fakultät zu Basel 1460–1900*. Basel: Rheinhardt, 1917, pp. 178, 181, 191, 192, 196, 213, 244, 466–469.

## BERLIN ACADEMY OF SCIENCES

Founded as the Society of Sciences in 1700; reorganized as the Academy of Sciences and Fine Literature in 1744. Under the jurisdiction of the Elector of Brandenburg, who became the King in Prussia in 1701.

### Chemical Position

Society of Sciences: Despite the hopes of G. W. Leibniz, the Society had neither a salaried position for chemistry nor a laboratory. The chemists among the ordinary members (see below) received as chemists no more than modest funds for their research. However, the Society did pay *Neumann* and *Pott* as teachers in the Medical-Surgical College and *Neumann* as Treasurer (1731–1737).

| Chemist | Years | Institutional salary | Chemical publications | |
|---|---|---|---|---|
| | | | Before | During |
| F. Hoffmann | 1709–1712 | None | Yes | No |
| C. Neumann | 1721–1737 | Yes (see profile) | No | Yes |
| J. H. Pott | 1722–1744 | None | Yes | Yes |
| A. S. Marggraf | 1738–1744 | None | No | Yes |

Academy of Sciences: When King Frederick II reorganized the Society into the Academy, he established a few salaried positions. Both *Pott* and *Marggraf* began receiving salaries as academicians for chemistry. In 1753 a laboratory with residence (cost: 2,708 Rtl) was completed and *Marggraf* was made its director. The next year *Pott's* salary as academician was added to his salary as teacher in the Medical-Surgical College, probably so that there would be no more than one salaried academician for chemistry. He was retired in 1760. After the Seven Years' War, the laboratory complex was remodeled and a budget (250 Rtl) provided. During the next decade *Marggraf*'s pupils C. A. Gerhard and F. C. Achard became members. Then in 1777, with *Pott* dead and *Marggraf* ailing, Frederick decided to recruit a new chemist, approaching first T. Bergman (proposed salary: 1,500 Rtl), then J. J. Ferber. Ferber was willing to come, but *Achard*'s allies were sufficiently organized to undermine his candidacy. In 1782 *Marggraf* was succeeded by *Achard*, who obtained a larger budget (400 Rtl) four years later. *Achard* withdrew from active participation in the Academy in 1800 to concentrate on developing methods for large-scale production of sugar from beets in Silesia.

## Institutional Histories / 231

| Chemical representative | Years | Institutional salary | Chemical publications | |
|---|---|---|---|---|
| | | | Before | During |
| J. H. Pott | 1744–1760 | Yes (see profile) | Yes | Yes |
| A. S. Marggraf | 1744–1782 | Yes (see profile) | Yes | Yes |
| J. G. Lehmann | 1754–1761 | None (but see profile) | Yes | Yes |
| C. P. Brandes | 1760–1776 | None | No | No |
| C. A. Gerhard | 1768–1821 | Yes (see profile) | Yes | Yes |
| F. C. Achard | 1776–1800 | Yes, from 1779 (see profile) | Yes | Yes |
| M. H. Klaproth | 1788–1817 | Yes, from 1800 (see profile) | Yes | Yes |
| S. F. Hermbstaedt | 1800–1833 | Yes, from 1812 (see profile) | Yes | Yes |

### Bibliography

Ferber, J. J. "Briefe an Friedrich Nicolai aus Mitau und St. Petersburg," edited by H. Ischreyt. *Schriftenreihe Nordost-Archiv*, 7 (1974), 32, 42, 44–45.

Harnack, A. *Geschichte der Königlich Preussischen Akademie der Wissenschaften zu Berlin*. 3 vols. Berlin: Reichsdruckerei, 1900.

Kirsten, C. (Archiv, Academy of Sciences, Berlin). Letter regarding the financial support for chemistry and chemists in the Berlin Society and Academy (30 April 1971).

Müller, H.-H. "Akademie und Wirtschaft im 18. Jahrhundert: Agrarökonomische Preisaufgaben und Preisschriften der Preussischen Akademie der Wissenschaften." *Studien zur Geschichte der Akademie der Wissenschaften der DDR*, 3 (1975), 91–92, 129–132.

## BERLIN MEDICAL-SURGICAL COLLEGE

Founded in 1723; closed in 1809 with the transfer of many professors to Berlin University and/or the Medical-Surgical Military Academy. Under the jurisdiction of the King in Prussia.

### Chemical Position

Probably thanks to *G. E. Stahl*, the College opened with *J. H. Pott* as OP of theoretical chemistry and *C. Neumann* as OP of practical chemistry. Upon *Neumann*'s death in 1737, *Pott* took over his responsibilities and salary. He had access to the Court Apothecary Shop for instructional purposes. In 1753 *Pott* tried to arrange for the appointment of his new son-in-law, E. G. Kurella, as AOP of chemistry with rights of succession for the ordinary chair. However, *J. T. Eller* managed to get his own protégé, C. P. Brandes, named AOP. Thwarted, *Pott* launched a literary campaign against *Eller* that soon degenerated into an angry feud. Upon *Pott*'s death in 1777, the Academy of Sciences, which had been paying his salary, recommended that a pharmacist be appointed to succeed him. The academicians apparently hoped that the salary savings could be used for their own purposes. Court Apothecary, C. H. Pein, was soon appointed OP of chemistry and pharmacy. He was dismissed for incompetence in 1792 and *S. F. Hermbstaedt*, who had been seeking the post for three years, was named his replacement. In 1798 D. L. Bourguet was appointed AOP of chemistry. When the College closed, *Hermbstaedt* was transferred to Berlin University's Philosophical Faculty as AOP of technical chemistry.

| Chemical representative | Years | Institutional salary | Chemical publications | |
|---|---|---|---|---|
| | | | Before | During |
| J. H. Pott | 1723–1777 | Yes (see profile) | Yes | Yes |
| C. Neumann | 1723–1737 | Yes (see profile) | No | Yes |
| C. P. Brandes | 1754–1776 | None? | No | Yes |
| C. H. Pein | 1777–1792 | Yes | No | No |
| S. F. Hermbstaedt | 1792–1809 | Yes (see profile) | Yes | Yes |
| D. L. Bourguet | 1798–? | None | Yes | Yes |

## Bibliography

Dorwart, R. A. *The Prussian Welfare State before 1740.* Cambridge, Mass.: Harvard University Press, 1971, pp. 253, 263–266.

Lehmann, H. *Das Collegium medico-chirurgicum in Berlin als Lehrstätte der Botanik und der Pharmacie.* Diss., Berlin University; Berlin: Triltsch and Huther, 1936.

"Die Matrikel des preussischen Collegium medico-chirurgicum in Berlin 1730 bis 1797," edited by A. von Lyncker. *Archiv für Sippenforschung,* 11 (1934), 129–158; 12 (1935), 97–135. (Cross-referenced as "Matrikel.")

Stürzbecher, M. "Die Anfänge des medizinischen Unterrichts am Collegium medico-chirurgicum in Berlin." In *Verhandlungen des XX. Internationalen Kongresses für Geschichte der Medizin.* Berlin: Olms, 1968, p. 774.

"Ueber die Medizinalanstalten in Berlin." *Berlinisches Jahrbuch der Pharmacie,* 2 (1796), 198.

Also see Berlin Academy (Harnack).

## BERLIN MINING SCHOOL

Founded in 1770. Under the jurisdiction of the King in Prussia.

### Chemical Position

C. A. Gerhard, the School's chief planner, arranged for the appointment of the apothecary and minor chemist, V. Rose, Sr., as the teacher of chemistry. Rose never actually lectured—first because too few students enrolled, then because he became sick. His successor, E. G. Kurella, did not actually begin teaching until 1774. Soon afterwards, Gerhard, who was responsible for mineralogy and the mining sciences, took over chemistry as well. He obtained some government support for experiments. In 1779 he relinquished chemistry to F. C. Achard, who passed it on to M. H. Klaproth five years later.

| Chemical representative | Years | Institutional salary | Chemical publications | |
|---|---|---|---|---|
| | | | Before | During |
| V. Rose, Sr. | 1770–1771 | 100 Tl | Yes | No |
| E. G. Kurella | 1771–1775? | 100 Tl | Yes | No |
| C. A. Gerhard | 1775?–1779 | Yes (see profile) | Yes | Yes |
| F. C. Achard | 1779–1784 | Yes (see profile) | Yes | Yes |
| M. H. Klaproth | 1784–1816 | Yes (see profile) | Yes | Yes |
| S. F. Hermbstaedt | 1817–1833 | Yes (see profile) | Yes | Yes |

### Bibliography

Krusch, P. *Die Geschichte der Bergakademie zu Berlin von ihrer Gründung bis zur Neueinrichtung im Jahre 1860.* Berlin: Geologische Landesanstalt and Bergakademie, 1904, pp. iv–xxv.

## BERLIN VETERINARY MEDICINE SCHOOL

Founded in 1790. Under the jurisdiction of the King in Prussia.

### Chemical Position

In 1790, after studying chemistry and botany in Leipzig at state expense, C. Ratzeburg was appointed as the School's apothecary and charged with teaching these and related subjects.

| Chemical representative | Years | Institutional salary | Chemical publications | |
|---|---|---|---|---|
| | | | Before | During |
| C. Ratzeburg | 1790–1808 | 1806: 500 Rtl | No | No |
| F. W. Christ | 1808–1817 | 1815: 500 Rtl | No | No |

*Bibliography*

Boetzkes, H. "Ueber die Rolle und die Aufgaben der Apotheke der Veterinärmedizinischen Fakultät in den vergangenen 170 Jahren." *Pharmazie, Beilage Pharmazeutische Praxis* (1960), 257–261.

Rieck, W. "Zur ältesten Geschichte der Tierärztlichen Hochschule Berlin." *Veterinärhistorisches Jahrbuch*, 4 (1928), 118–119, 135–136.

## BERLIN ARTILLERY ACADEMY

Founded in 1791; incorporated in the General War School in 1810. Under the jurisdiction of the King in Prussia.

*Chemical Position*

The Academy's founder, General G. F. von Tempelhoff, arranged for *M. H. Klaproth*, who had been giving chemical instruction to the artillery corps as an unaffiliated and unsalaried "professor," to be appointed to a salaried chair for chemistry. Klaproth was transferred to the staff of the General War School in 1810, but his salary was attached to his chair at Berlin University.

| Chemical representative | Years | Institutional salary | Chemical publications | |
|---|---|---|---|---|
| | | | Before | During |
| M. H. Klaproth | 1791–1812? | Yes (see profile) | Yes | Yes |
| S. F. Hermbstaedt | 1820–1833 | Yes (see profile) | Yes | Yes |

*Bibliography*

*Adres-Calender der Königl.-Preussischen Haupt- und Residenz-Städte Berlin und Potsdam.* Berlin: Akademie, 1793, p. 76; 1795, pp. 76–77; 1800, pp. 49–50.

Scharfenort, L. von. *Die königlich Preussische Kriegsakademie, 1810–1910.* Berlin: Mittler, 1910, pp. 9–10, 322.

## BERLIN SURGICAL PÉPINIÈRE

Founded in 1795; renamed the Medical-Surgical Pépinière in 1804 and the Medical-Surgical Military Academy in 1811. Under the jurisdiction of the King in Prussia.

*First Chemical Instruction*

At first the students were required to make their own arrangements for taking chemistry at the Medical-Surgical College.

*Chemical Position*

In 1797, when the school was reorganized and its budget enlarged, the professors at the Medical-Surgical College were granted small salaries in recompense for allowing 26 students to enroll in their courses without paying the regular fees. *S. F. Hermbstaedt* was assigned experimental physics, chemistry, and pharmacy and provided with access to the laboratory and supplies of the Court Apothecary Shop for teaching purposes. From 1804 he was responsible for chemistry and pharmacy, while C. D. Turte took on physics and chemistry.

Institutional Histories / 234

| Chemical representative | Years | Institutional salary | Chemical publications | |
|---|---|---|---|---|
| | | | Before | During |
| S. F. Hermbstaedt | 1797–1833 | Yes (see profile) | Yes | Yes |
| C. D. Turte | 1804–1832? | 1811: 300–390 Tl | No | Yes |

*Bibliography*
Preuss, J. D. E. *Das Königlich Preussische medizinisch-chirurgische Friedrich-Wilhelm's Institut (ursprünglich chirurgische Pépinière) zu Berlin: Ein geschichtlicher Versuch.* Berlin: Unger, 1819, pp. 59–60, 68–69, 164–165.
Schickert, W. *Die Militärärztlichen Bildungsanstalten von ihrer Gründung bis zur Gegenwart.* Berlin: Mittler, 1895, pp. 41–43, 88–91, 264.

## BONN UNIVERSITY
Founded in mid-1770s as the "Maxische Akademie"; raised to university status in 1784. Under the jurisdiction of the Elector of Cologne until 1794. Closed by France in 1797 and replaced by a central school, which, in turn, was closed in 1804.

*First Chemical Instruction*
In 1789 A. N. Arndt, the new OP of mining, began teaching smelting and assaying in the mint.

*Chemical Position*
Medical Faculty: From 1784 the government planned to establish a chair of chemistry. In 1789 J. A. Kracht, a cameralistics student, was selected for the post and sent to Würzburg and Vienna to study chemistry. However, he died while in Vienna. The chair was then promised to the physician and amateur chemist, *F. Wurzer*, on the condition that he undertake the necessary training at his own expense. He did so and was named OP of chemistry and materia medica in 1793.

| Chemical representative | Years | Institutional salary | Chemical publications | |
|---|---|---|---|---|
| | | | Before | During |
| F. Wurzer | 1793–1797 | Yes (see profile) | Yes | Yes |

Central School: Though *Wurzer* refused to take an oath of loyalty to the French Republic in 1797, he was more flexible the following year. Consequently, he was appointed professor of chemistry, physics, natural history, and botany in the new Central School. From 1800 to 1804 he was responsible for physics and chemistry.

| Chemical representative | Years | Institutional salary | Chemical publications | |
|---|---|---|---|---|
| | | | Before | During |
| F. Wurzer | 1798–1804 | Yes (see profile) | Yes | Yes |

*Bibliography*
Braubach, M. *Die erste Bonner Hochschule: Maxische Akademie und kurfürstliche Universität, 1774/77 bis 1798.* Bonn: Bouvier and Röhrscheid, 1966, pp. 175–179.
"Quellen zur Geschichte des Rheinlandes im Zeitalter der französischen Revolution . . . 1797–1801," edited by J. Hansen. Gesellschaft für rheinische Geschichtskunde: *Publikationen,* 42, no. 4 (1938), 393–395, 886–887, 953, 1087, 1263–1264. (Cross-referenced as "Quellen".)

## BRUNSWICK COLLEGIUM CAROLINUM

Founded in 1745. Under the jurisdiction of the Duke of Brunswick until 1806. Eventually became the Technische Hochschule.

### First Chemical Instruction
Around 1747 H. M. Kaulitz, a mining official without a formal appointment at the Collegium, was asked to teach chemistry. Although his classes were small, he continued offering chemistry courses until 1770, when he was transferred to Blankenburg.

### Chemical Position
Kaulitz's successor, L. Crell, was named OP of metallurgy. His attempts to get a laboratory were unsuccessful. Upon his departure for Helmstedt University in 1774, the chair was abolished. In 1789 G. F. Hildebrandt, OP of anatomy at Brunswick Anatomical-Surgical College, was given permission to teach chemistry in the Collegium. Partly because his requests for the title of OP of chemistry fell on deaf ears, he accepted a call to Erlangen University in 1793.

| Chemical representative | Years | Institutional salary | Chemical publications | |
|---|---|---|---|---|
| | | | Before | During |
| L. Crell | 1771–1774 | Yes (see profile) | Yes | Yes |
| G. F. Hildebrandt | 1789–1793 | None | No | Yes |

### Bibliography
Bleek, W. "Von der Kameralausbildung zum Juristenprivileg: Studium, Prüfung und Ausbildung der höheren Beamten des allgemeinen Verwaltungsdienstes in Deutschland im 18. und 19. Jahrhundert." *Historische und Pädagogische Studien*, 3 (1972), 158, 515, 604.
Brunswick Collegium Carolinum. Lecture catalogues (1761–1800). In *Gelehrte Beyträge zu den Braunschweigischen Anzeigen* (1761–1787), then *Braunschweigisches Magazin* (1788–1800).
Eschenburg, J. J. *Entwurf einer Geschichte des Collegii Carolini in Braunschweig*. Berlin and Stettin: Nicolai, 1812, pp. 10–17.
Hickel, E., and M. Okrusch. "Die Mineralogie am Collegium Carolinum zu Braunschweig in der Zeit von 1745 bis 1900." Technische Universität Carolo-Wilhelmina, Braunschweig: *Mitteilungen*, 11 (1976), 40–41, 62, 73, 75.
Niedersächsisches Staatsarchiv, Wolfenbüttel (2 Alt 14703, 14812).
*Die Technische Hochschule Braunschweig*. Edited by W. Schneider. West Berlin: Länderdienst, 1963, pp. 10–17.

## BRUNSWICK ANATOMICAL-SURGICAL COLLEGE

Founded in 1750. Under the jurisdiction of the Duke of Brunswick until 1806. Closed in 1869.

### Chemical Position
In 1792 G. F. Hildebrandt, OP of anatomy and physiology since 1785, obtained responsibility and funds for teaching chemistry as well. However, when he went to Erlangen University, chemistry was not included in his successor's responsibilities.

| Chemical representative | Years | Institutional salary | Chemical publications | |
|---|---|---|---|---|
| | | | Before | During |
| G. F. Hildebrandt | 1792–1793 | Yes | Yes | Yes |

## Bibliography

Brunswick Anatomical-Surgical College. Lecture catalogues (1765–1800). In *Gelehrte Beyträge zu den braunschweigischen Anzeigen* (1765–1787), then *Braunschweigisches Magazin* (1788–1800).

Döhnel, K.-R. "Das Anatomisch-Chirurgische Institut in Braunschweig 1750–1869." *Braunschweiger Werkstücke*, 19 (1957), 35–36.

## BUDAPEST UNIVERSITY

Founded in Tyrnau (Trnava) in 1635; provided with a medical faculty in 1770; relocated to Buda in 1777 and across the Danube to Pest in 1784. Under the jurisdiction of the Hapsburg Emperor.

### Chemical Position

Medical Faculty: Modeled after Vienna University's Medical Faculty, the new Faculty had a chair of botany and chemistry. Initially *J. J. Winterl*, the chair's first occupant, was unsuccessful in his efforts to get budgetary support. In 1773, for instance, he was admonished that prospective physicians and pharmacists had no need to see "vague, useless, and curious experiments." With the move to Pest in 1784, however, he obtained a budget for experiments (1784: 300 fl; 1795: 900 fl for botany and chemistry) and perhaps an official laboratory. That same year, P. Kitaibel was appointed adjunct professor of botany and chemistry (budget, 1795: 200 fl).

| Chemical representative | Years | Institutional salary | Chemical publications | |
|---|---|---|---|---|
| | | | Before | During |
| J. J. Winterl | 1770–1809 | Yes (see profile) | Yes | Yes |
| P. Kitaibel | 1784–1816 | 1795: 400 fl | No | Yes |
| J. Schuster | 1809–1838 | Yes | Yes | Yes |

### Bibliography

Antall, J., V. R. Harko, and T. Vida. "Die Ofener Jahre der Medizinischen Fakultät nach der Uebersiedlung der Tyrnauer Universität 1777–1784." *Orvostörténeti Közlemények*, 57–59 (1971), 143, 145, 152.

Endre, H. *Emlékkönyv a Budapesti Királyi Magyar Tudomány Egyetem Orvosi Karának Multjáról és Jelenéről*. Budapest: Athenaeum, 1896, pp. 70–71, 86–98, 365–366.

Györy, T. "Die ersten Jahre der medizinischen Fakultät in Nagyszombat (Tyrnau)." *Sudhoffs Archiv*, 25 (1932), 230, 234–235, 238.

Weh-von Prockl, M. *Personalbibliographien der Professoren der medizinischen Fakultät der ungarischen Universität in Tyrnau und Ofen 1769–1784*. Diss., Erlangen University; Erlangen: Hogl, 1974, pp. 70–80.

## COLMAR CENTRAL SCHOOL

Founded in 1795. Under the jurisdiction of the French Directory of the Upper Rhine. Closed in 1803 and replaced by a secondary school.

### Chemical Position

As the result of a competition held in the spring of 1796, J. C. Bartholdi was appointed professor of physics and chemistry. He continued teaching these subjects in Colmar until 1808.

| Chemical representative | Years | Institutional salary | Chemical publications | |
|---|---|---|---|---|
| | | | Before | During |
| J. C. Bartholdi | 1796–1803 | Yes | No | Yes |

*Bibliography*
See Strasbourg University (Berger-Levrault, pp. cxcv–cxcvii).

## COLOGNE UNIVERSITY

Founded in 1388. Under the jurisdiction of Cologne's Town Council until 1796. Closed by France in 1797 and replaced with a central school, which, in turn, was closed in 1804.

*First Chemical Instruction*
In 1698 J. B. Lambswerde, a member of the Medical Faculty, requested that a chemical laboratory be constructed. Though no action was taken, he may have offered courses on chemistry.

*Chemical Position*
Medical Faculty: Sometime before 1774, J. G. Menn, first OP of medicine from 1761 until his death in 1781, assumed responsibility for chemistry and materia medica. A chemical laboratory was completed for him in 1777. His successor, the apothecary J. M. Müller, was dismissed in 1797 when he refused to swear allegiance to the French Republic. The following year P. Best recommended the establishment of a special medical school to succeed the Medical Faculty and nominated A. N. Scherer for a chair of chemistry and materia medica. For want of funds, nothing came of this proposal.

| Chemical representative | Years | Institutional salary | Chemical publications | |
|---|---|---|---|---|
| | | | Before | During |
| J. G. Menn | 1770?–1781 | Yes | No | No |
| J. H. Müller | 1782–1797 | 1790: 200 fl | No | No |

Central School: Best was made OP of chemistry and experimental physics at this school.

| Chemical representative | Years | Institutional salary | Chemical publications | |
|---|---|---|---|---|
| | | | Before | During |
| P. Best | 1798–1804? | Yes | No | No |

*Bibliography*
Bianco, F. J. von. *Die alte Universität Köln*, vol. I. Cologne: Gehly, 1855, pp. 606, 608.
Bogumil (Stadtarchiv, Cologne). Letter containing evidence of Menn's responsibility for chemistry in 1774 (29 May 1970).
Keussen, H. "Die alte Universität Köln." Kölnischer Geschichtsverein: *Veröffentlichungen*, 10 (1934), 489.
*Medizinische National-Zeitung für Deutschland*, 1 (17 December 1798), 811.
Menn, J. G. *Rede von der Nothwendigkeit der Chemie*. Cologne: Universitätsdruckerey, 1777.
Moritz, F. "Aus der medizinischen Fakultät der alten Universität Köln." In *Festschrift zur Errinerung an die Gründung der alten Universität Köln in 1388*. Cologne, 1938, pp. 237–287.
Pribilla, W. "Die Geschichte der Anatomie an der Universität Köln von 1478 bis 1798." Deutsch-nordische Gesellschaft für Geschichte der Medizin, der Zahnheilkunde und der Naturwissenschaften: *Arbeiten*, 26 (1940), 18.
Schmidt, A. *Die Kölner Apotheken*. Bonn: Hanstein, 1918, pp. 86–87, 149.
"Ueber die innere Verfassung der Universität Kölln, in Rücksicht der Arzneywissenschaft." *Medicinisches und Physisches Journal*, no. 25 (1790), 26–28.
Also see Bonn University ("Quellen," pp. 400, 539–544, 879–883, 946–948).

## DUISBURG UNIVERSITY

Founded in 1655; closed in 1818. Under the jurisdiction of the Elector of Brandenburg, who became the King in Prussia in 1701.

### First Chemical Instruction

F. G. Barbeck, OP of medicine from 1671 to 1703, evidently lectured on chemistry from time to time.

### Chemical Position

Medical Faculty: In 1726 J. A. Timmermann, OP of medicine since 1724, received permission to establish an anatomical theater and chemical laboratory in a Lutheran building. (Duisburg was predominantly Calvinist.) In announcing his chemical lectures (1730–1737), he referred to this laboratory as the "Laboratorio Academico," implying that he had been assigned chemistry. His successor, J. G. Leidenfrost, probably assumed responsibility for chemistry, since he often offered courses on the subject. C. J. Carstanjan, AOP of medicine from 1788 to 1792, then OP until the University was closed, gave chemistry courses almost every year. In 1799 a report referred to him as the professor of chemistry and botany.

| Chemical representative | Years | Institutional salary | Chemical publications | |
|---|---|---|---|---|
| | | | Before | During |
| J. A. Timmermann | 1726?–1742? | 250 Tl | No | No |
| J. G. Leidenfrost? | 1743?–1788? | 1743–1770: 250–270 Tl<br>1770–1794?: 370 Tl | No | Yes |
| C. J. Carstanjan | 1788?–1818 | 1806: 484 Rtl | No | No |

### Bibliography

Ebbers (Stadtarchiv, Duisburg). Letter containing a transcription of all the chemistry offerings at Duisburg between 1727 and 1818 (7 April 1967).

Hesse, W. *Beiträge zur Geschichte der früheren Universität in Duisburg.* Duisburg: Nieten, 1879, pp. 66, 91, 98–99.

Ring, W. *Geschichte der Universität Duisburg.* Duisburg: Stadtverwaltung, 1920, pp. 171–172, 180.

Roden, G. von. "Die Universität Duisburg." *Duisburger Forschungen,* 12 (1968), 203, 210–211, 234–238, 269.

## ERFURT UNIVERSITY

Founded in 1392. Under the jurisdiction of the Elector of Mainz until 1802. Closed by Prussia in 1816.

### First Chemical Instruction

In the late sixteenth century, J. Gramman and A. Starck taught pharmaceutical chemistry at Erfurt.

### Chemical Position

Medical Faculty: The statutes of 1634 called for a chemical laboratory; but because of the Thirty Years' War this provision was never realized. Around 1673 A. Martini was named AOP for chemistry and botany. His successor, C. Cramer, received a token salary. From 1682 until the 1750s, unsalaried AOPs and OPs on their way up the seniority ladder to salaried chairs represented chemistry. In 1756, with the abandonment of seniority as the basis for assignments, H. Ludolf, Jr. rose to a salaried chair without relinquishing responsibility for chemistry—a responsibility that was fulfilled by substitutes, because he was in Mainz. The following year C. A. Mangold, AOP of chemistry since 1755, was named to a new salaried chair of chemistry and botany. Starting in 1777, the OP responsible for chemistry had access to the new laboratory of Erfurt's Academy of Useful Sciences. Nonetheless, chemistry went unrepresented from 1789

Institutional Histories / 239

to 1795. Then *J. B. Trommsdorff* was made a salaried AOP of chemistry and pharmacy. He remained responsible for these subjects, rising to OP in 1811, until the University was closed.

| Chemical representative | Years | Institutional salary | Chemical publications | |
|---|---|---|---|---|
| | | | Before | During |
| A. Martini | 1673?–1678 | None | No | No |
| C. Cramer | 1678–1682 | 50 fl | Yes | Yes |
| J. L. Körber | 1682?–1686 | None | No | Yes |
| L. F. Jacobi | 1689–1715 | None | No | Yes |
| H. C. Alberti | 1690–1692 | None | No | No |
| H. Ludolf, Sr. | 1719–1727 | None | No | Yes |
| H. P. Juch | 1727–1728 | None | No | No |
| I. J. Stahl | 1728–1729 | None | Yes | No |
| L. T. Luther | 1729–1737 | None | No | No |
| A. E. Büchner | 1737–1745 | None | Yes | Yes |
| *H. Ludolf, Jr.* | 1745–1764 | Yes (see profile) | Yes | Yes |
| Substitutes | | | | |
| J. C. Riedel | 1753–1756 | None | No | No |
| A. Nunn | 1757–1762? | None | No | No |
| *C. A. Mangold* | 1755–1767 | Yes (see profile) | Yes | Yes |
| J. P. Nonne | 1767–1771 | Yes | Yes | No |
| W. B. Trommsdorff | 1772–1782 | 1777: 40 Rtl + goods | Yes | Yes |
| J. J. Planer | 1783–1789 | Yes? | Yes | Yes |
| *J. B. Trommsdorff* | 1795–1816 | Yes | Yes | Yes |

Philosophical Faculty: In 1810 *C. F. Bucholz* was named AOP of chemistry in this Faculty and three years later promoted to OP.

| Chemical representative | Years | Institutional salary | Chemical publications | |
|---|---|---|---|---|
| | | | Before | During |
| *C. F. Bucholz* | 1810–1816 | Yes? | Yes | Yes |

**Bibliography**

Abe, H. R. (Lecturer, Erfurt Medical Academy). Letters regarding chemistry at Erfurt University (23 February, 19 May 1970).
Hampel, A. "Die beiden Lehrstühle für Chemie und für Pathologie an der Erfurter Medizinischen Fakultät während der ersten Hälfte des 18. Jahrhunderts." Diss., Erfurt Medical Academy, 1968.
Loth, R. "Die Dozenten der medizinischen Fakultät der Universität Erfurt in den Jahren 1646–1816." Akademie gemeinnütziger Wissenschaften zu Erfurt: *Jahrbücher*, N.F. 33 (1907), 179–250. (Cross-referenced as Loth.)
———. "Die Entwicklung der Anatomie, Chirurgie und Geburtshilfe auf der Universität Erfurt." Allgemeiner ärztlicher Verein von Thüringen: *Korrespondenz-Blätter*, 34 (1905), 161–169, 233–255. (Cross-referenced as Loth, "Entwicklung".)
Oergel, D. "Die Akademie nützlicher Wissenschaften zu Erfurt von ihrer Wiederbelebung durch Dalberg bis zu ihrer endgültigen Anerkennung durch die Krone Preussen (1776–1816)." Akademie gemeinnütziger Wissenschaften zu Erfurt: *Jahrbücher*, N.F. 30 (1904), 160, 165, 167, 206.
Stieda, W. "Erfurter Universitätsreformpläne im 18. Jahrhundert." Akademie gemeinnütziger Wissenschaften zu Erfurt: *Sonderschriften*, 5 (1934).
Wiegand, R. "Namensverzeichnis zur allgemeinen Studentenmatrikel der ehemaligen Univer-

sität Erfurt für die Zeit von 1637 bis 1816." *Beiträge zur Geschichte der Universität Erfurt*, 9 (1962), 9–161; 10 (1963), 13–165. (Cross-referenced as "Matrikel".) Also see Mainz University (Mathy, "Medizinhistorische Minaturen," pp. 96–105).

## ERLANGEN UNIVERSITY

Founded in 1743. Under the jurisdiction of the Margrave of Bayreuth until 1769, then the Margrave of Bayreuth-Ansbach until 1791, then the King in Prussia until 1806.

### Chemical Position

Medical Faculty: When the University was founded, J. F. Weissmann, a renowned local physician with a private laboratory, was made OP of medicine and given responsibility for chemistry. In 1753 the subject was transferred to H. F. Delius. Soon afterwards, space was set aside for a small laboratory in the new wing of the University's building. Finding it inadequate, Delius soon established a laboratory in his residence. From 1783 he received free charcoal for experiments. Upon his death in 1791, J. C. Schreber was temporarily given responsibility for chemistry, then *G. F. Hildebrandt* was called as Delius's successor from Brunswick. He soon transferred to the Philosophical Faculty.

| Chemical Representative | Years | Institutional salary | Chemical publications | |
|---|---|---|---|---|
| | | | Before | During |
| J. F. Weissmann | 1743–1753 | Yes | Yes | Yes |
| H. F. Delius | 1753–1791 | Yes | Yes | Yes |
| J. C. Schreber | 1791–1793 | Yes | Yes | Yes |
| G. F. Hildebrandt | 1793–1796 | Yes | Yes | Yes |

Philosophical Faculty: In 1796 *Hildebrandt* obtained a transfer to this Faculty on the grounds that it was the proper home of chemistry. He was provided with a new laboratory in 1799 and 300 Tl for a chemical-physical apparatus in 1800. Later, he had a budget (160 fl).

| Chemical representative | Years | Institutional salary | Chemical publications | |
|---|---|---|---|---|
| | | | Before | During |
| G. F. Hildebrandt | 1796–1816 | Yes | Yes | Yes |
| J. S. C. Schweigger | 1817–1819 | Yes | Yes | Yes |
| C. W. G. Kastner | 1821–1857 | Yes | Yes | Yes |

### Bibliography

Gastauer, T. *Die Personalbibliographien des Lehrkörpers der philosophischen Fakultät zu Erlangen von 1743 bis 1806.* Diss., Erlangen University; Erlangen: Hogl, 1968.

Glasser, L. *Personalbibliographien der Professoren der Medizinischen Fakultät der Universität Erlangen von 1743–1792.* Diss., Erlangen University; Erlangen: Hogl, 1967.

Henrich, F. "Aus Erlangens chemischer Vergangenheit." Physikalisch-medizinische Sozietät in Erlangen: *Sitzungsberichte*, 38 (1906), 103–139.

―――. "Ueber einen akademischen Lehrgang der Chemie zur Zeit der Phlogistontheorie." In *Beiträge aus der Geschichte der Chemie dem Gedächtnis von Georg W. A. Kahlbaum gewidmet*, edited by P. Diergart. Vienna and Leipzig: Deuticke, 1909, pp. 400–405.

Kaulbars-Sauer, B. *Personalbibliographien der Professoren der medizinischen Fakultät der Universität Erlangen von 1792–1850.* Diss., Erlangen University; Munich: Mikrokopie GmbH, 1969.

Schleebach, A. *Die Entwicklung der chemischen Forschung und Lehre an der Universität Erlangen von ihrer Gründung (1743) bis zum Jahre 1820.* Bayreuth: Seuffer, 1937.

## FRANKFURT AN DER ODER UNIVERSITY

Founded in 1506; closed in 1811 with the transfer of many professors to Breslau University. Under the jurisdiction of the Elector of Brandenburg, who became the King in Prussia in 1701.

*First Chemical Instruction*

In the 1690s both OPs of medicine, V. Behr and B. Albinus, were offering to teach chemistry.

*Chemical Position*

Medical Faculty: In 1740 J. F. Cartheuser, AOP of medicine at Halle University, was appointed to a supernumerary chair for chemistry, pharmacy, and materia medica. Four years later he kept these subjects when he rose to the second chair of anatomy and botany. His original chair was abolished. In 1759 he rose to the first chair of therapy and pathology. The second chair for theory, chemistry, anatomy, and botany remained vacant until 1763 when Helmstedt's chemical representative, P. I. Hartmann, was recruited with funds allocated to the University to make up for the neglect during the Seven Years' War. Hartmann remained responsible for chemistry when he moved to the first chair in 1777. His successor, B. C. Otto, former OP of natural history and economics at Greifswald University, was responsible for natural history, botany, chemistry, anatomy, and obstetrics. In late 1810 J. F. John, a Berlin chemist who had recently qualified as a PD of chemistry, was named to a new third chair for chemistry and pharmacy. However, when the University was closed a few months later, he was not among those transferred to Breslau University.

| Chemical representative | Years | Institutional salary | Chemical publications | |
|---|---|---|---|---|
| | | | Before | During |
| J. F. Cartheuser | 1740–1759 | Yes (see profile) | Yes | Yes |
| P. I. Hartmann | 1763–1788 | 1,000 Tl | Yes | Yes |
| B. C. Otto | 1788–1810 | 1,000 Tl | Yes | Yes |
| J. F. John | 1810–1811 | Yes | Yes | Yes? |

*Bibliography*

"Aeltere Universitätsmatrikeln der Universität Frankfurt a. d. O.: Die Matrikel, 1649–1811," edited by E. Friedländer. Archivverwaltung, Prussia: *Publikationen*, 36 (1888). (Cross-referenced as "Matrikel".)
Bardong, O. "Die Breslauer an der Universität Frankfurt (Oder): Ein Beitrag zur schlesischen Bildungsgeschichte 1648–1811." Historische Kommission für Schlesien: *Quellen und Darstellungen zur schlesischen Geschichte*, 14 (1970), 88, 96, 104, 113.
Frankfurt a. d. Oder University. Lecture catalogues (1679 . . . WS 1793/94). Bibliotheca, Wroclaw University.
Hausen, C. R. *Geschichte der Universität und Stadt Frankfurt an der Oder*. Berlin: Apitz, 1800, pp. 24–25, 113.
Hitzig, J. E. *Verzeichnis im Jahre 1825 in Berlin lebender Schriftsteller*. Berlin: Dümmler, 1826, pp. 126–128.
Lenz, M. *Geschichte der königlichen Friedrich-Wilhelms-Universität zu Berlin*. 4 vols. Halle: Waisenhaus, 1910, I: 41.
Also see Pfaff (*Sammlung*, p. 263) and Halle University (Conrad, p. 56).

## FREIBERG MINING ACADEMY

Founded in 1765. Under the jurisdiction of the Elector of Saxony.

*First Chemical Instruction*

In 1733, over three decades before the school was founded, J. F. Henckel evidently obtained permission to use his official facilities for teaching chemistry in Freiberg.

## Institutional Histories / 242

### Chemical Position

Upon the Academy's founding in 1765, C. E. Gellert, who as Henckel's successor had been teaching chemistry in Freiberg since 1749, was named professor of metallurgical chemistry. He outlived his first intended successor, C. F. Wenzel. His actual successor, W. A. Lampadius, was AOP of chemistry for a year, then OP of chemistry and smelting. In 1797 Lampadius obtained a laboratory (cost: 3,257 Tl) and budget (100 Rtl).

| Chemical representative | Years | Institutional salary | Chemical publications | |
|---|---|---|---|---|
| | | | Before | During |
| C. E. Gellert | 1765–1795 | Yes (see profile) | Yes | Yes |
| W. A. Lampadius | 1794–1842 | Yes (see profile) | Yes | Yes |

### Bibliography

Baumgärtel, H. "Bergbau und Absolutismus: Der sächsische Bergbau in der zweiten Hälfte des 18. Jahrhunderts und Massnahmen zu seiner Verbesserung nach dem Siebenjährigen Kriege." *Freiberger Forschungshefte: Kultur und Technik*, D44 (1963), 64, 82, 95.
Herrmann, W. "Die Entstehung der Freiberger Bergakademie." Ibid., D2 (1953), 23–42.
Lampadius, W. A. *Beyträge zur Erweiterung der Chemie.* Vol. I. Freiberg: Craz and Gerlach, 1804, frontispiece [picture of the laboratory] and "Erklärung der Titelvignette."
Weber, W. "Innovationen im frühindustriellen deutschen Bergbau und Hüttenwesen: Friedrich Anton von Heynitz." *Studien zu Naturwissenschaft, Technik und Wirtschaft im Neunzehnten Jahrhundert*, 6 (1976), 152–167.

## FREIBURG IM BREISGAU UNIVERSITY

Founded in 1457; relocated to Konstanz from 1686 to 1698 and again from 1713 to 1715 on account of wars. Under the jurisdiction of the Hapsburg Emperor until 1798.

### First Chemical Instruction

Around 1620 a laboratory was established at the University. According to the curriculum of 1624, the second OP of medicine was to use the laboratory to show the medical students how to prepare chemical medicines.

### Chemical Position

Medical Faculty: Though the curricula of 1624 and 1671 both assigned chemistry to one of the OPs, these plans were probably not followed for long. Regular teaching of the science began in 1759 when F. J. B. Baader, OP of materia medica, was assigned chemistry and botany as well. A decade later A. Lipp was appointed to a new chair for chemistry and botany. His successor, taking advantage of the funds made available by the confiscation of Jesuit properties, obtained a laboratory and auditorium, budget (75 fl), and laboratory assistant (wage: 150 fl) in 1779/80.

| Chemical representative | Years | Institutional salary | Chemical publications | |
|---|---|---|---|---|
| | | | Before | During |
| F. J. B. Baader | 1759–1768 | Yes | No | No |
| A. Lipp | 1768–1775 | Yes | No | No |
| F. Menzinger | 1775–1818 | Yes | No | No |

### Bibliography

Diepgen, P., and E. T. Nauck. "Die Freiburger medizinische Fakultät in der österreichischen Zeit." *Beiträge zur Freiburger Wissenschafts- und Universitätsgeschichte*, 16 (1957), 31, 36, 69–73.

Lüttringhaus, A., and C. Baumfelder. "Die Chemie an der Universität Freiburg i. Br. von den Anfängen bis 1920." Ibid., 18 (1957), 23–76.
Nauck, E. T. "Zur Geschichte des medizinischen Lehrplans und Unterrichts der Universität Freiburg i. Br." Ibid., 2 (1952), 27, 33–34, 36, 41–42, 44, 106–109, 116–117.
Wolz, W. "Pharmazeutische Ausbildung an der Universität Freiburg im Breisgau und im Oberrheingebiet." Ibid., 24 (1960), 94–95.

## FULDA UNIVERSITY

Founded in 1734. Under the jurisdiction of the Prince-Bishop of Fulda until 1802. Closed by Orange-Nassau in 1805.

### Chemical Position

Medical Faculty: According to the statutes of 1734, the professor of physiology would also be responsible for physics, chemistry, and botany. However, there is no evidence that this provision was followed when the Medical Faculty was established in 1738. At some later date, possibly in the 1760s, F. A. Schlereth, OP of medicine since 1759, became responsible for teaching medical practice and chemistry. In 1775 the pharmacist F. C. C. Lieblein was named AOP of practical chemistry and seven years later, OP of chemistry and botany.

| Chemical representative | Years | Institutional salary | Chemical publications | |
|---|---|---|---|---|
| | | | Before | During |
| F. A. Schlereth | 1760s?–1782 | Yes | No | No? |
| F. C. C. Lieblein | 1775–1805 | Yes, from 1782? | No | Yes |

### Bibliography

Mühl, W. A. *Die Aufklärung an der Universität Fulda mit besonderer Berücksichtigung der philosophischen und juristischen Fakultät (1734–1805)*. Fulda: Parzeller, 1961, p. 11.
Siemens, R. "Zur Geschichte der Apotheken Fuldas." Fuldaer Geschichtsverein: *Fuldaer Geschichtsblätter*, 9 (1910), 180–181.
"Die Studentenmatrikel der Adolphsuniversität zu Fulda (1734–1805)." Fuldaer Geschichtsverein: *Veröffentlichungen*, 15 (1936).

## GIESSEN UNIVERSITY

Founded in 1607. Under the jurisdiction of the Landgrave of Hesse-Darmstadt.

### First Chemical Instruction

Between 1610 and 1620, J. D. Mylius, a student of "Chymiatrie," and G. Horst and L. Jungermann, both OPs of medicine, were all deeply involved in chemical research and instruction.

### Chemical Position

Medical Faculty: From the early 1700s, AOP G. C. Möller evidently had responsibility for teaching chemistry. His successor, J. T. Hensing, had access to a laboratory in which to conduct his classes. Between 1723 and his death in 1726, Hensing was also OP of natural and chemical philosophy in the Philosophical Faculty. After Hensing's death, however, chemistry lacked an official representative for so long that the laboratory was diverted to other purposes. In 1765 J. W. Baumer, who had been called from Erfurt University, was given chemistry with a salary increase (200 fl) and a budget (30 fl). Soon afterwards, construction began on a laboratory; but it was not finished until 1783 (cost: 100 Rtl?). Upon Baumer's death in 1788, chemistry was assigned to K. W. Müller, an OP of medicine who had earlier quarreled with him about the right to use the laboratory. In 1818 Müller's successor, P. F. W. Vogt, led the way in arranging for the transfer of chemistry to the Philosophical Faculty, where it was represented first by L. Zimmermann, then J. Liebig.

| Chemical representative | Years | Institutional salary | Chemical publications | |
|---|---|---|---|---|
| | | | Before | During |
| G. C. Möller | 1700?–1717 | None | No | Yes |
| J. T. Hensing | 1717–1726 | None? | Yes | Yes |
| J. W. Baumer | 1765–1788 | 950 fl | Yes | Yes |
| K. W. Müller | 1788–1815 | 1795: 387 fl + goods<br>1803: 452 fl + goods | Yes | Yes |
| P. F. W. Vogt | 1817–1818 | Yes | No | No |

Economics Faculty: When this Faculty was established in 1777, Baumer was assigned to its chair of chemistry and mineralogy (salary increase: 30 fl plus 2 loads of charcoal). However, the government allowed the chair and Faculty to lapse after Baumer's death.

| Chemical representative | Years | Institutional salary | Chemical publications | |
|---|---|---|---|---|
| | | | Before | During |
| J. W. Baumer | 1777–1788 | 30 fl | Yes | Yes |

### Bibliography
Hessisches Staatsarchiv, Darmstadt (Abt. VI, 1 Giessen: Konv. 28, Fsc. 1a, 5, 13).
Hock, L. "Beitrag zur Geschichte der Chemie in Giessen." In *Ludwigs-Universität, Justus Liebig-Hochschule, 1607–1957: Festschrift zur 350-Jahrfeier*. Giessen, 1957, pp. 288–307.
Lehnert, G. "Wie Liebig Professor wurde." *Volk und Scholle*, 8 (1930), 50–54.
Leist, W. (Bibliothek, Giessen University). Letter concerning Müller's salary and budget (18 March 1971).
Schmitz, R. *Die deutschen pharmazeutisch-chemischen Hochschulinstitute*. Ingelheim am Rhein: Boehringer, 1969, pp. 129–131.
*Die Universität Giessen von 1607 bis 1907*. Vol. I. Giessen: Töpelmann, 1907, pp. 378, 380–381, 383, 388, 431, 444.
Weihrich, G. *Beiträge zur Geschichte des chemischen Unterrichts an der Universität Giessen*. Giessen: Münchow, 1891.
Also see Mangold (Schyra, "Das Reformwerk," pp. 25–27).

## GÖTTINGEN UNIVERSITY

Founded in 1734. Under the jurisdiction of the Elector of Hanover (King of England) until 1806.

### First Chemical Instruction

In 1735 the physician J. C. Cron was granted permission to offer private "collegia chymico-metallurgico-practica."

### Chemical Position

Medical Faculty: The University's planners frequently mentioned the desirability of having a chemistry professor and laboratory in the Medical Faculty. Although there were not enough funds for anything to come of this talk, one or more of the professors and private lecturers taught chemistry nearly every semester. In 1753 the science was formally assigned to *R. A. Vogel*, who was called from Erfurt to Göttingen as a salaried AOP of medicine. He declined to use space in the University Apothecary Shop for teaching the subject, probably because no budget was provided. He rose to OP of medicine and chemistry in 1760. His successor, *J. F. Gmelin*, started as an OP in the Philosophical Faculty and AOP in the Medical Faculty in 1775. But three years later, he moved completely into the Medical Faculty, most likely so that he could share in the M.D. fees. In 1783 a laboratory measuring forty feet by twenty feet, a lecture hall, and residence were constructed for him (cost: 7,406 Tl).

| Chemical representative | Years | Institutional salary | Chemical publications | |
|---|---|---|---|---|
| | | | Before | During |
| R. A. Vogel | 1753–1774 | Yes (see profile) | No | Yes |
| J. F. Gmelin | 1775–1804 | Yes, from 1778 (see profile) | Yes | Yes |
| F. Stromeyer | 1805–1835 | Yes | No | Yes |
| L. Crell | 1810–1816 | Yes (see profile) | Yes | No |

## Bibliography

Bärens, J. G. "Kurze Nachricht von Göttingen entworfen im Jahre 1754." Geschichtsverein für Göttingen und Umgebung: *Jahrbuch*, 1 (1908), 55–117.

Cron, J. C. *De praestantia et utilitate studii chymici auctoritate et consensu amplissimae facultatis medicae pauca disserit, ac simul ad collegia chymico-metallurgico-practica . . . invitat.* Göttingen: Hager, 1735.

Ganss, G.-A. *Geschichte der pharmazeutischen Chemie an der Universität Göttingen, dargestellt in ihrem Zusammenhang mit der allgemeinen und der medizinischen Chemie.* Diss., Göttingen University; Marburg: Euker, 1937.

Göttingen University. Lecture catalogues (1748–1800). In *Göttingische Anzeigen von gelehrten Sachen* (1748–1800).

Hamann (Niedersächsisches Staatsarchiv, Hanover). Letter with information about the new laboratory's estimated cost (3,300 Rtl) and Gmelin's salary in 1782 (27 November 1970).

"Die Matrikel der Georg-August-Universität zu Göttingen 1743–1837," edited by G. von Selle. Historische Kommission für Hannover, Oldenburg, Braunschweig, Schaumburg-Lippe und Bremen: *Veröffentlichungen*, 1–2 (1937). (Cross-referenced as "Matrikel".)

Pohlai (Archiv, Göttingen University). Letter bearing on the salaries of Vogel and Gmelin and on the actual cost of the laboratory (17 December 1970).

Pütter, J. S. *Versuch einer academischen Gelehrten-Geschichte von der Georg-Augustus-Universität zu Göttingen.* Vol. I. Göttingen: Vandenhoek, 1765, pp. 159, 291; Vol. II. Göttingen: Vandenhoeck and Ruprecht, 1788, pp. 45–46, 148, 258, 330–331; Vol. III, continuation by J. C. F. Saalfeld. Göttingen: Helwing, 1820, pp. 75–78, 330–331, 447–453.

Rössler, E. F. *Die Gründung der Universität Göttingen.* Göttingen: Vandenhoeck and Ruprecht, 1855.

## GRAZ LYCEUM

Founded in 1782 on the basis of the old incomplete university (founded 1585) in Graz; returned to university status in 1827. Under the jurisdiction of the Hapsburg Emperor.

### Chemical Position

Medical Faculty: In 1782 the Viennese authorities instructed the professor of internal medicine to teach botany and "some chemistry" as well.

| Chemical representative | Years | Institutional salary | Chemical publications | |
|---|---|---|---|---|
| | | | Before | During |
| von Sartori | 1782–1814? | Yes | No | No |

## Bibliography

Fischer, I. *Medizinische Lyzeen: Ein Beitrag zur Geschichte des medizinischen Unterrichts in Oesterreich.* Vienna and Leipzig: Braumüller, 1915, pp. 10, 18.

Krones, F. von. *Geschichte der Karl Franzens-Universität in Graz.* Graz: Universität Verlag, 1886, pp. 467, 508–509, 590.

## GREIFSWALD UNIVERSITY

Founded in 1456. Under the jurisdiction of the King of Sweden from 1637 to 1815, then Prussia.

*First Chemical Instruction*
Medical professors began teaching pharmaceutical chemistry there in the late sixteenth century.

*Chemical Position*
Medical Faculty: C. March, OP of mathematics, was AOP of chemistry from 1648 to 1655 when he went to Rostock University. A century later, various officials recommended the establishment of a chemical laboratory as one way of improving the University's attractiveness; but nothing came of these proposals. In 1771 the Medical Faculty requested that an adjunct be named for chemistry and anatomy. Three years later, C. E. Weigel, adjunct for botany, was appointed to a new ordinary chair of chemistry and pharmacy. In 1796 he obtained a budget for apparatus and experiments (1796–1800: 300 Rtl; 1801 on: 113 Rtl).

| Chemical representative | Years | Institutional salary | Chemical publications | |
|---|---|---|---|---|
| | | | Before | During |
| C. March | 1648–1655 | 100 Rtl as AOP | No | No |
| C. E. Weigel | 1774–1831 | Yes (see profile) | Yes | Yes |

*Bibliography*
Anselmino, O. "Nachrichten von früheren Lehrern der Chemie an der Universität Greifswald." Naturwissenschaftlicher Verein für Neuvorpommern und Rügen: *Mittheilungen*, 38 (1906), 105–140.
Boriss, H. "Die Entwicklung der Botanik und der botanischen Einrichtungen an der Universität Greifswald." In *Festschrift zur 500-Jahrfeier der Universität Greifswald, 17.10.1956*, vol. II. Greifswald, 1956, pp. 515–540.
Kosegarten, J. G. L. *Geschichte der Universität Greifswald*. Vol. I. Greifswald: Koch, 1856, p. 258.
Seth, I. *Die Universität Greifswald und ihre Stellung in der schwedischen Kulturpolitik 1637–1815*. Berlin: VEB, 1956, pp. 145–146, 176–177, 234, 238.
Valentin, J. "Die Entwicklung der pharmazeutischen Chemie an der Ernst-Moritz-Arndt-Universität in Greifswald." In *Festschrift*, II:468–479.

## HALLE UNIVERSITY

Founded in 1694. Under the jurisdiction of the Elector of Brandenburg, who became the King in Prussia in 1701.

*Chemical Position*
Medical Faculty: In writing the Faculty's statutes in 1694, F. Hoffmann made the senior professor (himself) responsible for practical medicine, plus anatomy, surgery, and chemistry. While he was in Berlin from 1709 to 1712, his former student, AOP H. Henrici, acted as his substitute. Hoffmann probably remained responsible for chemistry until 1718 when the addition of a third chair to the Faculty rendered the original distribution of subjects obsolete. From then until the 1780s, the only official chemical representative was C. C. Strumpf, who served as AOP of chemistry and botany from 1747 to 1754. However, one or more men, including J. Juncker, M. Alberti, P. Gericke, J. F. Cartheuser, J. H. Schulze, F. Hoffmann, Jr., J. J. Lange (OP of mathematics), F. C. Juncker, A. Nietzki, P. Müller, and F. A. Richter, were usually ready to offer courses on chemistry during these decades. Such courses could bring in as much as 10 Rtl per student. Finally in 1787, F. A. C. Gren, who had been serving as a PD for chemistry, was appointed AOP of chemistry. A year later, after a brief period as OP in the Philosophical Faculty (see below), he returned to the Medical Faculty as OP of chemistry and experimental physics. Although he could not persuade the authorities to equip a laboratory in the space set aside for this purpose around 1790, he did obtain a budget for experiments (1791:

203 Tl for chemistry and physics). When Gren died in 1798, H. F. Link and A. N. Scherer were considered as possible successors. The call went to Link because he had an M.D.; but when he turned it down, Scherer was called to the Philosophical Faculty (see below).

| Chemical representative | Years | Institutional salary | Chemical publications | |
|---|---|---|---|---|
| | | | Before | During |
| F. Hoffmann | 1694–1718? | Yes (see profile) | Yes | Yes |
| Substitute | | | | |
| H. Henrici | 1709–1712 | No | No | No |
| C. C. Strumpf | 1747–1754 | No | Yes | No |
| F. A. C. Gren | 1787–1798 | Yes, from 1788 (see profile) | Yes | Yes |

Philosophical Faculty: In 1788 Gren was promoted to OP for natural philosophy, including chemistry, in the Philosophical Faculty. He probably transferred to the Medical Faculty in hopes of eventually sharing in the M.D. fees. With the appointment of his successor Scherer, chemistry was definitively transferred to the Philosophical Faculty. In 1801 Scherer's successor, L. W. Gilbert, sold his physical-chemical apparatus to the University for 1,000 Tl and two years later became Director of the Physical Cabinet and Chemical Laboratory.

| Chemical representative | Years | Institutional salary | Chemical publications | |
|---|---|---|---|---|
| | | | Before | During |
| F. A. C. Gren | 1788 | Yes (see profile) | Yes | Yes |
| A. N. Scherer | 1799–1800 | Yes (see profile) | Yes | Yes |
| L. W. Gilbert | 1801–1811 | Yes | No | No |
| C. W. G. Kastner | 1812–1818 | Yes | Yes | Yes |
| J. S. C. Schweigger | 1819–1857 | Yes | Yes | Yes |

## Bibliography

Bergner, E., and H. Goerke. "Der schwedische Arzt Nils Rosén über die Medizin an der Universität Halle im Jahre 1729." *Sudhoffs Archiv*, 27 (1953), 219–223.
Conrad, J. "Die Statistik der Universität Halle während der 200 Jahre ihres Bestehens." In *Festschriften der vier Fakultäten zum zweihundertjährigen Jubiläum der vereinigten Friedrichs-Universität Halle-Wittenberg, den 3 August 1894: Festschrift der Philosophischen Fakultät*. Halle: Waisenhaus, 1895, pp. 51, 70.
Förster, J. C. *Uebersicht der Geschichte der Universität zu Halle in ihrem ersten Jahrhundert*. Halle: Waisenhaus, 1799, pp. 200, 202, 224–225.
Friedländer, H. "Zur Geschichte der medicinischen Fakultät Halle." *Archiv für die gesamte Medicin*, 3 (1842), 1–21.
Halle University. Lecture catalogues (1694 . . . 1800). Archiv, Halle University.
Kaiser, W., and K.-H. Krosch. "Zur Geschichte der Medizinischen Fakultät der Universität Halle." Halle University: *Wissenschaftliche Zeitschrift: Mathematisch-naturwissenschaftliche Reihe*, 13 (1964), 141–180, 363–430, 583–616, 797–936; 14 (1965), 1–48, 269–302, 357–432, 581–676; 15 (1966), 193–345, 1011–1124; 16 (1967), 603–644. (Cross-referenced as Kaiser and Krosch.)
———. "Die Statuten der Medizinischen Fakultät im 18. Jahrhundert." Halle University: *Wissenschaftliche Beiträge der Martin-Luther-Universität Halle-Wittenberg*, no. 3 (1967), 77–103. (Cross-referenced as Kaiser and Krosch, "Statuten".)
Kaiser, W., K.-H. Krosch, and W. Piechocki. "Collegium clinicum Halense." Ibid., pp. 9–76. (Cross-referenced as Kaiser and Krosch, "Collegium".)
Kaiser, W., and W. Piechocki. "Hallesches Druck- und Verlagswesen des 18. und des frühen 19. Jahrhunderts im Dienste der medizinisch-naturwissenschaftliche Publizistik." Halle Uni-

versity: *Wissenschaftliche Zeitschrift: Mathematisch-naturwissenschaftliche Reihe*, 21 (1972), 64–65, 76–77, 79–84.
Kaiser, W., and A. Skrobacki. "Adam Nietzki (1714–1780)." Ibid., 28 (1979), 135–149.
"Die Matrikel der Martin Luther-Universität, Halle-Wittenberg (1690–1730)," edited by F. Zimmermann and F. Juntke. Halle: Universitäts- und Landesbibliothek Sachsen-Anhalt: *Arbeiten*, 2 (1960). (Cross-referenced as "Matrikel".)
Schimank, H. "Ludwig Wilhelm Gilbert und die Anfänge der 'Annalen der Physik.'" *Sudhoffs Archiv*, 47 (1963), 361.
Schrader, W. *Geschichte der Friedrichs-Universität zu Halle*. 2 vols. Berlin: Dümmler, 1894–1895, I: 575, 578; II: 20, 427–428, 432, 496, 528.

## HEIDELBERG UNIVERSITY

Founded in 1386; closed from 1632 to 1652 and from 1688 to 1704 on account of wars. Under the jurisdiction of the Palatine Elector from 1648 to 1802, then Baden.

### First Chemical Instruction
In 1655 both OPs of medicine were teaching chemistry.

### Chemical Position
Medical Faculty: In 1727 the Medical Faculty proposed that one of its members be made responsible for chemistry. Evidently this proposal was accepted, because four years later, apothecary D. Nebel, OP of medicine since 1708, was described as the professor "in chemistry and botany." He had charge of the University Apothecary, who, as "chymicus," was responsible for demonstrating chemical operations. Nebel's successor as chemical representative was probably his son W. B. Nebel, who was assigned medical institutions, pharmacy, and chemistry by the statutes of 1743. These statutes specified that chairs be assigned according to seniority. Accordingly, a succession of men represented chemistry until 1767 when D. W. Nebel was made AOP of anatomy, surgery, and chemistry. In 1771 he was promoted to a new ordinary chair of chemistry and pharmacy. The Faculty's efforts to get a laboratory in 1786 were as fruitless as prior requests in 1680, 1687, and 1743 had been.

| Chemical representative | Years | Institutional salary | Chemical publications | |
|---|---|---|---|---|
| | | | Before | During |
| D. Nebel | 1727?–1733 | Yes | Yes | Yes? |
| W. B. Nebel | 1733?–1746 | 1746: 536 fl + goods | Yes | Yes? |
| J. C. Möller? | 1746?–1749? | 536 fl + goods | No | No |
| F. J. von Oberkamp? | 1749?–1750? | 536 fl + goods | No | No |
| J. J. Moers | 1750?–1758 | 536 fl + goods | No | No |
| M. Gattenhof | 1758–1767 | 600 fl | No | No |
| D. W. Nebel | 1767–1804 | 1771: 250 fl<br>1773: 400 fl<br>1788–1801: 600 fl + goods<br>1804: 719 fl | No | Yes |
| C. W. G. Kastner | 1805–1812 | Yes | Yes | Yes |
| L. Gmelin | 1814–1851 | Yes, from 1816? | Yes | Yes |

### Bibliography
*Acta sacrorum secularium quum anno MDCCLXXXVI . . . festum seculare quartum pio solemnique ritu celebravit Academia heidelbergensis*. Heidelberg: Wiesen, 1787, pp. 254, 551–556.
*Die Matrikel der Universität Heidelberg*. Edited by G. Toepke. Vol. IV. Heidelberg: Winter, 1903, pp. 77, 143, 292.
Merkel, G. "Wirtschaftsgeschichte der Universität Heidelberg im 18. Jahrhundert." Kommission für geschichtliche Landeskunde in Baden-Württemberg: *Veröffentlichungen*, series B, 73 (1973), 263–264.

Stübler, E. *Geschichte der medizinischen Fakultät der Universität Heidelberg, 1386–1925.* Heidelberg: Winter, 1926, pp. 107, 117–123, 125–126, 129, 135–139.

*Urkundenbuch der Universität Heidelberg.* Edited by E. Winkelmann. 2 vols. Heidelberg: Winter, 1886, I: 389–390, 408–409; II: 211, 225, 254, 260, 266. (Cross-referenced as Winkelmann.)

Weisert (Archiv, Heidelberg University). Letter regarding the salaries of the chemical representatives from the 1760s to 1804 and the State Economics High School (4 May 1977).

## HEIDELBERG STATE ECONOMICS HIGH SCHOOL

Founded in Kaiserslautern in 1774; relocated to Heidelberg in 1784. Under the jurisdiction of the Palatine Elector until 1802. Merged into Heidelberg University as the State Economics Section of the Philosophical Faculty in 1803.

### First Chemical Instruction

See Kaiserslautern Cameral High School.

### Chemical Position

G. A. Suckow, the OP of chemistry in Kaiserslautern, remained responsible for the subject in Heidelberg, where he soon obtained a laboratory. He was transferred to the University's Philosophical Faculty in 1803. But after his death in 1813, his chair was allowed to lapse, partly because the Medical Faculty still seemed the most appropriate place for chemistry.

| Chemical representative | Years | Institutional salary | Chemical publications | |
|---|---|---|---|---|
| | | | Before | During |
| G. A. Suckow | 1784–1803 | Yes (see profile) | Yes | Yes |

### Bibliography

Stieda, W. "Die Nationalökonomie als Universitätswissenschaft." Sächsische Gessellschaft der Wissenschaften: Philologisch-Historische Klasse: *Abhandlungen,* 25, no. 2 (1906), 126–130, 332–334, 376, 389–390.

Also Heidelberg University (Winkelmann, II: 281, 316–317; Weisert) and Kaiserslautern Cameral School (Funk, pp. 117–121; Webler).

## HELMSTEDT UNIVERSITY

Founded in 1576. Under the jurisdiction of the Dukes of Brunswick-Celle, Brunswick-Hanover, and Brunswick-Wolfenbüttel from the mid-seventeenth century until 1705; the Elector of Hanover and Duke of Brunswick-Wolfenbüttel until 1745; and then the Duke of Brunswick-Wolfenbüttel until 1806. Closed by Westphalia in 1809.

### First Chemical Instruction

In the late 1660s, the medical OPs H. Conring and H. Meibom were both offering courses on chemistry.

### Chemical Position

Medical Faculty: In 1688 J. A. Stisser was appointed AOP of chemistry. Three years later he rose to OP of botany, chemistry, anatomy, and physiology. Soon afterwards he established an "academic laboratory" at his own expense. However, his successor neither received nor assumed responsibility for chemistry. In 1726 AOP J. A. Schmidt was named to a new fourth chair for the subject. After his death two years later, the chair was abolished. Continuous representation of chemistry began in 1730 when P. Gericke, AOP at Halle University, was called to the vacant third chair as OP of anatomy, pharmacy, and chemistry. Although he could not obtain funding for a laboratory, his successor J. G. Krüger did persuade the government to finance a public chemical course with experiments (budget: 50 Rtl). Krüger's second successor, G. C. Beireis, remained responsible for chemistry until his death in 1809, even though the chemist *L. Crell* was a member of the Faculty from 1774.

| Chemical representative | Years | Institutional salary | Chemical publications | |
|---|---|---|---|---|
| | | | Before | During |
| J. A. Stisser | 1688–1700 | 1688: 250 Rtl<br>1700: 416 Rtl | Yes | Yes |
| J. A. Schmidt | 1726–1728 | 100 Rtl | No | Yes |
| P. Gericke | 1730–1750 | Yes (see profile) | No | Yes |
| J. G. Krüger | 1751–1759 | 400 Rtl | Yes | Yes |
| P. Hartmann | 1760–1763 | 400 Rtl | Yes | No |
| G. C. Beireis | 1763–1809 | over 400 Rtl | No | Yes |

*Bibliography*
Böhmer, J. C. *Memoriae professorum Helmstadiensium in medicorum ordine qui diem suum obierunt prolusionibus binis descriptae.* Wolfenbüttel: Freytag, 1719, pp. 59–62.
Helmstedt University. Lecture catalogues (WS 1613/14, WS 1667/68 . . . WS 1809/10). Herzogsbibliothek, Wolfenbüttel.
―――. "Die Matrikel." Niedersächsisches Staatsarchiv, Wolfenbüttel. (Cross-referenced as "Matrikel".)
Niedersächsisches Staatsarchiv, Wolfenbüttel (37 Alt 257, 259, 386, 437, 438, 442, 446, 447, 448, 644; 4 Alt 19 nr. 50).

## HERBORN HIGH SCHOOL

Founded in 1584. Under the jurisdiction of the Prince of Nassau until 1743, then the Dutch Crown until 1806. Closed by Nassau in 1817.

*First Chemical Instruction*
J. A. Hoffmann, OP of medicine, offered chemistry in 1767.

*Chemical Position*
Medical Faculty: S. J. L. Döring, AOP from 1794 to 1798 and then OP, was evidently responsible for chemistry. He tried to shift the science into the Philosophical Faculty in 1798 so that he could more legitimately claim the right to teach the subjects covered by his senior colleague. However, it was decided that chemistry should remain in the Medical Faculty.

| Chemical representative | Years | Institutional salary | Chemical publications | |
|---|---|---|---|---|
| | | | Before | During |
| S. J. L. Döring | 1794–1817 | 1794: 300 Tl +<br>firewood | No | No |

*Bibliography*
Grün, H. "Die Medizinische Fakultät der Hohen Schule Herborn." Verein für nassauische Altertumskunde und Geschichtsforschung, Wiesbaden: *Nassauische Annalen*, 70 (1959), 137–138, 141–143.

## INGOLSTADT UNIVERSITY

Founded in 1472; moved to Landshut in 1800, then Munich in 1825. Under the jurisdiction of the Elector of Bavaria.

*First Chemical Instruction*
In 1666 J. Tilemann, former OP of medicine at Marburg, lectured on "the alchemical art" for 2 fl a week.

## Chemical Position

Medical Faculty: In 1700 the Faculty failed in its efforts to secure the appointment of an AOP of chemistry. During the 1730s, the University erected a new building that included a small laboratory. However, chemistry seems not to have been assigned to an OP until 1754 when J. A. Carl was appointed to a new chair of chemistry, botany, and materia medica. Five years later he surrendered chemistry to J. P. Spring, who received a new chair for chemistry and practical medicine. Upon Spring's departure in 1760, Carl resumed responsibility for chemistry, using the pharmacist G. L. C. Rousseau as "chemical demonstrator." In 1772, against the wishes of the Medical Faculty, Rousseau was appointed AOP of chemistry and experimental physics. The Faculty managed to push him into the Philosophical Faculty the following year. But in 1776, he returned to the Medical Faculty as OP of chemistry, natural history, and materia medica. Three years later he obtained a new laboratory (cost: ca. 1000 fl). By 1799 Rousseau's successor, G. A. Bertelle, had a budget for a laboratory assistant and supplies (272 fl).

| Chemical representative | Years | Institutional salary | Chemical publications | |
|---|---|---|---|---|
| | | | Before | During |
| J. A. Carl | 1754–1759 | 600 fl | Yes | Yes |
| J. P. Spring | 1759–1760 | 800 fl | No | Yes |
| J. A. Carl | 1760–1772? | 800 fl | Yes | Yes |
| G. L. Rousseau | 1760–1794 | 1760: 300 fl<br>1767: 400 fl<br>1794: 1000fl | No | Yes |
| G. A. Bertelle | 1794–1818 | 1794: 600 fl | No | No |

### Bibliography

Kallinich, G. *Das Vermächtnis Georg Ludwig Claudius Rousseaus an die Pharmazie: Zweihundert Jahre Pharmazie an der Universität Ingolstadt-Landshut-München 1760–1960.* Munich: Govi, 1960.

Prantl, C. *Geschichte der Ludwig-Maximilians-Universität in Ingolstadt, Landshut, München.* Vol. I. Munich: Kaiser, 1872, pp. 494–495, 497, 501, 530–531, 599–602, 608–610, 676–678, 683.

## INNSBRUCK UNIVERSITY

Founded in the 1670s; reduced to lyceum status in 1782; restored to university status in 1791. Under the jurisdiction of the Hapsburg Emperor until 1805. Reduced to lyceum status by Bavaria in 1810.

### Chemical Position

Medical Faculty: In 1735 chemistry was assigned to the OP of medical institutions. Four decades later, S. B. Schiverek was appointed to a new chair of botany and chemistry. In 1777 he persuaded the government to use funds from confiscated Jesuit properties for a new laboratory (cost: 994 fl) and budget (232 fl). However, when the University was downgraded five years later, he and the laboratory's apparatus were moved to Lemberg where the Hapsburg authorities were trying to establish a major university. Chemistry was then assigned to J. M. von Menghin, the Lyceum's teacher of medical theory and therapy. After the school was restored to university status, the chair of botany and chemistry was reinstated. Its second occupant was the Court Apothecary, M. M. Schoepfer, who was obliged to use his shop's garden and laboratory for his teaching.

| Chemical representative | Years | Institutional salary | Chemical publications | |
|---|---|---|---|---|
| | | | Before | During |
| J. F. von Payr | 1735–1741 | Yes | No | No |
| N. A. von Sterzinger | 1742–1764 | Yes | No | Yes |

| Chemical representative | Years | Institutional salary | Chemical publications | |
|---|---|---|---|---|
| | | | Before | During |
| J. M. von Menghin | 1764–1775 | Yes | No | No? |
| S. B. Schiverek | 1775–1782 | 900 fl | No | No |
| J. M. von Menghin | 1782–1789 | 1,200 fl | No | No |
| J. M. Luzenberg | 1790–1791 | Yes | No | No |
| M. M. Schoepfer | 1792–1805 | Yes | No | No |
| F. X. Schoepfer | 1805–1810 | Yes | No | No |

*Bibliography*

Ganzinger, K. "Oesterreichische Pharmazeuten als Hochschullehrer: Die Innsbrucker Apotheker und Professoren Mathäus Michael Schöpfer und Franz Xaver Schöpfer." *Oesterreichische Apotheker-Zeitung*, 32 (1978), 575–578.

Österreichisches Staatsarchiv, Vienna. Letter bearing on Schiverek, Menghin, and the new laboratory (13 October 1972).

Schadelbauer, K. "Zur Geschichte der Innsbrucker Medizinischen Fakultät." Innsbruck University: Medizinische Fakultät: *Forschungen und Forscher der tiroler Aerzteschule*, 2 (1950), 49–50, 281.

## JENA UNIVERSITY

Founded in 1558. Under the jurisdiction of the Dukes of Saxe-Weimar, Saxe-Gotha, Saxe-Coburg, and Saxe-Meiningen.

### First Chemical Instruction

In 1613 Z. Brendel, Sr., an OP of medicine, offered to teach an "exercitium chymicus."

### Chemical Position

Medical Faculty: In 1615 private lecturer W. Rathmann was made "chemical director." This position was held by a succession of private lecturers until 1639 when it was given to W. Rolfinck, OP of anatomy, surgery, and botany. He carried chemistry with him when he rose to the first chair for practical medicine in 1641. From then until the early nineteenth century, the most senior OP was generally responsible for practical medicine and chemistry.

| Chemical representative | Years | Institutional salary | Chemical publications | |
|---|---|---|---|---|
| | | | Before | During |
| W. Rathmann | 1615–? | No | No | No |
| V. T. Clemens | 1620–1637 | No | No | No |
| J. L. König | 1637–1638 | No | No | No |
| W. Rolfinck | 1639–1673 | Yes | Yes | Yes |
| R. W. Krause Substitute | 1673–1718 | 1713: 280 fl | No | Yes |
| J. A. Wedel | 1713–1718 | 1718: 300 fl | No | Yes |
| G. W. Wedel | 1719–1721 | Yes (see profile) | Yes | Yes |
| J. A. Slevogt | 1722–1726 | 290 Rtl + goods | Yes | No |
| J. A. Wedel | 1727–1747 | 290 Rtl + goods | Yes | Yes |
| S. P. Hilscher | 1747–1748 | 290 Rtl + goods | Yes | Yes |
| G. E. Hamberger | 1748–1755 | 290 Rtl + goods | Yes | Yes |
| J. C. Stock | 1756–1758 | 290 Rtl + goods | No | No |
| E. A. Nicolai | 1759–1802 | 1759: 290 Rtl + goods 1768–1802: 411 Rtl + goods | No | Yes |
| C. G. Gruner | 1802–1815 | Yes | No | No |

Philosophical Faculty: In 1789, after extensive training at state expense, *J. F. A. Göttling* was appointed to the Philosophical Faculty as a salaried AOP of chemistry, pharmacy, and technology. He was provided with space in the local castle for a laboratory and 100 Rtl for equipment. He eventually rose to OP of chemistry.

| Chemical representative | Years | Institutional salary | Chemical publications | |
|---|---|---|---|---|
| | | | Before | During |
| J. F. A. Göttling | 1789–1809 | Yes (see profile) | Yes | Yes |
| J. W. Döbereiner | 1810–1849 | Yes | Yes | Yes |

*Bibliography*

Chemnitius, F. *Die Chemie in Jena von Rolfinck bis Knorr (1629–1921)*. Jena: Fromann, 1929.
Giese, E., and B. von Hagen. *Geschichte der medizinischen Fakultät der Friedrich-Schiller-Universität Jena*. Jena: Fischer, 1958.
*Goethes amtliche Schriften*. Vol. I, edited by W. Flach. Weimar: Böhlaus, 1950, pp. 364–375.
Gutbier, A. "Goethe, Grossherzog Carl August und die Chemie in Jena," *Jenaer akademische Reden*, 2 (1926).
Jahn, I. "Geschichte der Botanik in Jena . . . (1558–1864)." Diss., Jena University, 1963, pp. 11, 34, 63, 65–69, 111, 115, 152, 154–158.
*Die Matrikel der Universität Jena*. Vol. II: *1652–1723*, edited by R. Jauernig and M. Steiger. Weimar: Böhlaus, 1961–1977. (Cross-referenced as *Matrikel*.)
"Modell-Buch . . . 1747." Archiv, Jena University. (Cross-referenced as "Modell-Buch".)
Staatsarchiv Weimar. Letter reporting Krause's salary in 1713 (11 November 1970).

## KAISERSLAUTERN CAMERAL SCHOOL

Founded in 1774; renamed the Cameral High School in 1779; relocated in Heidelberg as the State Economics High School in 1784. Under the jurisdiction of the Palatine Physical-Economic Society until 1777, then the Palatine Elector.

### Chemical Position

In 1774 G. A. Suckow, PD at Jena University, was called to Kaiserslautern as OP of physics, pure and applied mathematics, natural history, chemistry, and state economics. The following year he was provided with a laboratory in the Physical-Economic Society's linen factory. He evidently allowed the students to repeat his demonstration experiments. Not finding the small Catholic town to his liking, *Suckow* helped campaign for the move to Heidelberg.

| Chemical representative | Years | Institutional salary | Chemical publications | |
|---|---|---|---|---|
| | | | Before | During |
| G. A. Suckow | 1774–1784 | Yes (see profile) | Yes | Yes |

*Bibliography*

Freitag, W. "Die Entwicklung der Kaiserslauterer Textilindustrie seit dem 18. Jahrhundert." Institut für Landeskunde des Saarlandes, Saarbrücken: *Veröffentlichungen*, 8 (1963).
Funk, M. J. "Der Kampf der merkantilistischen mit der physiokratischen Doktrin in der Kurpfalz." *Neue Heidelberger Jahrbücher*, 18 (1914), 109–117.
Medicus, F. C. *Verzeichnis der Chymische Versuche so in dem Sommerhalben Jahre 1780 auf der Kameral Hohen Schule zu Lautern sind angestellt worden*. Mannheim and Kaiserslautern: Kameral Hohe Schule, 1781.
Webler, H. "Die Kameral-Hohe-Schule zu Lautern (1774–1784)." Historischer Verein der Pfalz: *Mitteilungen*, 43 (1927).

## KASSEL COLLEGIUM CAROLINUM

Founded in 1709; merged with the local Medical-Surgical College in 1764; closed in 1786. Under the jurisdiction of the Landgrave of Hesse-Kassel.

### First Chemical Instruction
Unknown.

### Chemical Position
Collegium Carolinum: In 1766 the government issued a curriculum that required medical students to take chemistry. Around the same time, C. Prizier, the Collegium's instructor of mathematics and physics since 1754, was charged with teaching chemistry and the mining sciences. In 1769 he was provided with a laboratory where he often experimented in the presence of the Landgrave.

| Chemical representative | Years | Institutional salary | Chemical publications | |
|---|---|---|---|---|
| | | | Before | During |
| C. Prizier | 1766?–1780 | 1776: 500 Rtl | No | No |

Medical Faculty: Prizier's successor, J. D. Ebert, belonged to the Medical Faculty, which was founded in 1773.

| Chemical representative | Years | Institutional salary | Chemical publications | |
|---|---|---|---|---|
| | | | Before | During |
| J. D. Ebert | 1780–1786 | Yes | No | No |

### Bibliography
Berge, O. "Die Innenpolitik des Landgrafen Friedrich II. von Hessen-Kassel: Ein Beitrag zur Geschichte des aufgeklärten Absolutismus in Deutschland." Diss., Mainz University, 1952, pp. 245–254.

Strieder, F. W. *Grundlage zu einer Hessischen Gelehrten und Schriftsteller Geschichte.* Vol. II. Kassel: Cramer, 1782, pp. 3–4; Vol. XI. Kassel: Griesbach, 1797, pp. 176–177.

Also see Marburg University (Hackenberg, pp. 53–54, 60–73).

## KIEL UNIVERSITY

Founded in 1665. Under the jurisdiction of the Duke of Holstein-Gottorp until 1773, then the Duke of Schleswig-Holstein.

### First Chemical Instruction
C. March, formerly of Greifswald University and Rostock University, offered to teach the preparation of chemical medicines in 1665.

### Chemical Position
Medical Faculty: In 1701 the Duke announced plans to establish a laboratory at the University; but he failed to follow through. During the 1770s, J. C. Kerstens, former OP of practical medicine and chemistry at Moscow University, was charged with teaching chemistry and pharmacy. His successor, *C. H. Pfaff*, received intensive training in Paris at state expense.

| Chemical representative | Years | Institutional salary | Chemical publications | |
|---|---|---|---|---|
| | | | Before | During |
| J. C. Kerstens | 1770/80–1801 | Yes | Yes | No |
| C. H. *Pfaff* | 1802–1846 | Yes (see profile) | Yes | Yes |

**Bibliography**

Kiel University. Lecture catalogues (WS 1665/1666 . . . WS 1799/1800). Bibliothek, Kiel University.

*Professoren und Dozenten der Christian-Albrechts-Universität zu Kiel 1665–1954.* Edited by R. Bülck and H.-J. Newiger. 4th ed. Kiel: Hirt, 1956, pp. 75–76.

Schipperges, H. "Geschichte der medizinischen Fakultät: Die Frühgeschichte 1665–1840." In *Geschichte der Christian-Albrechts-Universität Kiel 1665–1965.* Vol. IV. Kiel: Wachholtz, 1967, pp. 56, 82.

## KÖNIGSBERG UNIVERSITY

Founded in 1544. Under the jurisdiction of the Elector of Brandenburg, who became the King in Prussia in 1701.

### First Chemical Instruction

At least one of the medical OPs was teaching chemistry by the early seventeenth century.

### Chemical Position

Medical Faculty: The statutes of 1623 assigned practical medicine and chemistry to the most senior OP. There is no evidence, however, that this provision was in force for long. In 1725, to judge from the reform proposal advanced by the AOP of physics, C. G. Fischer, the Medical Faculty had a chair for chemistry and materia medica. Its occupant was probably C. L. Charisius, the fourth OP. His successor, C. D. Meltzer, was responsible for these subjects in 1737 when *J. T. Eller* and von Bülow in Berlin decreed that henceforth each OP would keep the same subjects as he progressed up the seniority ladder to a better income. In 1740 Meltzer was supplemented by Court Apothecary F. G. Haupt, who served as AOP of chemistry until his death two years later. Assuming that the 1737 decree was followed, Meltzer's various successors were responsible for chemistry throughout their careers. His third and fourth successors, A. J. Orlovius and *K. G. Hagen*, did have responsibility for chemistry.

| Chemical representative | Years | Institutional salary | Chemical publications | |
|---|---|---|---|---|
| | | | Before | During |
| G. Lothus | 1623–1635 | Yes | No | No |
| D. Beckher? | 1636–1655 | Yes | Yes | Yes |
| C. L. Charisius? | 1720–1728 | 80 fl | No | No |
| C. D. Meltzer | 1728–1747 | 1737: 80 fl<br>1741–1747: 800 fl + goods | Yes | Yes |
| J. G. Haupt | 1740–1742 | No | No | Yes |
| J. H. von Sanden? | 1749–1759 | No | No | No |
| J. C. Laubmeyer? | 1760–1765 | No | No | No |
| A. J. Orlovius | 1766?–1788 | 1766–1774: 80 fl<br>1774–?: 800 fl + goods<br>1787: 532 Tl | No | Yes |
| *K. G. Hagen* | 1788–1807 | Yes (see profile) | Yes | Yes |

Philosophical Faculty: In 1807, in response to a Senate recommendation, *Hagen* was transferred to the Philosophical Faculty as OP of physics, natural history, and chemistry.

| Chemical representative | Years | Institutional salary | Chemical publication | |
|---|---|---|---|---|
| | | | Before | During |
| K. G. Hagen | 1807–1829 | Yes (see profile) | Yes | Yes |

### Bibliography

Arnoldt, D. H. *Ausführliche, und mit Urkunden versehene Historie der Königsbergischen Universität.* Vol. II. Königsberg: Hartung, 1746, pp. 286, 318, 336; supp., pp. 64–68.

Goldbeck, J. F. *Nachrichten von der Königlichen Universität zu Königsberg in Preussen.* Königsberg: Goldbeck, 1782, pp. 17, 25–26, 40, 73, 92–93.

*Die Matrikel . . . der Albertus-Universität zu Königsberg.* Edited by G. Erler. 3 vols. Leipzig: Duncker and Humblot, 1908–1917. (Cross-referenced as *Matrikel*.)

Matthes, H. "150 Jahre pharmazeutische Chemie an der Universität Königsberg." *Pharmazeutische Zeitung,* 73 (1928), 1041–1052.

Predeek, A. "Ein verschollener Reorganisationsplan für die Universität Königsberg aus dem Jahre 1725." Historische Kommission für ost- und westpreussische Landesforschung: *Altpreussische Forschungen,* 4, no. 2 (1927), 74, 76, 78, 85–86.

Prutz, H. *Die Königliche Albertus-Universität zu Königsberg i. Pr. im neunzehnten Jahrhundert.* Königsberg: Hartung, 1894, pp. 8–9.

Also see Frankfurt a. d. Oder University (Lenz, I: 41; IV: 4–23) and Halle University (Conrad, p. 56).

## LEIPZIG UNIVERSITY

Founded in 1409. Under the jurisdiction of the Elector of Saxony.

### First Chemical Instruction

In 1558 the OP of therapy was assigned responsibility for pharmacy, including chemical medicines. By the early seventeenth century, one or more of the medical professors or private lecturers was offering chemistry nearly every semester.

### Chemical Position

Medical Faculty: In 1668 the Elector appointed M. H. Horn as AOP of chemistry against the Faculty's wishes. Horn remained responsible for the subject when he rose to the chair of pathology. However, chemistry was allowed to go unrepresented from his death in 1681 until 1699, when J. C. Schamberg was appointed AOP of chemistry. Like Horn, he kept the subject when he rose to an ordinary chair. His successor, M. Naboth, also began as an AOP of chemistry. Then in 1710 the Elector appointed J. C. Scheider to a new ordinary chair of chemistry, ordering the Faculty to cover the chair's salary by merging two other chairs as soon as one of the OPs died. He also ordered the University to provide Scheider with a laboratory. Using various forms of resistance, the Faculty managed to force Scheider to withdraw in 1714. Upon Naboth's death in 1721, the Faculty established an ordinary chair of chemistry with a meager salary and appointed A. F. Petzold to it. His second successor, C. G. Eschenbach, managed to get the chair's salary increased by 100 Rtl in 1790, gain entry to the council of professors in 1803, and after seventeen years of lobbying, obtain a small laboratory and budget (150 Rtl) in 1804.

| Chemical representative | Years | Institutional salary | Chemical publications | |
|---|---|---|---|---|
| | | | Before | During |
| M. H. Horn | 1669–1681 | Yes, from 1675 | No | No |
| J. C. Schamberg | 1699–1706 | Yes, from 1701 | No | Yes |
| M. Naboth | 1707–1721 | None | No | Yes |
| J. C. Scheider | 1710–1714 | None | No | No |
| A. F. Petzold | 1722–1761 | Yes | Yes | No |

| Chemical representative | Years | Institutional salary | Chemical publications | |
|---|---|---|---|---|
| | | | Before | During |
| A. Ridiger | 1762–1783 | Yes | Yes | Yes |
| C. G. Eschenbach | 1784–1831 | Yes | Yes | Yes |

**Bibliography**

*Allgemeine Literatur-Zeitung: Intelligenzblatt* (21 April 1790), 420.
*Die jüngere Matrikel der Universität Leipzig 1559–1809.* Edited by G. Erler. 3 vols. Leipzig: Giesecke and Devrient, 1909. (Cross-referenced as *Matrikel.*)
Mayr, E. "Die Entwicklung der Chemie und der pharmazeutischen Chemie an der Universität Leipzig." Diss., Leipzig University, 1965.

## MAINZ UNIVERSITY

Founded in 1477. Under the jurisdiction of the Elector of Mainz until 1797. Closed by France in 1798 and replaced by a central school, which, in turn, was closed in 1803.

*First Chemical Instruction*
J. J. Becher, OP of medicine from 1663 to 1664, may have taught chemistry at Mainz.

*Chemical Position*
Medical Faculty: In 1746 the Elector directed the Faculty to teach chemistry. J. Vogelmann, OP of practical medicine, took on this task. Upon his departure for Würzburg University in 1750, G. C. Schmidt was named OP of medical institutions, chemistry, and physics. From 1755 to 1764 H. *Ludolf*, Electoral Physician and absent OP of chemistry at Erfurt University, served as AOP of chemistry. Upon his departure, F. Holthof, AOP of materia medica, may have been assigned chemistry. The need for an OP of chemistry and a laboratory was frequently mentioned in the 1770s and early 1780s by the Elector's advisors. When the University was reformed in 1784, N. C. Molitor was appointed OP of chemistry, pharmacy, and materia medica and provided with a laboratory and lecture hall (cost: ca. 3,000 Rtl). In the late 1780s, extensive remodeling improved the laboratory's usefulness.

| Chemical representative | Years | Institutional salary | Chemical publications | |
|---|---|---|---|---|
| | | | Before | During |
| J. Vogelmann | 1746–1749 | 200 fl | No | No |
| G. C. Schmidt | 1750–1754 | Yes | No | No |
| H. Ludolf | 1755–1764 | None | Yes | No |
| F. Holthof? | 1764–1784 | None | No | No |
| N. C. Molitor | 1784–1798 | 1,800 fl | Yes | Yes |

Central School: When the Central School replaced the University, Molitor was assigned theoretical and practical chemistry and one Anschel, chemistry and experimental physics.

| Chemical representative | Years | Institutional salary | Chemical publications | |
|---|---|---|---|---|
| | | | Before | During |
| N. C. Molitor | 1798–1803 | 2,500 Fr | Yes | No |
| Anschel | 1798–? | 2,500 Fr | No | No |

## Institutional Histories / 258

*Bibliography*
Aumüller, G. "Zur Geschichte der Anatomischen Institut von Kassel und Mainz." *Medizinhistorisches Journal*, 5 (1970), 148–149, 156–158.
Brück, A. P. "Um die Reform der Mainzer juristischen und medizinischen Fakultäten im 18. Jahrhundert." In *Die alte Mainzer Universität*. Mainz: Kupferberg, 1946, pp. 59–65.
Dreyfus, F. G. *Société et mentalités à Mayence dans la seconde moitié du XVIIIe siècle*. Paris: Colin, 1968, pp. 146–148.
Jung, H.-W. "Anselm Franz von Bentzel im Dienste der Kurfürsten von Mainz." *Beiträge zur Geschichte der Universität Mainz*, 7 (1966), 96, 113.
Just, L., and H. Mathy. *Die Universität Mainz: Grundzüge ihrer Geschichte*. Trautheim and Mainz: Mushake and Franzmathes, 1965, p. 104.
Mainz. *Hofkalender*, 1740 . . . 1784.
Mathy, H. "Medizinhistorische Minaturen aus dem Bereich der Mainzer Universität vom Ende des 18. Jahrhunderts." Vereinigung "Freunde der Universität Mainz": *Jahrbuch*, 12 (1963), 82–92, 96–105.
———. "Neue Quellen zur Biographie des Mainzer Anatomen Franz George Ittner." *Medizinhistorisches Journal*, 13 (1978), 95.
*Medizinische National-Zeitung für Deutschland*, 1 (17 December 1798), 811.
Also see Bonn University ("Quellen," pp. 950–951).

## MARBURG UNIVERSITY

Founded in 1527. Under the jurisdiction of the Landgrave of Hesse-Kassel.

*First Chemical Instruction*
J. Hartmann, OP of mathematics, probably began offering private chemistry courses around 1605.

*Chemical Position*
Medical Faculty: In 1609 Landgrave Moritz appointed Hartmann OP of "chemiatriae" in the Medical Faculty. This was the first chair of chemistry in Europe. With government support, Hartmann soon established an "official chemical-medical laboratory." However, when he went to Kassel in 1621, his chair lapsed and the laboratory fell into disuse. The new statutes of 1653 assigned hygiene, therapy, dietetics, and pharmacy, including chemistry, to the seniormost OP. On account of incomplete staffing, these statutes seem not to have been followed for long. A new laboratory was constructed in 1685; but it soon fell into neglect, once again because of inadequate staffing. In 1731 the building housing it was destroyed to make room for the stables. Seventeen years later J. G. Duising, OP of medicine and physics, complained about the lack of various facilities, including a laboratory. But when the authorities asked for a specific proposal, he replied that the costs of construction, supplies, and an assistant would be so high that it would be cheaper for him to be provided with funds to establish a laboratory in his residence. Nothing came of this suggestion. The Medical Faculty evidently lacked an official chemical representative until 1795 when the subject was assigned to C. Mönch, OP of botany and, since 1789, of chemistry and natural history in the State Economics Institute (see below).

| Chemical representative | Years | Institutional salary | Chemical publications | |
|---|---|---|---|---|
| | | | Before | During |
| J. Hartmann | 1609–1621 | Yes | Yes | Yes |
| J. Tilemann | 1653–1655 | Yes | No | No |
| J. J. Waldschmiedt | 1685?–1689 | 240 fl | Yes | Yes |
| C. Mönch | 1795–1805 | 400 Rtl | Yes | Yes |
| F. Wurzer | 1805–1844 | Yes (see profile) | Yes | Yes |

State Economics Institute: When the Institute was established in 1789, C. Mönch was charged with teaching chemistry and natural history. In 1793 he was provided with a laboratory (cost: 1,236 Rtl).

| Chemical representative | Years | Institutional salary | Chemical publications | |
|---|---|---|---|---|
| | | | Before | During |
| C. Mönch | 1789–1805 | None | Yes | Yes |

*Bibliography*

Ganzenmüller, W. "Das chemische Laboratorium der Universität Marburg im Jahre 1615." In his *Beiträge zur Geschichte der Technologie und der Alchemie*. Weinheim: Verlag Chemie, 1956, pp. 314–322.

Gundlach, F. "Catalogus Professorum Academiae Marburgensis." Historische Kommission für Hessen und Waldeck: *Veröffentlichungen*, 15 (1927), 182, 185, 192, 462.

Hackenberg, R. *Die Entwicklung der Naturwissenschaften an der Universität Marburg von 1750 bis zur westfälischen Zeit*. Diss., Marburg University; Marburg: Mauersberger, 1972.

Hessisches Staatsarchiv. Letter with microfilms bearing on salaries at Marburg University from 1785 to 1820 (27 July 1976).

Hof, D. *Die Entwicklung der Naturwissenschaften an der Universität Marburg/Lahn zur Zeit des Cartesianismusstreites bis 1750*. Diss., Marburg University; Marburg: Gorich and Weiershauser, 1971.

Meinel, C. "Die Chemie an der Universität Marburg seit Beginn des 19. Jahrhunderts: Ein Beitrag zu ihrer Entwicklung als Hochschulfach." *Academia Marburgensis*, 3 (1978).

Schmitz, R. *Die Naturwissenschaften an der Philipps-Universität Marburg 1527–1977*. Marburg: Elwert, 1978, pp. 185–234, 331–361.

## MÜNSTER UNIVERSITY

Founded in 1773; provided with a medical faculty in 1783–1793. Under the jurisdiction of the Bishop of Münster until 1802. Closed by Prussia in 1818.

*Chemical Position*

Medical Faculty: In 1793, after training at state expense, B. Bodde was appointed to the new chair of chemistry and pharmacy.

| Chemical representative | Years | Institutional salary | Chemical publications | |
|---|---|---|---|---|
| | | | Before | During |
| B. Bodde | 1793–1818 | Yes | No | No |

*Bibliography*

Pieper, A. *Die alte Universität Münster 1773–1818*. Münster: Regensberg, 1902, pp. 72, 83, 94.

## MUNICH ACADEMY OF SCIENCES

Founded in 1759. Under the jurisdiction of the Elector of Bavaria.

*Chemical Position*

The initial statutes charged the Philosophical Class with, among other things, carrying out chemical investigations of Bavarian natural materials. J. P. Spring, OP of practical medicine and chemistry at Ingolstadt University, was evidently chosen to represent chemistry. He came to Munich in 1760 as a high mining official with responsibility for lecturing on the subject. However, lacking adequate support, he seems not to have continued his lectures for long. After the Seven Years' War, Count S. F. Haimhausen, chief of mining, and J. A. Wolter, Electoral Physician, failed in their efforts to arrange financing for a laboratory. In 1776 F. M. Baader, a new member and overseer of the natural collections, was made responsible for reporting on current scientific developments. Viewing chemistry as an essential key to nature, he often discussed chemical issues in his mineralogical reports and general addresses. In 1792 M. Imhof, a new member, was charged with lecturing on experimental physics, evidently including chemistry, and provided with a budget (100 fl). Upon Baader's death in 1797, he obtained a higher budget (150 fl). He succeeded Baader as director of the Philosophical Class in 1800 and continued lecturing on experimental physics and chemistry until he resigned in 1811. Meanwhile, the Academy was reorganized in 1807 and *A. F. Gehlen* was appointed to its new chair for chemistry.

| Chemical representative | Years | Institutional salary | Chemical publications | |
|---|---|---|---|---|
| | | | Before | During |
| J. P. Spring | ca. 1760 | None? | Yes | Yes |
| M. Imhof | 1792?–1811? | 1801: 500 fl<br>1802: 600 fl + residence<br>1811: 950 fl | No | Yes |
| A. F. Gehlen | 1807–1815 | Yes (see profile) | Yes | Yes |
| H. A. Vogel | 1816–1851 | Yes | Yes | Yes |

*Bibliography*

"Akademie der Wissenschaften zu München." *Jenaische Allgemeine Literatur Zeitung: Intelligenzblatt* (6 April 1805), 315–316.

*Electoralis academiae scientiarum boicae primordia: Briefe aus der Gründungszeit der bayerischen Akademie der Wissenschaften.* Edited by M. Spindler. Munich: Beck, 1959, pp. 4–5, 40, 145–146, 453, 462, 526.

Hammermayer, L. "Gründungs- und Frühgeschichte der Bayerischen Akademie der Wissenschaften." *Münchener historische Studien: Abteilung Bayerische Geschichte*, 4 (1959), 370, 380.

Prandtl, W. *Die Geschichte des chemischen Laboratoriums der bayerischen Akademie der Wissenschaften in München.* Weinheim: Verlag Chemie, 1952.

Rehlingen, von. (Archiv, Academy of Sciences, Munich). Letter summarizing the fragmentary archival holdings concerning Baader and Imhof (15 June 1971).

Schwertl (Bayerisches Hauptstaatsarchiv, Munich). Letter reporting Imhof's salary in 1801 and 1802 (28 October 1980).

Westenrieder, L. *Geschichte der baierischen Akademie der Wissenschaften.* Vol. II. Munich: Lindauer, 1807, pp. 239–240, 420.

Also see Ingolstadt University (Kallinich, pp. 30, 472).

## OLMÜTZ LYCEUM

Founded in 1782 on the basis of the old incomplete university (founded 1581) in Olmütz; restored to university status in 1827. Under the jurisdiction of the Hapsburg Emperor.

*Chemical Position*

Medical Faculty: In 1782 the Viennese authorities instructed the professor of internal medicine to teach botany and "some chemistry" as well.

| Chemical representative | Years | Institutional salary | Chemical publications | |
|---|---|---|---|---|
| | | | Before | During |
| A. Sebald | 1782–1785 | Yes | No | No |
| S. Dürer | 1785–1787 | Yes | Yes | No |
| A. Beüttl | 1787–1805 | 600 fl | No | No |

## Bibliography

Zapletal, V. "Počátky lékařského studia na Moravě (1753–1800)." *Scripta Medica*, 3 (1957), 237–282.
Also see Graz Lyceum (Fischer, pp. 10, 38–39).

## PRAGUE UNIVERSITY

Founded in 1349. Under the jurisdiction of the Hapsburg Emperor.

### First Chemical Instruction

The Medical Faculty's statutes of the 1650s(?) called for instruction on chemical medicines. However, due to inadequate staffing, this provision may not have been implemented.

### Chemical Position

Medical Faculty: In 1745 J. A. J. Scrinci, OP of medical institutions, was assigned chemistry and experimental physics as well. Two years later, the government denied the Faculty's request for a laboratory, ordering the students to observe chemical operations in nearby apothecary shops. Although Scrinci did little teaching after 1758 on account of poor health, he remained formally responsible for chemistry until 1772. In that year, the science was turned over to T. von Bayer, OP of pathology and therapy. From 1773 to 1775 he was assisted by J. G. Mikan, new AOP of chemistry and botany. Then Mikan was appointed to a new ordinary chair for these subjects. He obtained a laboratory in 1785 and his son, J. C. Mikan, was appointed adjunct professor of chemistry and botany in 1798. Upon the elder Mikan's death in 1811, the subjects were assigned to separate chairs.

| Chemical representative | Years | Institutional salary | Chemical publications | |
|---|---|---|---|---|
| | | | Before | During |
| J. A. J. Scrinci | 1745–1772 | 1772: 1,100 fl | Yes | Yes |
| T. von Bayer | 1772–1775 | Yes | No | Yes |
| J. G. Mikan | 1773–1811 | 1778: 1,100 fl | No | Yes |
| J. C. Mikan | 1798–1812 | None | No | No |
| J. J. von Freyssmuth | 1812–1819 | 1812: 1,200 fl | No | Yes |

Law Faculty: In 1763 J. T. Peithner, an official in Prague whose proposals concerning mining education led to the revitalization of the Schemnitz Mining School, was appointed to this Faculty to teach "theoretical chemistry, subterranean surveying, mineralogy, metallurgy, subterranean geography, and subterranean mechanics and architecture." Though he did not receive a salary as OP of metallurgy, he did have a budget (300 fl) for experiments. In 1772 he was transferred to Schemnitz and the chair of metallurgy was abolished.

| Chemical representative | Years | Institutional salary | Chemical publications | |
|---|---|---|---|---|
| | | | Before | During |
| J. T. Peithner | 1763–1772 | None | No | Yes |

### Bibliography

Egert, U. *Personalbibliographien von Professoren der philosophischen Fakultät zu Prag . . . von 1800 bis 1860*. Vol. I. Diss., Erlangen University; Erlangen: Hogl, 1970, pp. 63–65.
Kirndörfer, D. *Die Personalbibliographien der Professoren und Dekane (Nicht-Professoren) der medizinischen Fakultät der Karl-Ferdinands-Universität in Prag im Zeitraum von 1749–1800*. Diss., Erlangen University; Erlangen: Hogl, 1971, pp. 6–24, 39–43, 83–86.
Österreichisches Staatsarchiv, Vienna. Letter bearing on Scrinci, Peithner, the Mikans, and Freyssmuth (13 October 1972).

Tomek, W. *Geschichte der Prager Universität*. Prague: Haase, 1849, pp. 298–299, 310–317.
Vogel, H. *Personalbibliographien der Professoren und Dozenten der Botanik, Chemie, . . . an der Medizinischen Fakultät der Karl-Ferdinands-Universität in Prag im ungefähren Zeitraum von 1800–1850*. Diss., Erlangen University; Erlangen: Hogl, 1972.
Wraný, A. *Geschichte der Chemie und der auf chemischer Grundlage beruhenden Betriebe in Böhmen bis zur Mitte des 19. Jahrhunderts*. Prague: Rivnác, 1902, pp. 173–183, 195–196.
Also see Born (Deutsch, p. 155), Schemnitz Mining Academy (Hornoch, pp. 33–36), and Vienna University (Oberhummer, pp. 145–148).

## ROSTOCK UNIVERSITY

Founded in 1419. Under the jurisdiction of the Duke of Mecklenburg and Rostock's Town Council until 1760, the Council until 1789, the Duke and Council until 1827, and then the Duke.

### First Chemical Instruction

The mathematical and medical professors were teaching Paracelsian doctrines at Rostock from the 1560s.

### Chemical Position

Medical Faculty: The statutes of 1623 and 1669 emphasized the importance of chemistry to medicine. Accordingly, from at least the mid-1640s, the second of the two OPs paid by the Duke was responsible for chemistry. In 1655 S. Wirdig gave his inaugural address on "chemistry's dignity and preeminence." A century later when the Duke withdrew his support from the University and founded Bützow University, the Medical Faculty was reduced to one OP who, of necessity, focused on therapy.

| Chemical representative | Years | Institutional salary | Chemical publications | |
|---|---|---|---|---|
| | | | Before | During |
| S. Schultetus | 1645?–1654 | Yes | No | No |
| S. Wirdig | 1654–1687 | Yes | No | Yes |
| J. E. Schaper | 1692–1721 | Yes | No | Yes |
| G. C. Handtwig | 1738–1759? | Yes | No | Yes? |

Philosophical Faculty: In 1792, three years after the Duke resumed his support of the University, H. F. Link, a PD at Göttingen University, was called to the Philosophical Faculty as OP of chemistry, botany, and natural history. By 1811, when he left for Breslau University, *Link* had obtained a budget for experiments (50 Tl). His successor, G. P. Mähl, was AOP of chemistry and pharmacy in the Medical Faculty from 1812 and OP of these same subjects in the Philosophical Faculty from 1817.

| Chemical representative | Years | Institutional salary | Chemical publications | |
|---|---|---|---|---|
| | | | Before | During |
| H. F. Link | 1792–1811 | Yes (see profile) | Yes | Yes |
| G. P. Mähl | 1817–1833 | Yes | Yes | Yes |

### Bibliography

*Die Matrikel der Universität Rostock*. Vol. V: *Ost. 1789–1830*, edited by E. Schäfer. Rostock: Stiller, 1912. (Cross-referenced as *Matrikel*.)
Ruser, I. "Die Chemie und ihr Einfluss auf die Medizin an der Rostocker Medizinischen Fakultät in der Zeit von 1419 bis 1959." Diss., Rostock University, 1964.

Schott, G. "Zur Geschichte der Chemie an der Universität Rostock (bis 1945)." Rostock University: *Wissenschaftliche Zeitschrift: Mathematisch-naturwissenschaftliche Reihe*, 18 (1969), 986–1017.

## SCHEMNITZ MINING ACADEMY

Founded as a mining school in 1735; reorganized as a "practical school" in 1763; elevated to the status of a mining academy in 1770. Under the jurisdiction of the Hapsburg Emperor.

### First Chemical Instruction

From 1735 to 1749?, Assaying Master Schmidt taught practical assaying and smelting.

### Chemical Position

When the school was reorganized, Imperial Physician G. von Swieten arranged for the appointment of N. J. Jacquin as chemistry teacher. Jacquin was provided with a residence, laboratory, and supplies. In 1768 he was called to Vienna University, and J. A. Scopoli, who had been giving a metallurgical-chemical course in Idria, was appointed his successor. In the 1780s Scopoli's successor, A. von Ruprecht, inaugurated laboratory instruction, making this one of the first schools in Europe where students could receive practical training in chemistry.

| Chemical representative | Years | Institutional salary | Chemical publications | |
|---|---|---|---|---|
| | | | Before | During |
| N. J. Jacquin | 1763–1768 | Yes (see profile) | No | No |
| J. A. Scopoli | 1769–1777 | Yes (see profile) | Yes | Yes |
| A. von Ruprecht | 1777–1792 | 1777–1780: 1,000 fl | No | Yes |
| | | 1780–1792: 1,900 fl | | |
| M. Patzier | 1792–1810 | 1792–1810: 1,200 fl | No | No |

### Bibliography

*Gedenkbuch zur hundertjährigen Gründung der königl. ungarischen Berg-und Forst-Akademie in Schemnitz 1770–1870*. Edited by G. Faller. Schemnitz: Joerges, 1871, pp. 5–8, 10–11, 32.

Hornoch, A. T. "Zu den Anfängen des höheren bergtechnischen Unterrichtes in Mitteleuropa." *Berg- und Hüttenmännische Monatshefte*, 89 (1941), 16–22, 33–37, 49–51.

Österreichisches Staatsarchiv, Vienna. Letter bearing on the salaries of Ruprecht and Patzier (13 October 1972).

Proszt, J. "Die Schemnitzer Bergakademie als Geburtsstätte chemisch-wissenschaftlicher Forschung in Ungarn." *Historia Eruditionis Superioris Rerum Metallicarum et Saltuariarum in Hungaria 1735–1935*. Vol. III. Sopron, 1938.

Townson, R. *Travels in Hungary with a Short Account of Vienna in the Year 1793*. London: Robinson, 1797, pp. 422–23.

Also see Vienna University (Oberhummer, pp. 145–150).

## STRASBOURG UNIVERSITY

Founded in 1621. Under the jurisdiction of Strasbourg's Town Council until 1789. Closed in 1792 and replaced by a medical school (1794–1870) and a central school (1796–1802).

### First Chemical Instruction

J. V. Scheid, AOP of medicine from 1680, was the first to teach chemistry at Strasbourg.

### Chemical Position

Medical Faculty: In 1685 J. B. Boecler was appointed to a new third chair for chemistry and materia medica. Henceforth, chemistry was represented by one of the three OPs. During the 1730s, while J. J. Sachs was responsible for the subject, he was provided with an assistant. In 1790 J. Hermann also had a paid assistant (salary: 200 livres).

## Institutional Histories / 264

| Chemical representative | Years | Institutional salary[a] | Chemical publications | |
|---|---|---|---|---|
| | | | Before | During |
| J. B. Boecler, Sr. | 1685–1701 | Yes | No | No |
| J. S. Henninger | 1703–1719 | Yes | No | No |
| J. B. Boecler, Jr. | 1719–1733 | Yes | Yes | No |
| J. J. Sachs | 1733–1738 | Yes | Yes | No |
| J. P. Boecler | 1738–1759 | Yes | No | Yes |
| J. R. Spielmann | 1759–1783 | Yes | Yes | Yes |
| J. Hermann | 1783–1792 | Yes | Yes | Yes |

[a]According to a report dated 1790, the thirteen most senior OPs each received 600 livres plus goods and the four remaining OPs each received 562½ livres. Earlier in the century, the junior OPs received 300 livres. When attendance was good, the total income of the senior OPs ranged upwards from 2,000 livres.

Medical School: P. F. Nicolas, professor of medical chemistry and pharmacy, resigned shortly after the School opened in the spring of 1795. He was succeeded by F. L. Ehrmann, then G. Masuyer. However, as of spring 1799, no chemical instruction had yet been offered.

| Chemical representative | Years | Institutional salary | Chemical publications | |
|---|---|---|---|---|
| | | | Before | During |
| P. F. Nicolas | 1794–1795 | Yes? | Yes | Yes? |
| F. L. Ehrmann | 1796–1798 | Yes? | Yes | No |
| G. Masuyer | 1798–1838 | 1803: 3000 fr | No | Yes |

Central School: After an open competition, a jury named F. L. Ehrmann the first professor of chemistry and physics. He was succeeded by J. L. A. Herrenschneider.

| Chemical representative | Years | Institutional salary | Chemical publications | |
|---|---|---|---|---|
| | | | Before | During |
| F. L. Ehrmann | 1796–1800 | Yes? | Yes | No |
| Herrenschneider | 1800–1802 | Yes? | No | No |

### Bibliography

*Die alten Matrikeln der Universität Strassburg 1621 bis 1793.* Edited by G. C. Knod. 3 vols. Strasbourg: Trübner, 1897–1902. (Cross-referenced as *Matrikel.*)

Berger-Levrault, O. *Annales des Professeurs des Académies et Universitaires Alsaciennes, 1523–1871.* Nancy: Berger-Levrault, 1892.

"Vermischte Bemerkungen über Strassburg und Paris." *Medizinische National-Zeitung für Deutschland,* 2 (24 June 1799), 604–608.

Wieger, F. *Geschichte der Medicin und ihrer Lehranstalten in Strassburg vom Jahre 1497 bis zum Jahre 1872.* Strasbourg: Trübner, 1885, pp. 56, 61–65, 164.

## STUTTGART UNIVERSITY

Founded as a military school at a castle between Stuttgart and Ludwigsburg in 1770; relocated to Stuttgart in 1776; raised to university status in 1781; closed in 1794. Under the jurisdiction of the Duke of Württemberg.

## Chemical Position

Medical Faculty: In 1775 the school physician, C. G. Reuss, was charged with teaching natural history, chemistry and materia medica. When the Medical Faculty was established the next year, he was appointed to the chair for these subjects. He successfully defended his job against critics who were concerned about his unwillingness to publish. In 1792 he gave chemistry to the OP of zoology, C. F. Kielmeyer, who soon obtained space for a laboratory and a budget (50 to 60 fl).

| Chemical representative | Years | Institutional salary | Chemical publications | |
|---|---|---|---|---|
| | | | Before | During |
| C. G. Reuss | 1775–1792 | 1792: 600 fl? | No | No |
| C. F. Kielmeyer | 1792–1794 | 1793: 575 fl | Yes | No |

### Bibliography

Uhland, R. "Geschichte der hohen Karlsschule in Stuttgart." *Darstellungen aus der Württembergischen Geschichte*, 36 (1953), 219, 329, 358.

Wagner, H. *Geschichte der Hohen Carls-Schule*. Vol. III. Würzburg: Etlinger, 1858, pp. 87–92.

Also see Pfaff (Pfaff, pp. 39–45).

## TÜBINGEN UNIVERSITY

Founded in 1477. Under the jurisdiction of the Duke of Württemberg.

### First Chemical Instruction

In 1664 J. C. Brotbeck, an OP of medicine, offered to teach chemistry.

### Chemical Position

Medical Faculty: E. Camerarius, AOP of medicine from 1693, was listed as responsible for chemistry in the lecture catalogue for 1700. In 1732 the Duke allocated 300 fl for a chemistry laboratory. But objections from Tübingen's Town Council and the deaths of the Duke in 1733 and Camerarius in 1734 brought an end to the project. Camerarius's successor, J. Bacmeister, may have been responsible for chemistry. In any case, Bacmeister's successor, J. G. Gmelin, was OP of chemistry and botany. In 1753 he was provided with a laboratory (cost: 653 fl). Henceforth, the Faculty had a chemical representative. In 1796 C. F. Kielmeyer was assigned to a new fifth chair for chemistry alone. But in 1801 he assumed botany and natural history as well and in 1809 he converted the laboratory into a dissecting room.

| Chemical representative | Years | Institutional salary | Chemical publications | |
|---|---|---|---|---|
| | | | Before | During |
| E. Camerarius | 1693?–1734 | Yes, from 1710 | No? | Yes |
| J. Bacmeister? | 1736–1748 | Yes | No | No |
| J. G. Gmelin | 1748–1755 | Yes | Yes | Yes |
| P. F. Gmelin | 1755–1768 | Yes | Yes | Yes |
| C. F. Jäger | 1768–1774 | Yes | No | Yes |
| G. C. C. Storr | 1774–1796 | Yes | No | Yes |
| C. F. Kielmeyer | 1796–1817 | Yes | Yes | Yes |
| C. G. Gmelin | 1817–1860 | Yes | Yes | Yes |

### Bibliography

Beese, M. *Die medizinischen Promotionen in Tübingen 1750–1799*. Diss., Tübingen University; 1 Tübingen: Spengler, 1977, pp. 25–26.

Bök, A. F. *Geschichte der herzoglich Würtenbergischen Eberhard Carls Universität zu Tübingen.* Tübingen: Cotta, 1774, pp. 159–160, 185, 195–196, 197–198, 247–248, 289.

Klüpfel, K. *Geschichte und Beschreibung der Universität Tübingen.* Tübingen: Fues, 1849, pp. 255–256, 478–479.

*Die Matrikeln der Universität Tübingen.* Edited by A. Bürk and W. Wille. Vols. II–III. Tübingen: Universitätsbibliothek, 1953–1954. (Cross-referenced as *Matrikel.*)

Pfaff, B. H. *Johann Georg Gmelin, Philipp Friedrich Gmelin, Georg Friedrich Sigwart, Karl Friedrich Clossius und ihre Tätigkeit in Lehre und Forschung auf dem Gebiet der gerichtlichen Medizin in Tübingen.* Diss., Tübingen University; 1 Tübingen: Spangenherg, 1976, pp. 4–40.

Tübingen University. Lecture catalogues (1664 . . . SS 1800). Bibliothek, Tübingen University.

Thümmel, H.-W. "Die Tübinger Universitätsverfassung im Zeitalter des Absolutismus." *Contubernium: Beiträge zur Geschichte der Eberhard-Karls-Universität Tübingen,* 7 (1975), 216, 229, 751.

Wankmüller, A. "Zur Geschichte des chemischen und pharmazeutischen Laboratoriums in Tübingen—Apotheker und Professor Christian Gottlob Gmelin." *Pharmazeutische Centralhalle für Deutschland,* 89 (1950), 8–12.

## VIENNA UNIVERSITY

Founded in 1365. Under the jurisdiction of the Hapsburg Emperor.

### First Chemical Instruction

Since the Medical Faculty was still claiming as late as 1696 that chemistry and alchemy had nothing to do with medical practice, the earliest chemical instruction was probably in the eighteenth century.

### Chemical Position

Medical Faculty: The Faculty failed in its efforts to get a laboratory in 1718 and to obtain a new chair for botany and chemistry in 1726. However, aided by Imperial Physician G. van Swieten, it acquired new chairs for surgery and for botany and chemistry in 1749. R. Laugier was named to the second chair, and in 1756 a laboratory and residence were completed for him in the University's new buildings. In 1768 he was replaced, because he had neglected his duties, by N. J. Jacquin. In 1792 Jacquin's son took over the chemical lectures, having been prepared for the post at the government's expense (the cost of his travels to Western Europe and the instruments he bought there was 8,650 fl). In 1797 he was given full responsibility for the chair.

| Chemical representative | Years | Institutional salary | Chemical publications | |
|---|---|---|---|---|
| | | | Before | During |
| R. Laugier | 1749–1768 | 1749–1755: 1500 fl<br>1756–1768: 2000 fl | No | No |
| N. J. Jacquin | 1768–1791 | Yes (see profile) | No | Yes |
| J. F. von Jacquin | 1792–1838 | 1792–1796: none<br>1796: 2000 fl | No | Yes |

### Bibliography

*Acta Facultatis Medicae Universitatis Vindobonensis 1399–1724.* Edited by K. Schrauf and L. Senfelder. Vol. VI. Vienna: Verlag des Wiener Medizinischen Doktorenkollegiums, 1912, pp. 139, 399.

Baresel, W. *Personalbibliographien von Professoren der medizinischen Fakultät der Universität Wien im ungefähren Zeitraum von 1745–1790 und der Josephs-Akademie in Wien von 1780–1790.* Diss., Erlangen University; Erlangen: Hogl, 1971, pp. 47–48, 87–96.

Kink, R. *Geschichte der kaiserlichen Universität zu Wien.* Vol. I, pts. 1–2. Vienna: Gerold, 1854, 1: 444, 453–454; 2: 254–271.

Klingenstein, G. "Vorstufen der theresianischen Studienreform in der Regierungszeit Karls

VI." Institut für österreichische Geschichtsforschung: *Mitteilungen*, 76 (1968), 370–371.
Müller, W. *Gerard van Swieten.* Vienna: Braumüller, 1883.
Oberhummer, W. "Die Chemie an der Universität Wien in der Zeit von 1749 bis 1848 und die Inhaber des Lehrstuhls für Chemie und Botanik." *Studien zur Geschichte der Universität Wien,* 3 (1965), 126–202.
Österreichisches Staatsarchiv, Vienna. Letter bearing on the salaries of the two Jacquins (13 October 1972).

## VIENNA THERESIANUM

Founded in 1746; merged with the Savoy Noble School (opened 1749) in two stages between 1776 and 1778; closed in 1784; reopened in 1797. Under the jurisdiction of the Hapsburg Emperor.

### First Chemical Instruction

J. H. G. Justi, professor of the cameral, commercial, and mining sciences from 1752 to 1753, may have taught chemistry there.

### Chemical Position

Sometime between 1773 and 1779, the professor of mineralogy, F. X. Eder, was assigned chemistry and metallurgy as well. He held these subjects until the school was closed. When it was reopened in 1797, the minor chemist, J. A. Scherer, was appointed professor of chemistry.

| Chemical representative | Years | Institutional salary | Chemical publications | |
|---|---|---|---|---|
| | | | Before | During |
| F. X. Eder | 1773/79–1784 | Yes | Yes | No |
| J. A. Scherer | 1797–1803 | Yes | Yes | Yes? |

### Bibliography

*Album des kaiserl. königl. Theresianums (1746–1880).* Edited by M. von Gemmell-Flischbach. Vienna: Perles, 1880, pp. 16, 76.
Ekkard, R. *Litterarisches Handbuch der bekannten höhern Lehranstalten in und ausser Teutschland.* Vol. I. Erlangen: Schleich, 1780, p. 3; Vol. II. Erlangen: Palm, 1782, p. 173.
Österreichisches Staatsarchiv, Vienna. Letter bearing on Eder (13 October 1972).

## VIENNA MEDICAL-SURGICAL ACADEMY

Founded in 1781; elevated to the status of an academy able to grant the doctorate in surgery in 1786. Under the jurisdiction of the Hapsburg Emperor. Closed in 1874.

### Chemical Position

When the school was organized in 1781, botany and chemistry may have been assigned to a Professor Kanker. Upon his death, J. J. Plenck, OP of surgery and obstetrics at Budapest University, was called to Vienna as professor of botany, chemistry, and pharmacy. The large building constructed for the school in 1785 probably included a laboratory.

| Chemical representative | Years | Institutional salary | Chemical publications | |
|---|---|---|---|---|
| | | | Before | During |
| Kanker? | 1781–1783 | Yes | No | No |
| J. J. Plenck | 1783–1806 | 1,900 fl | Yes | Yes |

## Institutional Histories / 268

*Bibliography*

*Allgemeine deutsche Bibliothek*, 57 (1784), 632.
Brambilla, J. A. *Instruktion für Professoren der K. K. chirurgischen Militärakademie*. Vol. I. Vienna: Trattner, 1784, pp. 6, 8, 109–114.
"Nachricht von der neuen Kayserlich-Königlich Josephinischen medicinisch chirurgischen Academie zu Wien." *Medicinisches Journal*, no. 9 (1786), 17, 19.
Also see Budapest University (Endre, pp. 86–88, 139) and Vienna University (Baresel, pp. 173–188).

### WITTENBERG UNIVERSITY

Founded in 1503. Under the jurisdiction of the Elector of Saxony until 1815. Closed by Prussia in 1816, with the transfer of several professors to Halle University.

*First Chemical Instruction*

D. Sennert, an OP of medicine and renowned scholar, taught chemistry at Wittenberg in the early seventeenth century.

*Chemical Position*

Medical Faculty: In 1723 M. G. Loescher, OP of natural philosophy and AOP of medicine, requested appointment to a new chair for chemistry and pharmacy. Soon afterwards, T. Neukranz, a PD who had established a laboratory and auditorium in his residence, also petitioned for such an appointment. While Neukranz was soon named AOP of chemistry and pharmacy, his subsequent efforts to get a salaried chair for chemistry and pharmacy or mining failed as did Loescher's concurrent efforts. Not until the early nineteenth century were further attempts made to institute a chair of chemistry. In 1804 the Faculty requested that the apothecary, A. F. L. Dörffurt, be appointed as AOP of chemistry and pharmacy. The government, which had failed to follow through on announced plans to establish a laboratory, denied the request. However, in 1810 it finally appointed C. H. T. Schreger as OP of chemistry and materia medica. When the University closed, he was transferred to Halle University as an OP of medicine.

| Chemical representative | Years | Institutional salary | Chemical publications | |
|---|---|---|---|---|
| | | | Before | During |
| T. Neukranz | 1724–1732? | No | No | Yes |
| C. H. T. Schreger | 1810–1816 | Yes | Yes | No |

*Bibliography*

"Album Academice Vitebergensis . . . (1602–1660)," edited by B. Weissenborn. *Geschichtsquellen der Provinz Sachsen und des Freistaates Anhalt*, N.R. 14–15 (1934). (Cross-referenced as "Matrikel," 14).
"Album Academicae Vitebergensis . . . (1660–1812)," edited by F. Juntke. Universitäts- und Landesbibliothek Sachsen-Anhalt, Halle: *Arbeiten*, 1 (1952) and 5 (1966). (Cross-referenced as "Matrikel," 1 or 5).
Friedensburg, W. *Geschichte der Universität Wittenberg*. Halle: Niemeyer, 1917, pp. 458, 467, 494, 575, 577–579, 582, 587.
Stolz, R. "Johann Theodor Neukranz und sein chemiches Laboratorium an der Universität Wittenberg im ersten Drittel des 18. Jahrhuṅderts," *NTM*, 16 (1979), 72–79.

### WÜRZBURG UNIVERSITY

Founded in 1582. Under the jurisdiction of the Prince-Bishop of Würzburg until 1803.

*First Chemical Instruction*

J. B. A. Beringer, OP of medicine, taught chemistry in Würzburg around 1714.

## Chemical Position

Medical Faculty: In 1738 L. A. Dercum, OP of botany, was assigned chemistry and materia medica as well. Three years later these subjects were transferred to J. V. Scheidler, who was appointed to a new fifth chair. Upon his death in 1745, Dercum probably resumed responsibility for chemistry. In 1750 the subject was assigned to J. Vogelmann, who had just been called from Mainz University as OP of practical medicine. His successor, F. H. M. Wilhelm, had the same subjects until 1782 when chemistry was turned over to G. Pickel, who obtained a laboratory a few years later and an ordinary chair in 1795.

| Chemical representative | Years | Institutional salary | Chemical publications | |
|---|---|---|---|---|
| | | | Before | During |
| L. A. Dercum | 1738–1741 | 300 fl | No | No |
| J. V. Scheidler | 1741–1745 | Yes | No | No |
| L. A. Dercum? | 1745–1749 | 300 fl | No | No |
| J. Vogelmann | 1750–1764 | 1750: 280 Rtl | No | No |
| F. H. M. Wilhelm | 1767–1782 | Yes | No | Yes |
| G. Pickel | 1782–1836 | 1782: 200 Tl | Yes | Yes |

### Bibliography

Sticker, G. "Entwicklungsgeschichte der Medizinischen Fakultät an der Alma Mater Julia." In *Aus der Vergangenheit der Universität Würzburg*, edited by M. Buchner. Berlin: Springer, 1932, pp. 481–484, 490–491, 498–499, 503, 512.

Wegele, F. X. von. *Geschichte der Universität Würzburg*. 2 vols. Würzburg: Stahel, 1882, I: 411–412, 441–442, 450, 481; II: 373, 394–397, 415–416.

# Appendix III

# CRELL'S SUBSCRIBERS (1784–1791)

This appendix is devoted to the *Chemische Annalen*'s subscribers, whose names, titles, and residences appeared annually in the journal from 1784 through 1791. Based on Crell's annual listings, it first presents an alphabetical master list of the subscribers that provides biographers with an easy way to ascertain whether or not a given person was a subscriber and, if so, the years in which he subscribed. The master list is followed by a geographical index that points local historians to the subscribers living in any given town.

The master list presents the following information about each subscriber:

*NAME:* Each entry begins with the subscriber's surname. In spelling surnames, I have usually followed the orthography used by Crell. However, when biographical dictionaries indicate a different orthography, I have followed standard usage. The subscriber's surname is followed by his initials, whenever possible. In some instances Crell's lists are the source for the initials. In others, I have relied on biographical dictionaries, matriculation lists, other subscription lists, and information provided by local archivists. Some names appear in italics and/or are followed by numbers. The italics indicate that the person has a profile in Appendix I. The numbers are used as follows:

1 = among the subscribers listed in C. F. Wenzel, *Lehre von der Verwandschaft der Körper* (Dresden: Gerlach, 1777).
2 = among the contributors to Crell's journals between 1778 and 1791.
3 = among the subscribers listed in C. W. Scheele, *Sämmtliche Physische und Chemische Werke*, ed. S. F. Hermbstaedt, 2 vols. (Berlin: Rottmann, 1793).
4 = among the subscribers listed in L. B. Guyton de Morveau et al., *Methode der chemischen Nomenklatur für das antiphlogistische System*, trans. K. von Meidinger, facsimile of the 1793 ed. (Hildesheim: Olms, 1978).

*AGE IN 1784:* The next column indicates, when biographical information has been available, the subscriber's age at the start of the *Chemische Annalen*.

*OCCUPATION:* Following the subscriber's name and age in 1784 is his occupation upon first subscribing. In most instances, my source of information has been Crell's listing. However, when he used honorific titles or did not indicate the subscriber's occupation, I have drawn on biographical dictionaries and local archivists for supplemental information. If a subscriber changed his occupation, the new occupation is indicated in the next row.

*TOWN:* Next I identify the town from which the person first subscribed. I have used modern German orthography, except for a few German towns whose names are customarily anglicized in English works and for towns outside Germany. If a subscriber changed his residence, the new residence is indicated in the row below. If a subscriber's hometown was given instead of his current residence, the town's name is preceded by "ex"—e.g., "ex Copenhagen."

*YEARS SUBSCRIBING IN GIVEN OCCUPATION AND TOWN:* Each entry ends with the dates that the person subscribed while in the occupation and town indicated in the same row. If the subscriber is known to have died between 1784 and 1791, the date of death is also given.

I am indebted to Dr. Werner (Barmen-Elberfeld, now Wuppertal), Dr. W. Biber and Dr. H. Haeberli (Bern), Dr. Querfurth (Brunswick), J. A. Rust (Clausthal-Zellerfeld), Dr. Weidenhaupt (Düsseldorf), Dr. Andernacht (Frankfurt a. M.), Barabas and Harms (Hamburg), Dr. Mundhenke (Hanover), R. Volkmann (Helmstedt), Klaube (Kassel), and Dr. Reinhardt (Lüneburg) for information regarding subscribers who resided in these towns.

TABLE A3: Crell's Subscribers 1784–1791

| Name | Age in 1784 | Occupation | Town | Years subscribing in given occupation and town |
|---|---|---|---|---|
| Anonymous | | Sanitation official | Brunswick | 1784–90 |
| Anonymous | | Medical official | Düsseldorf | 1785–91 |
| Anonymous | | Academy librarian | St. Petersburg | 1786–91 |
| Anonymous | | Academy librarian | Stockholm | 1788–91 |
| Anonymous | | Electoral librarian | Dresden | 1789–91 |
| Anonymous | | Magliabecchi librarian | Florence | 1789–91 |
| Anonymous | | Reading Society librarian | Oehringen | 1789 |
| Anonymous | | Economic Society librarian | St. Petersburg | 1790–91 |
| Abich, R. A.(2) | | Mining councillor | Schöningen | 1784–91 |
| Abildgaard, P. C. | 44 | Professor of medicine-science | Copenhagen | 1787–91 |
| Achard, F. C. (1, 2) | 31 | Academician | Berlin | 1784–91 |
| Adami, J. A. | | Physician | Quackenbrück | 1784–85 |
| Allis | | Mining administrator | Přibram | 1786 |
| Amburger, J. A. A.(2) | 34 | Physician-apothecary | Offenbach | 1784 |
| Ameldung | | Apothecary | Osnabrück | 1784 |
| Amelung, A.(2) | 49 | Glaziery administrator | Grünenplan | 1784–91 |
| Andreae, J. G. R.(1–3) | 60 | Court apothecary | Hanover | 1784–91 |
| Andreae | | Apothecary | St. Petersburg | 1788–90 |

TABLE A3: Crell's Subscribers 1784–1791 (cont.)

| Name | Age in 1784 | Occupation | Town | Years subscribing in given occupation and town |
|---|---|---|---|---|
| Arndt | | Pharmacy provisor | Königsberg | 1784 |
| | | Apothecary | Königsberg | 1785 |
| Aschenborn, L. A.(1) | | Apothecary | Berlin | 1784–91 |
| Aschoff, L. P. | 26 | Pharmacy learner | Halle | 1785 |
| | | Pharmacy learner | Bielefeld | 1786–87 |
| Aschoff | | Apothecary | Bielefeld | 1786–87 |
| Bachmann | | Pharmacy learner | Berlin | 1787 |
| | | Pharmacy learner | Brunswick | 1788 |
| | | Pharmacy provisor | Obernkirchen | 1789–91 |
| Backhaus(3) | | Pharmacy learner | Hameln | 1784–85 |
| | | Pharmacy provisor | Hameln | 1787–88 |
| | | Pharmacy provisor | Berlin | 1789–91 |
| Bähr, J. J. Z. | | Smelting administrator | Unterharz | 1784 |
| Bärensprung, H. S.(1,3) | | Apothecary | Berlin | 1784–91 |
| Bärensprung | | Pharmacy learner | Langensalza | 1785, 1787–89 |
| Baldinger, E. G.(1) | 46 | Professor of medicine | Kassel | 1784–86 |
| | | Professor of medicine | Marburg | 1787–90 |
| Ballas, J. | | Mining student | Bohemia | 1785 |
| Balz | | Apothecary | Frankfurt a. M. | 1786–91 |
| Bang | | Pharmacy provisor | St. Petersburg | 1790–91 |
| Banks, J.(2) | 41 | President of the Royal Society | London | 1784–91 |
| Bassegli, von | | Count | ex Ragusa | 1784–86 |
| Becker(1) | | Apothecary | Brunswick | 1784–91 |
| Becker | | Pharmacy learner | ex Copenhagen | 1786 |
| | | Pharmacy learner | Langensalza | 1787–88 |
| Beckerhinn, C. P. D.(1,2) | | Pharmacy learner | Strasbourg | 1787–88 |
| | | Medical student | Strasbourg | 1789–91 |
| Beireis, G. C.(2) | 54 | Professor of medicine-science | Helmstedt | 1784–91 |
| Bell, J. F.(3) | | Apothecary | Berlin | 1784–91 |
| Bender, C. B. | | Medical student | Erlangen | 1785 |
| | | Physician | Kochendorf | 1788 |
| Benteli, S.(1) | | Apothecary | Bern | 1784–91 |
| Berend, A. W. | | Medical student | Berlin | 1784 |
| | | Apothecary | Berlin | 1785 |
| Berendt | | Apothecary | St. Petersburg | 1784–90 |
| Bergman, T. O.(2) | 49 | Professor of chemistry | Uppsala | 1784 (d. 1784) |
| Bergmann, C. W. | | Medical student | Berlin | 1784 |
| | | Apothecary | Berlin | 1785 |
| Berndt, J. | | Unknown | Moscow | 1784–85 |
| Bernhard | | Physician | Jüterbog | 1784–90 |
| Beroldingen, F. C. von(2) | 44 | Canon | Hildesheim | 1785–91 |
| Bertram, A. F. | | Pharmacy learner | Kassel | 1785 |
| Beseke, J. M. G. | 38 | Professor of science | Mitau | 1784 |
| Beuth, H. J. F. | 50 | Treasury councillor | Düsseldorf | 1784 |
| Beyer, A.(2) | 42 | Mining secretary | Schneeberg | 1784–91 |
| Beyer, S. F. | | Pharmacy learner | Berlin | 1784 |
| | | Apothecary | Berlin | 1785–91 |

TABLE A3: Crell's Subscribers 1784-1791 (cont.)

| Name | Age in 1784 | Occupation | Town | Years subscribing in given occupation and town |
|---|---|---|---|---|
| Bibo, J. H. | | Unknown | Moscow | 1784-85 |
| Bietzker | | Unknown | Oebisfelde | 1790-91 |
| Bindheim, J. J.(1,2) | 34 | Pharmacy learner | Berlin | 1784 |
| | | Apothecary | Moscow | 1786-91 |
| Bischoff, H. | | Pharmacy learner | Hameln | 1791 |
| Blancherie, M. C. de la | 32 | Science popularizer | Paris | 1786-88 |
| Blankennagel, G. von | | Water engineer | Moscow | 1784-86, 1789-91 |
| Blume | | Apothecary | Schlawe | 1785-91 |
| Blumenbach, J. F. | 32 | Professor of medicine-science | Göttingen | 1791 |
| Bode | | Pharmacy learner | Copenhagen | 1785-86 |
| Börnicke, J. | | Official | Moscow | 1788, 1791 |
| Bollmann | | Pharmacy learner | Minden | 1785 |
| Borges, W. H. L. | 27 | Medical student | Helmstedt | 1787-89 |
| | | Medical student | Erlangen | 1790 |
| | | Physician | Brunswick | 1791 |
| Bornemann | | Army surgeon | Werchoturje | 1784-85, 1787-89 |
| Bose, F. W. A. von | 31 | Saxon ambassador | Stockholm | 1784 |
| Bothmer, C. von | 48 | Mining director | Bayreuth | 1784-87 |
| Boving | | Pharmacy learner | Copenhagen | 1785-88 |
| | | Apothecary | Jutland | 1789-91 |
| Brande, J. K.(1,3) | 30 | Court apothecary | Hanover | 1784-91 |
| Brandis, J. D. | 22 | Medical student | Göttingen | 1784-86 |
| Brandt, B. P. | | Pharmacy learner | Allendorf | 1784 |
| | | Pharmacy learner | Kassel | 1785-89 |
| Braths | | Unknown | Herrnhut | 1784 |
| Brauer, Jr. | | Apothecary | Diepholz | 1785 |
| Braun, A. | | Medical student | Prague | 1785 |
| Braun | | Pharmacy learner | Langensalza | 1788 |
| | | Pharmacy learner | Augsburg | 1789-90 |
| | | Pharmacy learner | Nuremberg | 1791 |
| Brinckmann, J. P. B.(2) | 38 | Physician | Düsseldorf | 1784 (d. 1785) |
| Brockmann(2) | | University-town apothecary | Rinteln | 1786, 1789-91 |
| Brückmann, U. F. B.(2) | 56 | Ducal physician | Brunswick | 1784-91 |
| Brüel(2) | 22 | Smelting administrator | Zellerfeld | 1784-91 |
| Brugnatelli, L. V.(2) | 23 | Physician | Pavia | 1789-91 |
| Brun, W. C. A. | 40 | Apothecary | Güstrow | 1785-91 |
| Bucholtz, W. H. S.(2,4) | 50 | Physician and apothecary | Weimar | 1784-91 |
| Bühring, J. H. | | Apothecary | Burgdorf i. Br. | 1784-87, 1791 |
| Bühring | | Apothecary | Baruth | 1788-90 |
| Bülow, J. J. C. F. von | | Forestry administrator | Zellerfeld | 1784 |
| Bürger | | Surgeon | Burgdorf, Switzerland | 1785-91 |

TABLE A3: Crell's Subscribers 1784–1791 (cont.)

| Name | Age in 1784 | Occupation | Town | Years subscribing in given occupation and town |
|---|---|---|---|---|
| Büsing(3) | | Court apothecary | Schwerin | 1784–87 |
| Büttner, C. W. | 68 | Physician | Jena | 1784–91 |
| Burchhardt | | Apothecary | Blankenburg | 1784–88 |
| Butter, J. G. | | Apothecary | Moscow | 1784–91 |
| Byhan, F. | | Apothecary | St. Petersburg | 1784–91 |
| Cäser | | Town official | Magdeburg | 1784–85 |
| Cahlo | | Physician | Ovelgönne | 1785–90 |
| Camman, J. G. | | Medical student | Helmstedt | 1790 |
| Carita, G. L.(1) | | Apothecary | Berlin | 1784–86 |
| Carnap, von | | Unknown | Elberfeld | 1791 |
| Catherine II | 55 | Empress | St. Petersburg | 1787–91 |
| Cavendish, H.(2) | 53 | Gentleman | London | 1786–91 |
| Christiani, | | Pharmacy learner | ex Kiel | 1784 |
| O. W. C.(2) | | Pharmacy learner | Langensalza | 1785 |
| | | Pharmacy learner | Strasbourg | 1786–88 |
| | | Pharmacy learner | Kiel | 1789–91 |
| Claus(2) | | Mining administrator | Holzminden | 1784–87 |
| Claussen | | Law student | Halle | 1784–85 |
| Corvinus, J. H. | | Apothecary | Schöppenstedt | 1784–90 |
| Cothenius, C. A. von | 76 | Royal physician | Berlin | 1784–88 (d. 1789) |
| Couret, P. | | Pharmacy learner | Kempten | 1785–91 |
| Cretschmar, G. L. | | Apothecary | Elberfeld | 1784–91 |
| Cruse, G. F. | | Medical student | Berlin | 1784 |
| Curtius | | Physician | Lübeck | 1784, 1787–91 |
| Daniel, C. F. | 31 | Physician | Halle | 1784–85 |
| Daniels, W. | | Mining administrator | Eschweiler | 1790 |
| Decker | | Pharmacy learner | Oker | 1784 |
| | | Pharmacy learner | Hildesheim | 1785 |
| Deggeler | | Official | Hager | 1785–87 |
| Dehfloff, B. P. J. | | Unknown | Moscow | 1784–85 |
| Dehne, J. C. C.(2) | 34 | Physician | Schöningen | 1784–91 (d. 1791) |
| Delius, H. F.(1,2) | 64 | Professor of medicine-chemistry | Erlangen | 1784–91 (d. 1791) |
| Delius | | Pharmacy provisor | Yersmold | 1785 |
| Demler | | Medical student | Tübingen | 1785–91 |
| Dempfwolf, J. F. | 30 | Apothecary | Lüneburg | 1785–91 |
| Deutsch | | Apothecary | Neudam | 1784 |
| Devely | | Unknown | ex Yverdon | 1789–91 |
| Didrichson, D. | 32 | Economic society secretary | Copenhagen | 1789–91 |
| Dieterich, J. C. | 62 | Publisher | Göttingen | 1784 |
| Dietrich, P. H. von | 61 | Baron | Paris | 1784 |
| Dilly | | Apothecary | Herrnhut | 1784 |
| Dinckler | | Physician | Elberfeld | 1784 |
| Dönch(3) | | Pharmacy learner | Hanover | 1791 |
| Dörffurt, A. F. L. | 17 | Pharmacy learner | Genthin | 1786–87 |
| Döring, J. P. F. | | Mining councillor | Düsseldorf | 1784–91 |

TABLE A3: Crell's Subscribers 1784–1791 (cont.)

| Name | Age in 1784 | Occupation | Town | Years subscribing in given occupation and town |
|---|---|---|---|---|
| Döring | | Apothecary | Freienwalde | 1784 |
| | | Apothecary | Berlin | 1785 |
| Döring | | Mining administrator | Oberkaltenberg | 1785–91 |
| Döring | | Apothecary | Vladimir | 1786–87 |
| Dolhof | | Mayor | Magdeburg | 1784–89 |
| Dollfuss, J. C.(2) | | Pharmacy learner | London | 1786 |
| | | Physician | London | 1787–90 |
| Dorse | | Surgeon | Moscow | 1784–85 |
| Drechsler, C. G.(1) | | Apothecary | Zellerfeld | 1784 |
| Drewer | | Pharmacy learner | Hanover | 1784–85 |
| Ebeling, J. T. P. C. | 31 | Physician | Lüneburg | 1785–91 |
| Ebenberger, J. | | Apothecary | Prague | 1785–91 |
| Ebermeier, H. C. | 49 | Apothecary | Melle | 1784–87 |
| Ebert, J. H. | | Mine manager | Unterharz | 1784 |
| Eggers | | Pharmacy provisor | Moscow | 1790 |
| Ehmke | | Apothecary | Stolp | 1785–91 |
| Ehmsen(1) | | Court apothecary | Osnabrück | 1784–87 |
| Ehmsen | | Pharmacy learner | Hanover | 1784 |
| Ehrenfriedstein, J. | | Army apothecary | Lubny | 1784–85 |
| Ehrhart, F. | 42 | Royal botanist | Herrenhausen | 1784–86 |
| Ehrlich | | Pharmacy learner | Berlin | 1784 |
| | | Apothecary | Berlin | 1785 |
| Ehrmann, F. L.(2) | 43 | Science teacher | Strasbourg | 1784–91 |
| Eller, J. L. F. | | Navy surgeon | Hintbergen | 1785–86 |
| Elwert, J. K. P. | 24 | Medical student | Erlangen | 1784–85 |
| Engel, K. C. | 32 | Physician | Schwerin | 1785 |
| Erlach, J. von | 60 | Colonel | Bern | 1784–91 |
| Essmann | | Unknown | Moscow | 1784–85 |
| Euler, J. A.(2) | 50 | Academician | St. Petersburg | 1784–88, 1790 |
| Evenius | | Apothecary | Nizhni-Novgorod | 1784–91 |
| Felsch, C. | | Unknown | Moscow | 1784–85 |
| Felsch, D. | | Apothecary | Kazan | 1784–90 |
| Ferber, J. J.(2) | 41 | Academician | St. Petersburg | 1784–85 |
| Fiedler, C. W.(1,2) | 26 | Apothecary | Kassel | 1784–91 |
| Fiedler, J. G.(1) | 56 | Apothecary | Kassel | 1784, 1786–87 |
| Fischer(1) | | Smelting administrator | Oker | 1784–86 |
| Fischer | | Pharmacy provisor | Stargard | 1784–85 |
| | | Pharmacy provisor | Berlin | 1786–89 |
| Flach | | Pharmacy learner | Königsberg | 1785 |
| Flemming | | Physician | Jüterbog | 1784–90 |
| Florencourt, C. C. von(2) | 27 | Treasury councillor | Blankenburg | 1784 |
| Flügger, J. H.(1,2) | | Apothecary | Kassel | 1784 |
| Förchtl, J. O. S. | | Apothecary | Cologne | 1784–91 |
| Forcke, J. J. W. | | Medical student | Göttingen | 1786–87 |
| Forster, G. | 30 | Professor of science | Wilna | 1785–88 |

TABLE A3: Crell's Subscribers 1784–1791 (cont.)

| Name | Age in 1784 | Occupation | Town | Years subscribing in given occupation and town |
|---|---|---|---|---|
| Fränkel, G. | | Physician | Moscow | 1787–88 |
| Frenzel | | Apothecary | Erfurt | 1784–90 |
| Friedland | | Pharmacy learner | Hamburg | 1784–85 |
| | | Pharmacy learner | Osterode | 1786 |
| Friedrich-Carl(4) | | Prince | Rudolstadt | 1784–91 |
| Frischmann | | Apothecary | Erlangen | 1790 |
| Fuchs, G. F. C.(2) | 24 | Docent in science-medicine | Jena | 1784–91 |
| Fuhrmann | | Merchant | Moscow | 1786 |
| | | Official | Orel | 1790 |
| Fulda | | Mining councillor | Kassel | 1784 |
| Gadolin, J.(2) | 24 | Professor of science | Åbo | 1786, 1789–91 |
| Gärtner, Sr. | | Apothecary | Hanau | 1789–91 |
| Gärtner, Jr. | | Apothecary | Hanau | 1789–91 |
| Gäuke, J. C. | | Pharmacy learner | Berlin | 1784–85 |
| | | Pharmacy learner | Lissa | 1786–91 |
| Gahlen, von | | Apothecary | Barmen | 1784 |
| Gahn, J. G. | 39 | Mining official | Stockholm | 1789–91 |
| Gardner, F. | | Merchant | Moscow | 1786–91 |
| Gasser | | Pharmacy learner | Regensburg | 1787 |
| Gaupp(1) | | Apothecary | Kirchheim | 1784–91 |
| Gebeler | | Apothecary | Walsrode | 1784–91 |
| Gebhard | | Apothecary | Schafstädt | 1784 |
| Geelhaar, von | | Artillery captain | Berlin | 1785–89 |
| Gehler, J. C. | 52 | Professor of science | Leipzig | 1784 |
| Gehrt, C. L.(1,3) | 34 | Apothecary | Altona | 1791 |
| Gemmingen, von | | Official | Ansbach | 1784–89 |
| Gempt, J. H. | | Medical student | Helmstedt | 1787–88 |
| Georg | | Prince | Waldeck | 1791 |
| Georgi, J. G.(1,2) | 55 | Academician | St. Petersburg | 1784–91 |
| Gesler, von | | Count | Berlin | 1785–91 |
| Geutner | | Pharmacy learner | ex Königstein | 1786–91 |
| Geyr, von | | Magistrate | Düsseldorf | 1785–91 |
| Glawet, J. C. von | | Manufacturer | Moscow | 1786 |
| Gleditsch, J. G. | 70 | Professor of botany | Berlin | 1785–86 (d. 1786) |
| Glendenberg (1-3) | | Pharmacy learner | London | 1784–86 |
| | | Pharmacy provisor | Schwerin | 1787 |
| | | Apothecary | Schwerin | 1788–91 |
| Glenk | | Saltworks director | Weisbach | 1789–91 |
| Gmelin, C. G.(1) | 35 | Court apothecary and physician | Tübingen | 1784–91 |
| Gmelin, C. G.(2) | | Court apothecary | Stuttgart | 1784–91 |
| Gmelin, E. | 33 | Physician | Heilbronn | 1784–91 |
| *Gmelin*, J. F.(1,2) | 36 | Professor of chemistry | Göttingen | 1784–91 |
| Gönner | | Physician | Berlin | 1784–91 |
| Goethe, J. W. von | 35 | Official | Weimar | 1784 |

TABLE A3: Crell's Subscribers 1784–1791 (cont.)

| Name | Age in 1784 | Occupation | Town | Years subscribing in given occupation and town |
|---|---|---|---|---|
| Göttling, J. F. A.(2) | 31 | Medicine-science student | Göttingen | 1784–88 |
| | | Professor of chemistry-technology | Jena | 1789–91 |
| Goldhagen, J. F. G. | 42 | Professor of medicine | Halle | 1784–85 (d. 1788) |
| Gottschalk(3) | | Pharmacy learner | Hanover | 1784–85 |
| Graberg, A. L.(1-3) | | Apothecary | Brunswick | 1784–91 |
| Graf | | Town apothecary | Bayreuth | 1784–85 |
| Grattenauer(4) | | Bookdealer | Nuremberg | 1784 |
| Gravenhorst, C. J. | 53 | Chemical manufacturer | Brunswick | 1788–91 |
| Gren, F. A. C.(2) | 24 | Medical student | Halle | 1784–86 |
| | | Professor of science | Halle | 1787–91 |
| Grewe | | Court apothecary | St. Petersburg | 1784–91 |
| Grim, L. | | Pharmacy learner | Kiel | 1784 |
| Grönlund | | Apothecary | Copenhagen | 1787–91 |
| Groschke, J. G.(2) | 24 | Professor of science | Mitau | 1789–91 |
| Grossmann(1) | | Physician | Boizenburg | 1784–87, 1789 |
| Grote, von | | Official | Cologne | 1785 |
| Grudtner | | Pharmacy learner | Langensalza | 1785 |
| Gruner, C. G. | 40 | Professor of medicine | Jena | 1784–91 |
| Guckenberger, L. J. | | Physician | Hanover | 1786–90 |
| Günther(1,2) | | Apothecary | Copenhagen | 1785–87 |
| Guyton de Morveau(2) | 47 | General attorney | Dijon | 1785–91 |
| Hacken | | Apothecary | Kostroma | 1784–90 |
| Hacquet, B.(2) | 45 | Professor of medicine | Laibach | 1784 |
| Hänle, G. F.(2) | 21 | Apothecary | Lahr i. B. | 1786, 1790 |
| Hagen, K. G.(1,2) | 35 | Professor of science and apothecary | Königsberg | 1784–86 |
| Hagen | | Apothecary | Copenhagen | 1785–91 |
| Haim, G. | | Mining official | Salzburg | 1785 |
| Hammer | | Pharmacy learner | Hanover | 1788–90 |
| Hamstein, L. A. F. | | Mining official | Clausthal | 1784 |
| Hannemann, J. B. | | Pharmacy learner | Berlin | 1784 |
| | | Apothecary | Berlin | 1785 |
| Hannemann, N. D. | | Apothecary | Moscow | 1784–90 |
| Hannesmann | | Bookdealer | Kleve | 1786–91 |
| Harbordt, B. | | Medical student | Brunswick | 1784–86 |
| Hardenberg, C. A. von | 34 | Official | Brunswick | 1785–90 |
| Hardenberg, G. A. G. von | 19 | Official | Ansbach | 1791 |
| | | Mining councillor | Berlin | 1790–91 |
| Harsleben(2) | | Apothecary | Potsdam | 1786, 1788–91 |
| Haselberg, L. W. | 20 | Medical student | Göttingen | 1784 |
| Hasse, C. F.(3) | 39 | Apothecary | Hamburg | 1784–91 |

TABLE A3: Crell's Subscribers 1784–1791 (cont.)

| Name | Age in 1784 | Occupation | Town | Years subscribing in given occupation and town |
|---|---|---|---|---|
| Hasse, J. F. B.(2,3) | | Apothecary | Hamburg | 1784–91 |
| Hassenfratz, J. H.(2) | 29 | Mining official | Paris | 1790 |
| Hauch, A. W. | 29 | Stable master | Copenhagen | 1788–91 |
| Haukohl | | Unknown | Moscow | 1784–85 |
| Hauptvogel | | Official | Moscow | 1786 |
| Hausenbaum, C. | | Hospital apothecary | Moscow | 1784–85 |
| Hausmann, J. M.(2) | 35 | Dye-works owner | Colmar | 1786–91 |
| Hawkins, J.(2) | 26 | Gentleman | Brunswick | 1785 |
| | | Gentleman | London | 1789 |
| Hecht(2) | | Apothecary | Strasbourg | 1784–91 |
| Heerbrandt | | Bookdealer | Tübingen | 1784, 1786–88 |
| Hees, von | | Apothecary | Barmen | 1784 |
| Heinemann(2) | | Mining councillor | Brunswick | 1784–85 |
| | | Treasury councillor | Brunswick | 1786–88 |
| Held, J. N. | 54 | Physician | Wetzlar | 1784–86 (d. 1786) |
| Hellwig | | Physician | Fürstenburg | 1784–88 |
| Helvig, N. U.(2) | 46 | Apothecary | Stralsund | 1784–91 |
| Hemmelmann(1) | | Apothecary | Wolfenbüttel | 1784–85 |
| Hempel, J. G.(2) | 32 | Medical student | Helmstedt | 1784–85 |
| | | Physician | Helmstedt | 1786–91 |
| Hengstenberg, J. H. | | Apothecary | Elberfeld | 1784–85 |
| Henk, M. C. | | Unknown | ex Gladbach | 1784 |
| Henkel, C. | | Unknown | Moscow | 1784–85 |
| | | Unknown | St. Petersburg | 1786 |
| Hennemann, W. J. C. | 29 | Physician | Schwerin | 1784–91 |
| Herget, B. | | Physician and chemistry teacher | Prague | 1785–91 |
| Hermann, J. B.(2) | 46 | Professor of science | Strasbourg | 1784–91 |
| Hermann, K. S. L. | 19 | Pharmacy provisor | Zerbst | 1789–91 |
| Hermbstaedt, S. F.(2) | 24 | Pharmacy provisor | Berlin | 1784–85 |
| | | Science student | Berlin | 1786 |
| | | Chemistry teacher | Berlin | 1787–91 |
| Herrmann, J. | | Unknown | Kunzendorf | 1784 |
| Herrmann, B. F. J.(2) | 29 | Steelworks director | Catharinenburg | 1786–91 |
| Herrmann | | Court apothecary | Oehringen | 1790 |
| Heyer, J. C. H.(1-3) | 38 | Apothecary | Brunswick | 1784–91 |
| Hieppe, E. P. | | Medical student | Berlin | 1784 |
| | | Pharmacy learner | Strasbourg | 1785 |
| | | Apothecary | Wetzlar | 1787–91 |
| Hildebrand, G. | | Apothecary | Moscow | 1784–91 |
| Hildebrand, P. | | Army apothecary | Lubny | 1784–90 |
| Hildebrandt, G. F. | 20 | Professor of medicine | Brunswick | 1791 |

TABLE A3: Crell's Subscribers 1784–1791 (cont.)

| Name | Age in 1784 | Occupation | Town | Years subscribing in given occupation and town |
|---|---|---|---|---|
| Hink | | Apothecary | Einbeck | 1785–86 |
| Höfer, J. G. J. | | Apothecary | Gandersheim | 1784–91 |
| Höpfner, J. G. A.(2) | 25 | Physician | Bern | 1784–91 |
| Höpfner | | Apothecary | Bern | 1784–91 |
| Hoffmann, C. A.(2) | 28 | Pharmacy learner | Kassel | 1784–85 |
| | | Pharmacy provisor | Weimar | 1786–91 |
| Hoffmann, F. C.(1,2) | 40 | Apothecary | Leer | 1787–91 |
| Hohenthal, von | | Official | St. Petersburg | 1784 |
| Holtheuer | | Apothecary | St. Petersburg | 1784–85, 1788–89 |
| Holz, I. | | Apothecary | Prague | 1785–91 |
| Holze, C. F. | | Pharmacy learner | Berlin | 1784–86 |
| Hompesch, J. W. von | 23 | Canon | Düsseldorf | 1785–91 |
| Honig | | Official | Schöningen | 1784–91 |
| Honrich, L. W. | | Unknown | Moscow | 1784–88 |
| | | Unknown | Voronezh | 1789–90 |
| Hoppe, D. H.(2) | 24 | Pharmacy learner | Regensburg | 1787–90 |
| Hoppe, E. G.(1) | | Apothecary | Königsberg | 1784–85 |
| Hoym, von | | Treasury councillor | Brunswick | 1784–86 |
| Hummel | | Merchant | Helmstedt | 1786–91 |
| Huth(2) | | Teacher | Halle | 1785 |
| Ilisch, J. S.(3,4) | | Pharmacy learner | Horneburg | 1784–85 |
| | | Pharmacy learner | Riga | 1786 |
| | | Apothecary | Riga | 1787–91 |
| Illmann | | Official | St. Petersburg | 1786–87 |
| Ilsemann, J. C.(1,2) | 55 | Apothecary | Clausthal | 1784–91 |
| Ilsemann(1) | | Pharmacy learner | Hanover | 1784–86 |
| Ilsemann | | Pharmacy learner | Hanover | 1784–85 |
| Ilsemann(1) | | Apothecary | Hildesheim | 1785–89 |
| Ingwerfen | | Apothecary | Friederica | 1788–91 |
| Ischwald | | Physician | Schaffhausen | 1784–85 |
| Iser(3) | | Pharmacy provisor | Minden | 1785 |
| Jacobi, F. H. | 41 | Official | Düsseldorf | 1784–87 |
| Jahn, W. | | Medical student | Helmstedt | 1785 |
| | | Physician | Neumünster | 1786–87 |
| Jahn | | Apothecary | Meiningen | 1785 |
| Jenner, G. von | 50 | Official | Bern | 1789–91 |
| John | | Apothecary | Anklam | 1784–91 |
| Jordan | | Apothecary | Hoya | 1784–87 |
| Jordan(1) | | Pharmacy learner | Lüneburg | 1786 |
| Jürgenson, M. | | Imperial apothecary | Moscow | 1784–90 |
| Jugowitz | | Physician | Laibach | 1784 |
| Kähler, F. W. S. | | Pharmacy learner | Frankfurt a. M. | 1784–91 |
| Kaldewey(2) | | Physician and apothecary | Lünen | 1784–89 |
| Kalkau, J. | | Apothecary | Moscow | 1784–89 |
| Karsten, D. L. G.(2,3) | 16 | Mining student | Marburg | 1788 |
| | | Mining student | Halle | 1789 |
| | | Mining official | Berlin | 1790–91 |

TABLE A3: Crell's Subscribers 1784–1791 (cont.)

| Name | Age in 1784 | Occupation | Town | Years subscribing in given occupation and town |
|---|---|---|---|---|
| Karsten, W. J. G.(2) | 52 | Professor of science | Halle | 1784–87 (d. 1787) |
| Kast, J. H. | | Mining official | Zellerfeld | 1784 |
| Kasteleyn, P. J.(2) | 39 | Apothecary | Amsterdam | 1791 |
| Kastenbein, U. H. | | Smelting master | Clausthal | 1784 |
| Keber | | Merchant | Berlin | 1786–91 |
| Keidel | | Pharmacy learner | Halle | 1785–91 |
| Kelp | | Apothecary | Ovelgönne | 1785–90 |
| Kels, H. W.(2) | 25 | Pharmacy learner | Osnabrück | 1784–86 |
| | | Medical student | Göttingen | 1787–91 |
| Kern, C. W.(4) | | Treasury councillor | Ansbach | 1789–91 |
| Kessel | | Medical student | Königsberg | 1785 |
| | | Medical student | Berlin | 1786–91 |
| Kessler | | Physician | Magdeburg | 1784–88 |
| Keyser | | Apothecary | Detmold | 1786–91 |
| Kirsten | | Law student | Halle | 1784 |
| Kirwan, R.(2) | 49 | Gentleman | London | 1784–87 |
| | | Gentleman | Dublin | 1788–91 |
| Klaproth, M. H.(2,3) | 41 | Apothecary and science teacher | Berlin | 1784–91 |
| Klenze | | Magistrate | Schladen | 1784–87 |
| Klewitz, W. A. | | Treasury official | Magdeburg | 1785–91 |
| Klint, J. | | Army physician | Moscow | 1787–88 |
| Klockmann | | Apothecary | Schwerin | 1784–91 |
| Klötz | | Mining surveyor | Rothenburg a. d. S. | 1784–86 |
| Klügel, G. S. | 47 | Professor of science | Halle | 1788–91 |
| Knorre, J. G.(2) | 57 | Mint master | Hamburg | 1784–91 |
| Kobligk | | Physician | Elbing | 1784 |
| Koch, B. A.(4) | | Pharmacy learner | Bremen | 1785–91 |
| Kölbeke | | Official | Mikhailov | 1791 |
| Könnecke, A. L. | | Apothecary | Halberstadt | 1786 |
| Körber, J. F. | 19 | Medical student | Göttingen | 1784–85 |
| Köstelhön | | Unknown | Moscow | 1784–85 |
| Köster(1) | | Apothecary | Münden | 1785–91 |
| Köster, J. | | Apothecary | Orel | 1784–89, 1791 |
| Kohl, B. | | Unknown | Celle | 1787–90 |
| | | Unknown | Osterholz | 1791 |
| Kohl, C. E. F.(2) | 44 | Apothecary | Halle | 1784–86, 1788–91 |
| Kohl, J. E.(2) | | Mining councillor | Fürstenberg | 1784–90 |
| Kohlhass, J. J. | 37 | Physician | Regensburg | 1787–90 |
| Kohlrief, G. A. | 35 | Professor of medicine-science | St. Petersburg | 1788–89 |
| Kollowrath, A. N. von(4) | | Official | Milan | 1786–91 |
| Kopp | | Apothecary | Würzburg | 1791 |
| Koryzna, J. | | Police surgeon | Moscow | 1784–88 |
| Kraatz, D. F.(1) | | Pharmacy provisor | Berlin | 1784–89 |
| Kraft | | Medical student | Halle | 1784–85 |
| Kramer | | Physician | Halberstadt | 1787–91 |

TABLE A3: Crell's Subscribers 1784–1791 (cont.)

| Name | Age in 1784 | Occupation | Town | Years subscribing in given occupation and town |
|---|---|---|---|---|
| Kratz(1) | | Physician | Hildesheim | 1784 |
| Kratzenstein, C. G.(2) | 61 | Professor of science | Copenhagen | 1784–91 |
| Kretschmer | | Physician | Elberfeld | 1790 |
| Kreusel, J. C. | | Surgeon | Moscow | 1784–85 |
| | | Physician | Moscow | 1786–88 |
| Kron | | Hospital apothecary | Moscow | 1784–86 |
| Krüger, F. | | Bookdealer | Giessen | 1784 |
| Krüger | | Apothecary | Lüneburg | 1784–86, 1788–91 |
| Krüger | | Physician | Danzig | 1785 |
| Kücke | | Pharmacy learner | Brunswick | 1787 |
| | | Pharmacy learner | Lauterbach | 1788 |
| | | Pharmacy learner | Bremen | 1789–90 |
| | | Pharmacy learner | Hanover | 1791 |
| Küster, H. G. F. | | Medical student | Helmstedt | 1784 |
| Kummer, J. U. | | Medical student | Stettin | 1784 |
| Kundler, A. | | Physician | Essen | 1790 |
| Kunhardt, F. G. | 23 | Pharmacy learner | Frankfurt a. M. | 1784–91 |
| Kunsemüller, C. L.(2,3) | 23 | Pharmacy provisor | Hamburg | 1789 |
| | | Medical student | Berlin | 1790 |
| | | Physician | Hamburg | 1791 |
| Kurella, E. G. | 59 | Physician | Berlin | 1784–91 |
| Lammersdorf, J. A. | 26 | Physician | Hanover | 1784–91 |
| Lampe, P. A. | 30 | Physician | Danzig | 1785 |
| Lang | | Apothecary | Stuttgart | 1784–91 |
| Lange | | Apothecary | St. Petersburg | 1784–89, 1791 |
| Lange(3) | | Pharmacy learner | Frankfurt a. d. O. | 1786–91 |
| Langer, J. H. S. | 29 | Mining official | Weimar | 1785–87 (d. 1788) |
| Langguth, C. A. | 30 | Professor of medicine | Wittenberg | 1784–91 |
| Langguth, J. A. | | Physician | Köthen | 1784–91 |
| Langsdorf, C. C. von(2) | 27 | Saltworks administrator | Gerabronn | 1787–88 |
| Lasius, G. S. O.(2) | 32 | Engineer lieutenant | Hanover | 1785–90 |
| | | Engineer lieutenant | Hameln | 1791 |
| Lehnstubb | | Official | St. Petersburg | 1784–85 |
| Leipoldt, J. Z.(2) | | Apothecary | Augsburg | 1784 |
| Lembke | | Apothecary | Tula | 1784–85 |
| Lemke | | Pharmacy learner | Goslar | 1789–90 |
| *Leonhardi, J. G.*(2,3) | 38 | Professor of medicine | Wittenberg | 1789–91 |
| Leopold II | 37 | Grand Duke | Florence | 1784–89 |
| | | Emperor | Vienna | 1790–91 (d. 1791) |
| Leske, N. G. | 33 | Professor of science | Leipzig | 1784 (d. 1786) |
| Lichtenberg, G. C.(2) | 40 | Professor of science | Göttingen | 1784–91 |
| Lichtenstein, G. R.(2) | 39 | Professor of medicine | Helmstedt | 1784–91 |
| Lieber | | Pharmacy learner | Hamburg | 1791 |
| Linke | | Medical student | Leipzig | 1784 |
| Lober | | Apothecary | Erlangen | 1784 |

TABLE A3: Crell's Subscribers 1784-1791 (cont.)

| Name | Age in 1784 | Occupation | Town | Years subscribing in given occupation and town |
|---|---|---|---|---|
| Löwe, A. C. L.(2) | | Pharmacy learner | Quedlinburg | 1785 |
| Lowitz, J. T.(2) | 27 | Pharmacy learner | St. Petersburg | 1784, 1786 |
| | | Apothecary | St. Petersburg | 1787-91 |
| Lowson | | Unknown | Copenhagen | 1788-91 |
| Luck, H. G.(2) | | Pharmacy learner | Stade | 1784-85 |
| | | Pharmacy learner | Berlin | 1786-91 |
| Luckenbach | | Finance councillor | Bernburg | 1784-86 |
| Lütke, von | | Mayor | Moscow | 1787 |
| Lunde, J. W. | | Mining official | Clausthal | 1786-91 |
| Lystager | | Pharmacy learner | Copenhagen | 1785-88 |
| | | Apothecary | Jutland | 1789-91 |
| Mähl | | Town apothecary | Rostock | 1786-91 |
| Mätke, J. H. F. | | Mining official | Unterharz | 1784 |
| Mahlstedt, J. G. | | Unknown | Moscow | 1784-85 |
| Mandenberg, J. H.(2,3) | | Pharmacy learner | Berlin | 1786-89 |
| Manthey, J. G. L. | 15 | Medical student | Copenhagen | 1787-88 |
| | | Surgeon and docent in chemistry | Copenhagen | 1789-91 |
| Manuel, J. von | | General commissioner | Bern | 1784-90 |
| Markel | | Apothecary | Nuremberg | 1785 |
| Martin | | Apothecary | Strasbourg | 1784, 1786-91 |
| Martius, E. W.(2) | 28 | Pharmacy learner | Hermannstein | 1784 |
| | | Pharmacy learner | Strasbourg | 1786 |
| Masch, A. W. | | Castle apothecary | Stolp | 1784-91 |
| Masius | | Court physician | Schwerin | 1785 |
| Mayer, J. N. | 30 | Physician | Prague | 1786 |
| Mayer, J. T. | 32 | Professor of science | Erlangen | 1790 |
| Meder, F. von | | Waterworks administrator | Moscow | 1784-86 |
| Meding, F. A. L. von | 19 | Law student | Göttingen | 1787 |
| | | Unknown | Hanover | 1788-90 |
| | | Mining administrator | Clausthal | 1791 |
| Megenhard(1) | | Apothecary | Tuttlingen | 1786-91 |
| Meier, Jr.(1) | | Apothecary | Osnabrück | 1784-85 |
| Meineke, A. H. | 22 | Medical student | Helmstedt | 1786-90 |
| Meineke, D. H. G. | | Medical student | Hildesheim | 1791 |
| | | Pharmacy learner | Schwerin | 1784 |
| Meisner, G. N.(4) | | Pharmacy learner | Halle | 1784 |
| | | Pharmacy provisor | Halle | 1785 |
| Mejo, T.(1) | | Apothecary | Königsberg | 1784-85 |
| Melissino, J. von | | University official | Moscow | 1784-87 |
| Mensch, C. | | Lieutenant | Moscow | 1786-87 |
| Mertens | | Apothecary | Landsberg | 1784 |
| | | Apothecary | Berlin | 1785-91 |
| Merz | | Pharmacy learner | Weimar | 1787 |
| Métherie, J. C. de la(2) | 41 | Editor of science journal | Paris | 1786-91 |
| Mey, J. | | Army surgeon | Moscow | 1784-85 |
| Meyer, F. G. | 69 | Court physician | Hanover | 1784 (d. 1791) |

TABLE A3: Crell's Subscribers 1784-1791 (cont.)

| Name | Age in 1784 | Occupation | Town | Years subscribing in given occupation and town |
|---|---|---|---|---|
| Meyer, G. J. | | Pharmacy provisor | Oldenburg | 1785-86 |
| Meyer, J. C. F.(2,3) | 45 | Court apothecary | Stettin | 1784-91 |
| Meyer, J. S. F. | | Metallurgy-chemistry student | ex Clausthal | 1784 |
| | | Metallurgy-chemistry student | Helmstedt | 1785 |
| | | Metallurgy-chemistry student | Lautenthal | 1786 |
| Meyer(3) | | Apothecary | Neustadt a. R. | 1784-91 |
| Meyer | | Army surgeon | Scharnebeck | 1784-86 |
| Michaelis | | Law student | Halle | 1785 |
| Michaelson | | Apothecary | Demmin | 1785-91 |
| Miessl, J. N. | | Mining administrator | Joachimsthal | 1786 |
| Möhring, H. C.(1) | | Apothecary | Berlin | 1784-91 |
| Möllenhof | | Apothecary | Wetzlar | 1785 |
| Möller | | Official | Bielefeld | 1786-87 |
| Mönch, C.(2) | 40 | Professor of science and apothecary | Kassel | 1784-85 |
| | | Professor of science | Marburg | 1786-91 |
| Mönnicke, C. T.(1) | | Pharmacy provisor | Berlin | 1784-87 |
| Morell, C. F.(2) | 35 | Apothecary | Bern | 1784-91 |
| Moritz | | Law student | Halle | 1784-85 |
| Moser, A. | | Mining official | Salzburg | 1785 |
| Motz, von | | Baron | Salzburg | 1785-86 |
| Mües | | Goldworker | Osnabrück | 1784 |
| Mühlenbein | | Physician | Königslütter | 1789-91 |
| Mühlenstheth, J. A. | 38 | Apothecary | Copenhagen | 1785-91 |
| Müller, F. H. | 52 | Apothecary and porcelain producer | Copenhagen | 1785-88 |
| Müller | | Court apothecary | Gotha | 1784 |
| Müller | | Apothecary | Braunfels | 1785-91 |
| Müller | | Apothecary | Schöningen | 1791 |
| Müller | | Apothecary | Moscow | 1791 |
| Muhle, A. W. B.(2) | | Medical student | ex Fürstenau | 1784 |
| | | Pharmacy learner | Lüneburg | 1785 |
| | | Pharmacy provisor | Dannenberg | 1786 |
| | | Apothecary | Harburg | 1787-91 |
| Mukey, J. C. F. | | Apothecary | Bern | 1784-91 |
| Mumsen, J. | | Physician | Hamburg | 1784 |
| | | Physician | Copenhagen | 1785-91 |
| Murray, J. A.(2) | 44 | Professor of science | Göttingen | 1784-91 (d. 1791) |
| Murray, P. F. D. | 14 | Pharmacy learner | Hameln | 1791 |
| Mutzenbecher, S. D. | | Medical student | Göttingen | 1785-86 |
| Muus | | Rent secretary | Unknown | 1789-91 |
| Nagel, J. B. | | Apothecary | Berlin | 1784-86 |
| Nauwerk, C. L.(2) | 50 | Mining official | Dresden | 1786-91 (d. 1790) |

TABLE A3: Crell's Subscribers 1784–1791 (cont.)

| Name | Age in 1784 | Occupation | Town | Years subscribing in given occupation and town |
|---|---|---|---|---|
| Nesselrodt, C. von | | Count | Düsseldorf | 1784–91 |
| Nestler, C. G. | | Apothecary | Strasbourg | 1784–91 |
| Neuber | | Apothecary | Frankfurt a. M. | 1789–91 |
| Neumann | | Court apothecary | Kassel | 1784 |
| Neumann | | Castle apothecary | Bayreuth | 1784–85 |
| Nicolai, E. A. | 62 | Professor of medicine | Jena | 1784 |
| Nicolai, F. C. | 51 | Publisher | Berlin | 1786–91 |
| Niedner, C. F.(1) | | Apothecary | Stettin | 1784–91 |
| Niedt, J. W. | | Apothecary | Schwerin | 1784–86 |
| Nippa | | Apothecary | St. Petersburg | 1791 |
| Nissenius, C. F. | | Apothecary | Crossen | 1784 |
| Nitschmann, M. | | Apothecary | Sarepta | 1786–89 |
| Nittinger | | Pharmacy learner | Nuremberg | 1785 |
| Noble, J. le von | | Saltworks administrator | Hallein | 1785–86 |
| Nose, C. W.(1,2) | 31 | Physician | Elberfeld | 1784–91 |
| Oertel | | Court apothecary | Bayreuth | 1784–85 |
| Ofterdinger | | Apothecary | Ballingen | 1784–91 |
| Orlovius, A. J. | 49 | Professor of medicine | Königsberg | 1784 (d. 1788) |
| Osten, D. D. | | Unknown | Moscow | 1784–85 |
| Osterrode, J. T.(2) | | Apothecary | St. Petersburg | 1786–89 |
| Ottleben(2) | | Physician | Fallersleben | 1784–85 |
| Otto | | Law student | Halle | 1784–85 |
| Overcamp | | Lecturer | Greifswald | 1785 |
| Pabst, U. J. C.(2) | | Apothecary | Riga | 1784, 1787–91 |
| Palm | | Apothecary | Edingen | 1788–91 |
| Panke | | Apothecary | St. Petersburg | 1784–90 |
| Panzer, G. W. F. | 29 | Physician | Nuremberg | 1784–91 |
| Parisani | | Physician | Salzburg | 1785 |
| Pavonarius, J. S.(1-3) | | Apothecary | Stade | 1784–91 |
| Peckel | | Apothecary | Kongsberg | 1786–91 |
| Pepin | | Pharmacy learner | Hanover | 1784–85 |
| Pfaff | | Pharmacy learner | Hanover | 1784–90 |
| | | Apothecary | Steinfeld | 1791 |
| Pfister, J. F. | | Pharmacy learner | ex Schaffhausen | 1784 |
| Pflugmacher | | Pharmacy learner | Copenhagen | 1785–91 |
| Pfotenhauer | | Physician | Wittenberg | 1786–90 |
| Pfraumer | | Apothecary | Hallein | 1785 |
| Pickel, G.(2) | 33 | Professor of chemistry | Würzburg | 1791 |
| Pickhard | | Apothecary | Holzminden | 1784–87 |
| Piepenbring, G. H.(2) | 21 | Pharmacy learner | Pyrmont | 1785, 1789–91 |
| | | Pharmacy learner | Bückeburg | 1786–88 |
| Pirch, J. von | | Literature student | Halle | 1784 |
| | | Cameral student | Halle | 1785 |
| Pitiskus, M. C.(2) | | Pharmacy provisor | Oldenburg | 1784 |
| Planer, J. J. | 41 | Professor of medicine-science | Erfurt | 1784–90 (d. 1789) |

TABLE A3: Crell's Subscribers 1784–1791 (cont.)

| Name | Age in 1784 | Occupation | Town | Years subscribing in given occupation and town |
|---|---|---|---|---|
| Platenius, F. G. J. | | Apothecary | Elberfeld | 1784–85 |
| Plenciz, J. von | 32 | Professor of medicine | Prague | 1785 (d. 1785) |
| Ploucquet, W. G.(1) | 40 | Professor of medicine | Tübingen | 1784–91 |
| Pochodjaschin, von | | Major | St. Petersburg | 1788–89 |
| Podewillis, von | | Cameral student | Halle | 1785 |
| Polz | | Cameral student | Erlangen | 1790 |
| Poniatowsky, S. | 30 | Prince and Lord High Treasurer | Poland | 1784–91 |
| Pott | | Mathematics teacher | Salzburg | 1785 |
| Praetorius, B. G.(2) | 24 | Pharmacy learner | ex Riga | 1784 |
| | | Pharmacy learner | Langensalza | 1785–86 |
| Praun, C. von | | Mining official | Zellerfeld | 1784–89 |
| | | Official | Blankenburg | 1790 |
| | | Official | Brunswick | 1791 |
| Preick, J. S. G.(2) | | Apothecary | Coburg | 1784 |
| Quärnt | | Apothecary | Moscow | 1784–86 |
| Quittenbaum | | Pharmacy learner | Hanover | 1784 |
| Raspe, R. E.(2) | 48 | Mining consultant | Cornwall | 1787–91 |
| Rasumowsky, A. K. von | | Chamberlain | St. Petersburg | 1786–87 |
| Reden, F. W. von | 32 | Mining official | Breslau | 1791 |
| Reden, K. von | 48 | Mining director | Clausthal | 1784–91 (d. 1791) |
| Reiche, J. H. | | Smelting-mint administrator | Zellerfeld | 1784 |
| Reinhold, S. A. | | Physician | Barmen | 1784–91 |
| Reitz(4) | | Apothecary | Ansbach | 1787–91 |
| Reitz | | Apothecary | Erlangen | 1789 |
| Rennovanz | | Mining official | St. Petersburg | 1786–90 |
| Retz | | Pharmacy learner | Berlin | 1785 |
| | | Pharmacy provisor | Prague | 1786–91 |
| Reuss, A. C.(1) | 28 | Physician | Bruchsal | 1786–91 |
| Reuss, C. F.(2) | 39 | Professor of medicine | Tübingen | 1784–91 |
| Reuss, C. G. | 42 | Professor of medicine | Stuttgart | 1784–91 |
| Reuss, J. J. | 33 | Town physician | Stuttgart | 1784–91 |
| Richter, A. | | Army surgeon | Moscow | 1784–85 |
| Richter, F. A.(2) | 36 | Docent in medicine | Halle | 1784–90 |
| | | Professor of medicine | Halle | 1791 |
| Rieben, von | | Law student | Tübingen | 1784–88 |
| Riedel | | Official | Bayreuth | 1784–85 |
| Riehtsahl | | Tutor | Prague | 1785–91 |
| Rieken, J. F. | | Pharmacy learner | Halle | 1784 |
| | | Pharmacy learner | Jever | 1785–91 |
| Riess | | Mining official | Bibra | 1784–91 |
| Riess | | Merchant | Frankfurt a. M. | 1785 |
| Rissler | | Apothecary | Mühlhausen | 1784–91 |
| Rittenhouse, D. | 52 | Treasurer | Philadelphia | 1787–91 |
| Röhle, J. J. F. | | Apothecary | Vorsfelde | 1784 |

TABLE A3: Crell's Subscribers 1784–1791 (cont.)

| Name | Age in 1784 | Occupation | Town | Years subscribing in given occupation and town |
|---|---|---|---|---|
| Römer(1) | | Finance councillor | Brunswick | 1784–91 |
| Rönneberg | | Pharmacy learner | Dissen | 1784 |
| | | Pharmacy provisor | Steinfurt | 1785 |
| Röpke(2) | | Pharmacy provisor | Herrnhut | 1784 |
| Rössler, C. A.(2) | | Mining councillor | Prague | 1785–91 |
| Röstel | | Pharmacy learner | Copenhagen | 1785–86 |
| Roi, J. P. du(2) | 43 | Court physician | Brunswick | 1784–85 (d. 1785) |
| Rollmann | | Mining student | Königsborn b. U. | 1788–91 |
| Romanzow, N. von | 30 | Russian ambassador | Frankfurt a. M. | 1785–91 |
| Rose, V.(3) | 22 | Pharmacy learner | Stettin | 1784 |
| | | Apothecary | Berlin | 1785–91 |
| Rosner | | Physician | Augsburg | 1784 |
| Rottboell, C. F. | 57 | Professor of medicine | Copenhagen | 1785–91 |
| Rousseau, G. L. C.(2) | 60 | Professor of science | Ingolstadt | 1784 |
| Rowohl, J. A. | | Medical student | Berlin | 1784 |
| Rudolph | | Apothecary | Kaluga | 1784–85 |
| Rückert, G. C. A.(2) | 21 | Court apothecary | Ingelfingen | 1789–91 |
| Rüde, G. W. | 19 | Court apothecary | Kassel | 1784 |
| Ruge | | Pharmacy learner | Hanover | 1784–85 |
| | | Apothecary | Neuhaus | 1788–91 |
| Sallmuth | | Physician | Köthen | 1784–91 |
| Salzwedel, P.(1) | 32 | Apothecary | Frankfurt a. M. | 1784–91 |
| Sander(1) | | University apothecary | Göttingen | 1785–91 |
| Saussure, H. B. de | 44 | Professor of philosophy-science | Geneva | 1784–86 |
| | | Gentleman | Geneva | 1787–91 |
| Schaarup | | Merchant | Copenhagen | 1787–91 |
| Schacht | | Town apothecary | Quedlinburg | 1788–91 |
| Schadow | | Mining councillor | Berlin | 1786 |
| Schäffer, J. J. | | Surgeon | Moscow | 1784–85 |
| Schauer | | Merchant | Magdeburg | 1784–88 |
| Scheele, C. W.(2) | 42 | Apothecary | Köping | 1784–86 |
| Scheibel(1) | | Apothecary | Wahren | 1786 |
| Scherenberg | | Apothecary | Barmen | 1784 |
| Scherf, J. C. F.(2) | 34 | Court physician | Detmold | 1784–91 |
| Scheuchzer, C. | | Unknown | Zürich | 1784 |
| Schiller, J. M.(2) | 21 | Apothecary | Rothenburg o. d. T. | 1787–91 |
| Schindelmeiser | | Medical student | Königsberg | 1784 |
| Schlechtriem | | Pharmacy learner | Regensburg | 1787–88 |
| Schlenter | | Physician | Insterburg | 1785–91 |
| Schloenbach, C. L. | 17 | Official | Minden | 1788–91 |
| Schlüter, G. L. W. | 25 | Forest official | Zellerfeld | 1784 |
| Schmidt, E. A. | | Smelting administrator | Zellerfeld | 1784 |

TABLE A3: Crell's Subscribers 1784–1791 (cont.)

| Name | Age in 1784 | Occupation | Town | Years subscribing in given occupation and town |
|---|---|---|---|---|
| Schmidt, F.(4) | | Accountant | Prague | 1785–91 |
| Schmidt, J. F. | | Medical student | Helmstedt | 1786–87 |
| | | Medical student | Vienna | 1788 |
| | | Physician | Paderborn | 1789 |
| Schmidt, J. G. | 21 | Philosophy teacher | Hamburg | 1788–90 |
| Schnecker, J. E. D.(1) | | Physician | Hildesheim | 1784–86 |
| Schnizlein | | Official | Ansbach | 1790–91 |
| Schöller, M. L. | 22 | Pharmacy learner | ex Düsseldorf | 1784 |
| | | Physician | Düsseldorf | 1788–91 |
| Schönermark, von | | Captain | Berlin | 1785–91 |
| Schönwald, J. G.(2) | | Apothecary | Elbingen | 1784–91 |
| Schöpf, J. D.(4) | 32 | Physician | Erlangen | 1785–91 |
| Schrader, von | | Treasury councillor | Brunswick | 1785–91 |
| Schrader | | Pharmacy learner | Hameln | 1784–88 |
| Schrader | | Pharmacy learner | Brunswick | 1787 |
| Schreber, J. C. D.(1) | 45 | Professor of medicine | Erlangen | 1784–91 |
| Schreiber | | Apothecary | Melle | 1784 |
| Schroll, C. M. B. | 28 | Mining official | Salzburg | 1785 |
| Schubart | | Physician | Danzig | 1785 |
| Schulz | | Apothecary | Königsberg | 1785 |
| Schulze, J. G. F. | 32 | Physician and apothecary | Bernburg | 1784 |
| Schurer, F. L. | 20 | Science teacher | Strasbourg | 1790 |
| Schwabe | | Pharmacy learner | Hanover | 1784–85 |
| Schwarze, J. F.(4) | | Apothecary | Buttstädt | 1784–91 |
| Schwechten, J. H. | | Pharmacy learner | Berlin | 1784–86 |
| Schwenke, A. | | Pharmacy learner | Lich | 1784 |
| | | Pharmacy provisor | Wetzlar | 1785–86 |
| | | Apothecary | Wetzlar | 1787–88 |
| | | Apothecary | Lich | 1789–91 |
| Schwenkhart | | Professor | Vienna | 1785–86 |
| Scopoli, J. A.(2) | 61 | Professor of science | Pavia | 1784–88 (d. 1788) |
| Sebass | | Smelting administrator | Schornborn | 1784–90 |
| Seemann, A. F. C. | | Pharmacy learner | Helmstedt | 1784–85 |
| Seidenburg, J. G. | | Apothecary | Berlin | 1784–91 |
| Seidenburg(3) | | Apothecary | Ratzeburg | 1784–86 |
| Seidensticker | | Pharmacy learner | Wolfenbüttel | 1791 |
| Seidl, G. von | | Law student | Halle | 1784 |
| Selb | | Mining councillor | Hausach | 1787–91 |
| Seligmann | | Physician | Herrnhut | 1784 |
| Selle, C. G. | 36 | Professor of medicine | Berlin | 1784–91 |
| Sellmann | | Pharmacy learner | Nuremberg | 1785 |
| Seyler, A. J. G. | | Apothecary | Hanover | 1785–91 |

TABLE A3: Crell's Subscribers 1784–1791 (cont.)

| Name | Age in 1784 | Occupation | Town | Years subscribing in given occupation and town |
|---|---|---|---|---|
| Sickingen, C. von(2) | 47 | Bavarian ambassador | Paris | 1784 |
| Siecherer, F. | | Pharmacy learner | Frankfurt a. M. | 1784–91 |
| Siegel(1) | | Apothecary | Vaihingen | 1785–91 |
| Siemens, F. | | Smelting administrator | Unterharz | 1784 |
| Siemerling, C. F. | 32 | District physician | Aurich | 1787–91 |
| Sierstorpff, C. H. von | 34 | Court huntmaster | Brunswick | 1785–91 |
| Sievers, J. A. C.(1) | 22 | Pharmacy learner | Oker | 1784 |
| | | Pharmacy learner | London | 1785 |
| | | Unknown | Moscow | 1787 |
| Sievers, J. F. E. | 16 | Medical student | Helmstedt | 1786–87 |
| Silenz | | Pharmacy learner | ex Schleswig | 1786 |
| | | Pharmacy learner | Langensalza | 1787–88 |
| Simon | | Apothecary | Barr | 1784–85 |
| | | Apothecary | Colmar | 1786–91 |
| Sommer, J. C. | 43 | Professor of medicine | Brunswick | 1784–91 |
| Spalkhaver(3) | | Pharmacy learner | Langensalza | 1788 |
| | | Apothecary | Itzehoe | 1789–91 |
| Sperling | | Unknown | Moscow | 1784–85 |
| Spielmann, J. J. | 39 | Apothecary | Strasbourg | 1784–91 |
| Stallknecht | | Pharmacy provisor | Bayreuth | 1784–85 |
| | | Pharmacy provisor | Regensburg | 1787–90 |
| Stechow, von | | Baron | Berlin | 1785–86 |
| Stein, H. F. K. von | 27 | Mining councillor | Wetter | 1784–91 |
| Stein, J. E. | | Apothecary | Moscow | 1786–89, 1791 |
| Stephen | | Professor of medicine-science | Moscow | 1790–91 |
| Steudel | | Unknown | Esslingen | 1785, 1788–91 |
| Stockar, J. G. von(1) | 48 | Physician | Schaffhausen | 1784–85 |
| Stockse | | Army surgeon | Moscow | 1784–87 |
| Storr, G. C. C.(1,2) | 35 | Professor of science | Tübingen | 1784–91 |
| Stosch, C. W. | | Physician | ex Kiel | 1784 |
| | | Physician | Berlin | 1785–91 |
| Strougafschikow, von | | Lieutenant-colonel | St. Petersburg | 1787–91 |
| Struve, H.(1, 2) | 33 | Physician | Lausanne | 1784 |
| | | Professor of science | Lausanne | 1785–91 |
| Struve, O. W.(1) | | Physician | Lausanne | 1784 |
| Stucke, C. H.(2–3) | 21 | Pharmacy provisor | Arolsen | 1789–91 |
| Studer, S. | 27 | Hospital pastor | Bern | 1785–91 |
| Stüber | | Apothecary | Esslingen | 1784 |
| Stüber | | Apothecary | Neuchâtel | 1785 |
| Stutzer | | Pharmacy provisor | Bremen | 1786–87 |
| Suckow, G. A.(2) | 33 | Professor of science | Kaiserslautern | 1784 |
| | | Professor of science | Heidelberg | 1787–91 |
| Sües | | Apothecary | Zweibrücken | 1784–85 |

TABLE A3: Crell's Subscribers 1784–1791 (cont.)

| Name | Age in 1784 | Occupation | Town | Years subscribing in given occupation and town |
|---|---|---|---|---|
| Sundblad, J. | | Surgeon | Moscow | 1784–88 |
| Sylvestre | | Bookdealer | Paris | 1791 |
| Tannenberg, G. | | Apothecary | Moscow | 1784–91 |
| Taylor, C. | | Manufacturer | Manchester | 1785–86 |
| Tennant, S. | 23 | Gentleman | London | 1784–86 |
| Thiele | | Pharmacy learner | Bremen | 1786–91 |
| Thorn | | Pharmacy provisor | St. Petersburg | 1786–87 |
| | | Apothecary | St. Petersburg | 1788–89 |
| Thorspecken, G. C.(2) | | Pharmacy learner | ex Dessau | 1784 |
| Tichitz, J. | | Apothecary | Prague | 1785–91 |
| Tielebein, C. F.(2) | 31 | Apothecary | Schwerin | 1784–86 (d. 1786) |
| Tiemann | | Pharmacy learner | Stettin | 1784–91 |
| Tiemann | | Apothecary | Bielefeld | 1786–87 |
| Tiepolt | | Apothecary | Königsberg | 1785 |
| Tornesi | | Official | Bayreuth | 1784–85, 1787–91 |
| Trampel, J. E.(2) | 47 | Physician | Meinberg | 1784–88 |
| Trebra, F. W. H. von (2) | 44 | Mining director | Clausthal | 1784–91 |
| Treutel | | Bookdealer | Strasbourg | 1784 |
| Tychsen, N.(2) | 33 | Chemistry lecturer | Copenhagen | 1786–88 |
| | | Apothecary | Kongsberg | 1789–91 |
| Ufhausen | | Unknown | St. Petersburg | 1790 |
| Uhlendorf | | Pharmacy learner | Bentheim | 1786–91 |
| Uhthoff | | Bookdealer | Moscow | 1784–85 |
| Ungar, K. R. | 41 | University librarian | Prague | 1785–91 |
| Uslar, P. D. von(2) | 25 | Smelting administrator | Clausthal | 1784–91 |
| Vaughan, S. | | Unknown | Philadelphia | 1787–91 |
| Veltheim, A. F. von(1,2) | 43 | Mining director | Harbke | 1784–91 |
| Veltheim, C. C. von | 33 | Mining director | Rothenburg a. d. S. | 1784–91 |
| Versmann(1) | | Apothecary | Stade | 1784–87 |
| Versmann | | Pharmacy learner | Stade | 1785–87 |
| Vieweg, J. F. | 33 | Bookdealer | Berlin | 1786 |
| Vincent, F. M. | | Apothecary | Berlin | 1785 |
| Voigt, C. F. | | Apothecary | Erfurt | 1787–91 |
| Voigt, J. C. W.(2) | 32 | Mining secretary | Weimar | 1785–91 |
| Volkmar | | Mining secretary | Goslar | 1784–91 |
| Volta, A.(2) | 39 | Professor of science | Pavia | 1784–89 |
| Vorbrod | | Apothecary | St. Petersburg | 1788–91 |
| Voss, J. J. | 47 | Apothecary | Riga | 1784–86 |
| Vulpius, J. S. | 24 | Pharmacy learner | Stuttgart | 1788–89 |
| Wabst, A. J.(1) | | Apothecary | Brunswick | 1784–89, 1791 |
| Wachenhusen, S. W. | | Unknown | Schwerin | 1784 |
| Wachter | | Physician | Bernburg | 1787–91 |
| Wackenröder(1) | | Pharmacy learner | Hanover | 1784–90 |
| Wagenfeld | | Pharmacy learner | Brunswick | 1787 |

TABLE A3: Crell's Subscribers 1784–1791 (cont.)

| Name | Age in 1784 | Occupation | Town | Years subscribing in given occupation and town |
|---|---|---|---|---|
| Walz | | Apothecary | Stuttgart | 1784–91 |
| Wassidlo | | Pharmacy provisor | Herrnhut | 1784 |
| Wedderkop(2) | | Castle apothecary | Glückstadt | 1785–86 |
| Wegely, J. G. | | Manufacturer | Berlin | 1784–91 |
| Wegener, | | Law student | Helmstedt | 1785–86 |
| A. H. F. | | Law student | Göttingen | 1787–90 |
| | | Law student | Brunswick | 1791 |
| Wehrde, G. F. | 55 | Town apothecary | Hanover | 1788–90 |
| Weigel, B. N. | 63 | Town physician | Stralsund | 1784–89 |
| *Weigel,* C. E.(2.3) | 36 | Professor of science | Greifswald | 1784–91 |
| Weinl | | Court apothecary | Erlangen | 1785 |
| Weiss | | Medical student | Königsberg | 1784–85 |
| Well, J. J. von(1) | 59 | Professor of science | Vienna | 1784 |
| Wendelnstädt | | Physician | Wetzlar | 1789–91 |
| Wendland, G. H.(1) | | Apothecary | Berlin | 1786–91 |
| Wernberger, | 37 | Town physician | Erlangen | 1786–88 |
| E. L. | | Town physician | Windsheim | 1789–91 |
| Westendorff, J. C.(2) | 44 | Mine physician | Güstrow | 1784–91 |
| Westphal(4) | | Apothecary | Halle | 1784 |
| Westphalen, F. von | | Law student | Göttingen | 1786–87 |
| *Westrumb,* J. F.(2,3) | 33 | Apothecary | Hameln | 1784–91 |
| Wiedemann, J. F.(2) | 20 | Mining secretary | Stuttgart | 1789–91 |
| *Wiegleb,* J. C.(1–3) | 52 | Apothecary | Langensalza | 1784–91 |
| Wilcke, J. C.(2) | 52 | Professor of science and academician | Stockholm | 1787–91 |
| Wilkens, H. D.(2) | | Medical student | Helmstedt | 1784–87 |
| | | Chemistry-metallurgy student | Freiberg | 1788–89 |
| Wilkens | | Apothecary | Saarbrücken | 1787–91 |
| Willdenow, J. C.(1) | | Apothecary | Berlin | 1784–91 |
| Wille, C. L. A.(2) | | Mining administrator | Veckerhagen a. W. | 1784–86 |
| Winkler | | Mining administrator | Rothenburg a. d. S. | 1787–90 |
| Winterberger, B. G. | 35 | Apothecary | St. Petersburg | 1784–91 |
| *Winterl,* J. J.(2,4) | 45 | Professor of science | Budapest | 1787–91 |
| Wischnewsky, von | | Lieutenant | Moscow | 1786 |
| Witte | | Apothecary | Oldenburg | 1785–86 |
| Wittekop, J. H.(2) | | Law-cameral student | Helmstedt | 1784–85 |
| | | Law-cameral student | Göttingen | 1786–88 |
| | | Secretary | Magdeburg | 1789–91 |

TABLE A3: Crell's Subscribers 1784–1791 (cont.)

| Name | Age in 1784 | Occupation | Town | Years subscribing in given occupation and town |
|---|---|---|---|---|
| Witting | | Pharmacy learner | Gronau | 1786, 1788–91 |
| Wolf, J. | | Apothecary | Moscow | 1784–87, 1789–90 |
| Wolf | | Pharmacy provisor | Schöningen | 1784–90 |
| Wolf | | Pharmacy learner | Braunfels | 1785–91 |
| Wolfing | | Apothecary | Stuttgart | 1784–91 |
| Worms, D. | | Pharmacy learner | Altona | 1784 |
| Wurtz, G. C. | 28 | Physician | Strasbourg | 1784–90 |
| Wyss, J. A.(1) | 63 | Artillery lieutenant | Bern | 1784 |
| Zacharoff, J. | | Medical student | Göttingen | 1788–90 |
| Zacharoff | | Medical student | ex St. Petersburg | 1786–91 |
| Zacharoff | | Medical student | St. Petersburg | 1789 |
| | | Adjunct academician | St. Petersburg | 1790 |
| Zachert, Z. | | Apothecary | Moscow | 1784–86 |
| Zahm | | Apothecary | St. Petersburg. | 1790–91 |
| Zanders | | Physician | Solingen | 1785–86, 1788, 1790 |
| Zemisch | | Unknown | Moscow | 1786 |
| Zettel, A. B. | | Unknown | Moscow | 1784–86 |
| Zickner | | Pharmacy learner | Schöningen | 1784–91 |
| Ziegler, C. J. A. | 49 | Physician | Quedlinburg | 1784–91 |
| Zier | | Pharmacy learner | Zerbst | 1789–91 |
| Zimmermann, E. A. W.(2) | 41 | Professor of science | Brunswick | 1784–87 |
| Zois, von | | Baron | Laibach | 1784 |
| Zorn, J.(2,3) | 45 | Apothecary | Kempten | 1785–91 |
| Zybelin, S. | | Physician | Moscow | 1786–87, 1789 |

Table A4: Geographical Index to Crell's Subscribers

| Town | Persons subscribing | | | | |
|---|---|---|---|---|---|
| Åbo | Gadolin | | | | |
| Allendorf | Brandt | | | | |
| Altona | Gehrt | Worms | | | |
| Amsterdam | Kasteleyn | | | | |
| Anklam | John | | | | |
| Ansbach | Gemmingen | Hardenberg | Kern | Reitz | |
| | Schnizlein | | | | |
| Arolsen | Stucke | | | | |
| Augsburg | Braun | Leipoldt | Rosner | | |
| Aurich | Siemerling | | | | |
| Balingen | Ofterdinger | | | | |
| Barmen | Gahlen | Hees | Reinhold | Scherenberg | |
| Barr | Simon | | | | |
| Baruth | Bühring | | | | |
| Bayreuth | Bothmer | Graf | Neumann | Oertel | |
| | Riedel | Stallknecht | Tornesi | | |
| Bentheim | Uhlendorf | | | | |

TABLE A4: Geographical Index to Crell's Subscribers (cont.)

| Town | Persons subscribing | | | |
|---|---|---|---|---|
| Berlin | *Achard* | Aschenborn | Bachmann | Backhaus |
| | Bärensprung | Bell | Berend | Bergmann |
| | Beyer | Bindheim | Carita | Cothenius |
| | Cruse | Döring | Ehrlich | Fischer |
| | Gäuke | Geelhaar | Gesler | Gleditsch |
| | Gönner | Hannemann | Hardenberg | *Hermbstaedt* |
| | Hieppe | Holze | Karsten | Keber |
| | Kessel | *Klaproth* | Kraatz | Kunsemüller |
| | Kurella | Luck | Mandenberg | Mertens |
| | Möhring | Mönnicke | Nagel | Nicolai |
| | Retz | *Rose* | Rowohl | Schadow |
| | Schönermark | Schwechten | Seidenburg | Selle |
| | Stechow | Stosch | Vieweg | Vincent |
| | Wegely | Wendland | Willdenow | |
| Bern | Benteli | Erlach | Höpfner | Höpfner |
| | Jenner | Manuel | Morell | Mukey |
| | Studer | Wyss | | |
| Bernburg | Luckenbach | Schulze | Wachter | |
| Bibra | Riess | | | |
| Bielefeld | Aschoff | Aschoff | Möller | Tiemann |
| Blankenburg | Burchhardt | Florencourt | Praun | |
| Bohemia | Ballas | | | |
| Boizenburg | Grossmann | | | |
| Braunfels | Müller | Wolf | | |
| Bremen | Koch | Kücke | Stutzer | Thiele |
| Breslau | Reden | | | |
| Bruchsal | Reuss | | | |
| Brunswick | Anonymous | Bachmann | Becker | Borges |
| | Brückmann | Graberg | Gravenhorst | Harbordt |
| | Hardenberg | Hawkins | Heinemann | *Heyer* |
| | *Hildebrandt* | Hoym | Kücke | Praun |
| | Römer | Roi | Schrader | Schrader |
| | Sierstorpff | Sommer | Wabst | Wagenfeld |
| | Wegener | Zimmermann | | |
| Budapest | *Winterl* | | | |
| Bückeburg | Piepenbring | | | |
| Burgdorf, Br. | Bühring | | | |
| Burgdorf, Sw. | Bürger | | | |
| Buttstädt | Schwarze | | | |
| Catharinenburg | Herrmann | | | |
| Celle | Kohl | | | |
| Clausthal | Hamstein | Ilsemann | Kastenbein | Lunde |
| | Meding | Meyer | Reden | Trebra |
| | Uslar | | | |
| Coburg | Preick | | | |
| Colmar | Hausmann | Simon | | |
| Cologne | Förchtl | Grote | | |
| Copenhagen | Abildgaard | Becker | Bode | Boving |
| | Didrichson | Grönlund | Günther | Hagen |
| | Hauch | Kratzenstein | Lowson | Lystager |
| | Manthey | Mühlensteth | Müller | Mumsen |
| | Pflugmacher | Röstell | Rottboell | Schaarup |
| | Tychsen | | | |
| Cornwall | Raspe | | | |
| Crossen | Nissenius | | | |

TABLE A4: Geographical Index to Crell's Subscribers (cont.)

| Town | Persons subscribing | | | | |
|---|---|---|---|---|---|
| Dannenberg | Muhle | | | | |
| Danzig | Krüger | Lampe | Schubart | | |
| Demmin | Michaelson | | | | |
| Dessau | Thorspecken | | | | |
| Detmold | Keyser | Scherf | | | |
| Diepholz | Brauer | | | | |
| Dijon | Guyton de Morveau | | | | |
| Dissen | Rönneberg | | | | |
| Dresden | Anonymous | Nauwerk | | | |
| Dublin | Kirwan | | | | |
| Düsseldorf | Anonymous | Beuth | Brinckmann | Döring | |
| | Geyr | Hompesch | Jacobi | Nesselrodt | |
| | Schöller | | | | |
| Edingen | Palm | | | | |
| Einbeck | Hink | | | | |
| Elberfeld | Carnap | Cretschmar | Dinckler | Hengstenberg | |
| | Kretschmer | Nose | Platenius | | |
| Elbing | Kobligk | | | | |
| Elbingen | Schönwald | | | | |
| Erfurt | Frenzel | Planer | Voigt | | |
| Erlangen | Bender | Borges | Delius | Elwert | |
| | Frischmann | Lober | Mayer | Polz | |
| | Reitz | Schöpf | Schreber | Weinl | |
| | Wernberger | | | | |
| Eschweiler | Daniels | | | | |
| Essen | Kundler | | | | |
| Esslingen | Steudel | Stüber | | | |
| Fallersleben | Ottleben | | | | |
| Florence | Anonymous | Leopold | | | |
| Frankfurt a. M. | Balz | Kähler | Kunhardt | Neuber | |
| | Riess | Romanzow | Salzwedel | Siecherer | |
| Frankfurt a. d. O. | Lange | | | | |
| Freiberg | Wilkens | | | | |
| Freienwalde | Döring | | | | |
| Friederica | Ingwerfen | | | | |
| Fürstenau | Muhle | | | | |
| Fürstenberg | Hellwig | Kohl | | | |
| Gandersheim | Höfer | | | | |
| Geneva | Saussure | | | | |
| Genthin | Dörffurt | | | | |
| Gerabronn | Langsdorf | | | | |
| Giessen | Krüger | | | | |
| Gladbach | Henk | | | | |
| Glückstadt | Wedderkop | | | | |
| Göttingen | Blumenbach | Brandis | Dieterich | Forcke | |
| | *Gmelin* | *Göttling* | Haselberg | Kels | |
| | Körber | Lichtenberg | Meding | Murray | |
| | Mutzenbecher | Sander | Wegener | Westphalen | |
| | Wittekop | Zacharoff | | | |
| Goslar | Lemke | Volkmar | | | |
| Gotha | Müller | | | | |
| Greifswald | Overcamp | *Weigel* | | | |
| Gronau | Witting | | | | |
| Grünenplan | Amelung | | | | |

TABLE A4: Geographical Index to Crell's Subscribers (cont.)

| Town | Persons subscribing | | | |
|---|---|---|---|---|
| Güstrow | Brun | Westendorff | | |
| Hager | Deggeler | | | |
| Halberstadt | Könnecke | Kramer | | |
| Halle | Aschoff | Claussen | Daniel | Goldhagen |
| | *Gren* | Huth | Karsten | Karsten |
| | Keidel | Kirsten | Klügel | Kohl |
| | Kraft | Meisner | Michaelis | Moritz |
| | Otto | Pirch | Podewillis | Richter |
| | Rieken | Seidl | Westphal | |
| Hallein | Noble | Pfraumer | | |
| Hamburg | Friedland | Hasse | Hasse | Knorre |
| | Kunsemüller | Lieber | Mumsen | Schmidt |
| Hameln | Backhaus | Bischoff | Lasius | Murray |
| | Schrader | *Westrumb* | | |
| Hanau | Gärtner | Gärtner | | |
| Hanover | *Andreae* | Brande | Dönch | Drewer |
| | Ehmsen | Gottschalk | Guckenberger | Hammer |
| | Ilsemann | Ilsemann | Kücke | Lammersdorf |
| | Lasius | Meding | Meyer | Pepin |
| | Pfaff | Quittenbaum | Ruge | Schwabe |
| | Seyler | Wackenröder | Wehrde | |
| Harbke | Veltheim | | | |
| Harburg | Muhle | | | |
| Hausach | Selb | | | |
| Heidelberg | *Suckow* | | | |
| Heilbronn | Gmelin | | | |
| Helmstedt | Beireis | Borges | Camman | Gempt |
| | Hempel | Hummel | Jahn | Küster |
| | Lichtenstein | Meineke | Meyer | Schmidt |
| | Seemann | Sievers | Wegener | Wilkens |
| | Wittekop | | | |
| Hermannstein | Martius | | | |
| Herrenhausen | Ehrhart | | | |
| Herrnhut | Braths | Dilly | Röpke | Seligmann |
| | Wassidlo | | | |
| Hildesheim | Beroldingen | Decker | Ilsemann | Kratz |
| | Meineke | Schnecker | | |
| Hintbergen | Eller | | | |
| Holzminden | Claus | Pickhard | | |
| Horneburg | Ilisch | | | |
| Hoya | Jordan | | | |
| Ingelfingen | Rückert | | | |
| Ingolstadt | Rousseau | | | |
| Insterburg | Schlenter | | | |
| Itzehoe | Spalkhaver | | | |
| Jena | Büttner | Fuchs | *Göttling* | Gruner |
| | Nicolai | | | |
| Jever | Rieken | | | |
| Joachimsthal | Miessl | | | |
| Jüterbog | Bernhard | Flemming | | |
| Jutland | Boving | Lystager | | |
| Kaiserslautern | *Suckow* | | | |
| Kaluga | Rudolph | | | |
| Kazan | Felsch | | | |

TABLE A4: Geographical Index to Crell's Subscribers (cont.)

| Town | Persons subscribing | | | |
|---|---|---|---|---|
| Kassel | Baldinger | Bertram | Brandt | Fiedler |
| | Fiedler | Flügger | Fulda | Hoffmann |
| | Mönch | Neumann | Rüde | |
| Kempten | Couret | Zorn | | |
| Kiel | Christiani | Grim | Stosch | |
| Kirchheim | Gaupp | | | |
| Kleve | Hannesmann | | | |
| Kochendorf | Bender | | | |
| Königsberg | Arndt | Flach | *Hagen* | Hoppe |
| | Kessel | Mejo | Orlovius | Schindelmeiser |
| | Schulz | Tiepolt | Weiss | |
| Königsborn b. U. | Rollmann | | | |
| Königslutter | Mühlenbein | | | |
| Königstein | Geutner | | | |
| Köping | Scheele | | | |
| Köthen | Langguth | Sallmuth | | |
| Kongsberg | Peckel | Tychsen | | |
| Kostroma | Hacken | | | |
| Kunzendorf | Herrmann | | | |
| Lahr i. B. | Hänle | | | |
| Laibach | Hacquet | Jugowitz | Zois | |
| Landsberg | Mertens | | | |
| Langensalza | Bärensprung | Becker | Braun | Christiani |
| | Grudtner | Praetorius | Silenz | Spalkhaver |
| | *Wiegleb* | | | |
| Lausanne | Struve | Struve | | |
| Lautenthal | Meyer | | | |
| Lauterbach | Kücke | | | |
| Leer | Hoffmann | | | |
| Leipzig | Gehler | Leske | Linke | |
| Lich | Schwenke | | | |
| Lissa | Gäuke | | | |
| London | Banks | Cavendish | Dollfuss | Glendenberg |
| | Hawkins | Kirwan | Sievers | Tennant |
| Lubny | Ehrenfriedstein | Hildebrand | | |
| Lübeck | Curtius | | | |
| Lüneburg | Dempfwolf | Ebeling | Jordan | Krüger |
| | Muhle | | | |
| Lünen | Kaldewey | | | |
| Magdeburg | Cäsar | Dolhof | Kessler | Klewitz |
| | Schauer | Wittekop | | |
| Manchester | Taylor | | | |
| Marburg | Baldinger | Karsten | Mönch | |
| Meinberg | Trampel | | | |
| Meiningen | Jahn | | | |
| Melle | Ebermeier | Schreiber | | |
| Mikhailov | Kölbeke | | | |
| Milan | Kollowrath | | | |
| Minden | Bollmann | Iser | Schloebach | |
| Mitau | Beseke | Groschke | | |
| Moscow | Berndt | Bibo | Bindheim | Blankennagel |
| | Börnike | Butter | Dehfloff | Dorse |
| | Eggers | Essmann | Felsch | Fränkel |
| | Fuhrmann | Gardner | Glawet | Hannemann |
| | Haukohl | Hauptvogel | Hausenbaum | Henkel |

TABLE A4: Geographical Index to Crell's Subscribers (cont.)

| Town | Persons subscribing | | | |
|---|---|---|---|---|
| | Hildebrand | Honrich | Jürgenson | Kalkau |
| | Klint | Köstelhön | Koryzna | Kreusel |
| | Kron | Lütke | Mahlstedt | Meder |
| | Melissino | Mensch | Mey | Müller |
| | Osten | Quärnt | Richter | Schäffer |
| | Sievers | Sperling | Stein | Stephen |
| | Stockse | Sundblad | Tannenberg | Uhthoff |
| | Wischnewsky | Wolf | Zachert | Zemisch |
| | Zettel | Zybelin | | |
| Mühlhausen | Rissler | | | |
| Münden | Köster | | | |
| Neuchâtel | Stüber | | | |
| Neudamm | Deutsch | | | |
| Neuhaus | Ruge | | | |
| Neumünster | Jahn | | | |
| Neustadt a. R. | Meyer | | | |
| Nizhni-Novgorod | Evenius | | | |
| Nuremberg | Braun | Grattenauer | Markel | Nittinger |
| | Panzer | Sellmann | | |
| Oberkaltenberg | Döring | | | |
| Obernkirchen | Bachmann | | | |
| Oebisfelde | Bietzker | | | |
| Oehringen | Anonymous | Herrmann | | |
| Offenbach | Amburger | | | |
| Oker | Decker | Fischer | Sievers | |
| Oldenburg | Meyer | Pitiskus | Witte | |
| Orel | Fuhrmann | Köster | | |
| Osnabrück | Ameldung | Ehmsen | Kels | Meier |
| | Mües | | | |
| Osterholz | Kohl | | | |
| Osterode | Friedland | | | |
| Ovelgönne | Cahlo | Kelp | | |
| Paderborn | Schmidt | | | |
| Paris | Blancherie | Dietrich | Hassenfratz | Métherie |
| | Sickingen | Sylvestre | | |
| Pavia | Brugnatelli | Scopoli | Volta | |
| Philadelphia | Rittenhouse | Vaughan | | |
| Poland | Poniatowsky | | | |
| Potsdam | Harsleben | | | |
| Prague | Braun | Ebenberger | Herget | Holz |
| | Mayer | Plenciz | Retz | Riehtsahl |
| | Rössler | Schmidt | Tichitz | Ungar |
| Příbram | Allis | | | |
| Pyrmont | Piepenbring | | | |
| Quakenbrück | Adami | | | |
| Quedlinburg | Löwe | Schacht | Ziegler | |
| Ragusa | Bassegli | | | |
| Ratzeburg | Seidenburg | | | |
| Regensburg | Gasser | Hoppe | Kohlhass | Schlechtriem |
| | Stallknecht | | | |
| Riga | Ilisch | Pabst | Praetorius | Voss |
| Rinteln | Brockmann | | | |
| Rostock | Mähl | | | |
| Rothenburg a. d. S. | Klötz | Veltheim | Winkler | |
| Rothenburg o. d. T. | Schiller | | | |

TABLE A4: Geographical Index to Crell's Subscribers (cont.)

| Town | Persons subscribing | | | |
|---|---|---|---|---|
| Rudolstadt | Friedrich-Carl | | | |
| Saarbrücken | Wilkens | | | |
| St. Petersburg | Anonymous | Anonymous | Andreae | Bang |
| | Berendt | Byhan | Catharine | Euler |
| | Ferber | Georgi | Grewe | Henkel |
| | Hohenthal | Holtheuer | Illmann | Kohlrief |
| | Lange | Lehnstubb | Lowitz | Nippa |
| | Osterrode | Panke | Pochodjaschin | Rasumowsky |
| | Rennovanz | Strougafschikow | Thorn | Ufhausen |
| | | Vorbrod | Winterberger | Zacharoff |
| | Zacharoff | Zahm | | |
| Salzburg | Haim | Moser | Motz | Parisani |
| | Pott | Schroll | | |
| Sarepta | Nitschmann | | | |
| Schaffhausen | Ischwald | Pfister | Stockar | |
| Schafstädt | Gebhard | | | |
| Scharnebeck | Meyer | | | |
| Schladen | Klenze | | | |
| Schlawe | Blume | | | |
| Schleswig | Silenz | | | |
| Schneeberg | Beyer | | | |
| Schöningen | Abich | Dehne | Honig | Müller |
| | Wolf | Zickner | | |
| Schöppenstedt | Corvinus | | | |
| Schornborn | Sebass | | | |
| Schwerin | Büsing | Engel | Glendenberg | Hennemann |
| | Klockmann | Masius | Meineke | Niedt |
| | Tielebein | Wachenhusen | | |
| Solingen | Zanders | | | |
| Stade | Luck | Pavonarius | Versmann | Versmann |
| Stargard | Fischer | | | |
| Steinfeld | Pfaff | | | |
| Steinfurt | Rönneberg | | | |
| Stettin | Kummer | *Meyer* | Niedner | *Rose* |
| | Tiemann | | | |
| Stockholm | Anonymous | Bose | Gahn | Wilcke |
| Stolp | Ehmke | Masch | | |
| Stralsund | Helvig | Weigel | | |
| Strasbourg | Beckerhinn | Christiani | Ehrmann | Hecht |
| | Hermann | Hieppe | Martin | Martius |
| | Nestler | Schurer | Spielmann | Treutel |
| | Wurtz | | | |
| Stuttgart | Gmelin | Lang | Reuss | Reuss |
| | Vulpius | Walz | Wiedemann | Wolfing |
| Tübingen | Demler | Gmelin | Heerbrandt | Plouquet |
| | Reuss | Rieben | Storr | |
| Tula | Lembke | | | |
| Tuttlingen | Megenhard | | | |
| Unterharz | Bähr | Ebert | Mätke | Siemens |
| Uppsala | Bergman | | | |
| Vaihingen | Siegel | | | |
| Veckerhagen a. W. | Wille | | | |
| Vienna | Leopold | Schmidt | Schwenkhart | Well |
| Vladimir | Döring | | | |
| Vorsfelde | Röhle | | | |

TABLE A4: Geographical Index to Crell's Subscribers (cont.)

| Town | Persons subscribing | | | |
|---|---|---|---|---|
| Voronezh | Honrich | | | |
| Wahren | Scheibel | | | |
| Waldeck | Georg | | | |
| Walsrode | Gebeler | | | |
| Weimar | *Bucholtz* | Goethe | Hoffman | Langer |
| | Merz | Voigt | | |
| Weisbach | Glenk | | | |
| Werchoturje | Bornemann | | | |
| Wetter | Stein | | | |
| Wetzlar | Held | Hieppe | Möllenhof | Schwenke |
| | Wendelnstädt | | | |
| Wilna | Forster | | | |
| Windsheim | Wernberger | | | |
| Wittenberg | Langguth | *Leonhardi* | Pfotenhauer | |
| Wolfenbüttel | Hemmelmann | Seidensticker | | |
| Würzburg | Kopp | Pickel | | |
| Yersmold | Delius | | | |
| Yverdon | Devely | | | |
| Zellerfeld | Brüel | Bülow | Drechsler | Kast |
| | Praun | Reiche | Schlüter | Schmidt |
| Zerbst | Hermann | Zier | | |
| Zürich | Scheuchzer | | | |
| Zweibrücken | Sües | | | |

# SUBJECT INDEX

academicians for chemistry. *See* professors of chemistry
administrative schools, 24, 31, 42–45, 66–68, 70, 146, 196, 199, 204, 214, 226–227, 235, 249, 253, 267; laboratories, 44, 249
alchemy, alchemists, 6–7, 10–11, 14–16, 20, 25, 28, 66, 71, 164, 170, 179, 184, 190, 250, 266
anatomy, anatomists, 33–35, 38, 64, 108, 130, 165–166, 168–169, 174–175, 178–179, 181, 183–184, 187, 189, 191, 198, 200, 202, 214, 228, 235, 238, 241, 246, 248–249, 252
antiphlogistic revolution: in France, 96, 98–101, 103, 119–120; in Catholic Germany, 106–109; in the German chemical community, 3–5, 61, 84, 95–97, 100–107, 110–144, 146–147, 216, 220
assaying. *See* metallurgy

botany, botanists, 33–35, 37–38, 50, 55, 64–65, 72, 89, 106, 108, 166–167, 174, 178–179, 184–186, 188, 192, 195–196, 200, 202–205, 209, 211, 215–218, 228, 232, 234, 236, 238, 241–243, 245–246, 249, 251–252, 258, 260–262, 265–267, 269, 276–277

cameralism, mercantilism, state economics, 15–16, 24, 26, 28–29, 39–40, 44–45, 79, 204, 234, 241, 249, 253, 267, 275, 285–286
Catholic-Protestant comparisons, 5–6, 28, 32, 34, 36–37, 39–40, 48–49, 53, 58, 86–89, 106–112, 142
central schools, 220, 226–227, 234, 236–237, 257–258, 263–264
chemical concepts: affinity, 21, 129; ammonia, volatile alcali, 65–66; calcination and combustion, 98, 102, 123; carbon, 95, 101, 123; carbon dioxide, fixed air, 84, 94, 101, 118, 122–123; chlorine, dephlogisticated marine acid, 79, 105, 197, 205; heat 65–66; hydrogen, inflammable air, 98–100, 125, 127; light, 137; mercuric oxide, mercury calx per se, 106, 117–120, 122–139, 146; metallic oxides, calces, earths, 79, 99, 107–108, 119–120, 122–123, 128–129, 138, 149; metals, 98, 101–102, 105–108, 123; oxygen, dephlogisticated air, fire air, pure air, vital air, 98–99, 102, 105, 118–120, 122–139; phlogiston, 1, 66, 96, 99–106, 110, 112, 119–120, 125, 128, 135, 137, 140–141; phosphorus, 70, 98, 137, 140, 164; pinguic acid, 84, 118, 154; reduction, 106–107, 117–120, 122–123, 125–132, 134–139, 146; sal ammoniac, 193, 210; saltpeter, 24, 61, 65, 193; sulfur, 98, 101, 106; water, 98–99, 112, 120, 122–123, 125–130, 132–133, 136, 138; weight, 98–99, 126, 129, 131, 134–135
chemistry: images of, 2, 4, 8–11, 13–17, 20, 22–29, 70, 72, 149–150; social support for, 2, 13–61, 85–87, 130–131, 135, 145, 149–150, 153–299
———, uses: balloons, 191, 198, 200, 206, 209; beverage production, 8, 10–11, 23, 25, 47, 59, 195, 219; bleaching and dyeing, 10–11, 22–25, 191, 193, 197, 204–206, 211, 279; brick making, 24–26, 220; chemical manufacturing, 18, 24, 59, 65, 67, 78, 193, 208–211, 214, 218–219, 222, 277–279, 290; glass making, 10–11, 23–24, 164, 224,

chemistry, uses (cont.)
272; patent medicines, 7, 17, 60, 169, 174–175, 179, 202; porcelain making, 24, 47, 58–59, 176–177, 181, 183, 190, 198, 201, 212–213, 220; salt production, 24, 168, 193, 277, 282, 285; sugar production from beets, 193, 198, 200, 206–208, 222; water analysis, 60, 176, 178, 187, 204–205, 215. *See also* metallurgy; mining industry; pharmacy

chemists: social-religious backgrounds, 12, 57, 62, 64, 154, 164–224; educations, 38–39, 53–57, 64–66, 92, 97, 155, 164–224; foreign travels, 57, 65–66, 102, 111, 145, 164–168, 170, 172–173, 175, 181–190, 196, 199, 201–202, 208–209, 220, 222–223, 266; residences, 88, 106–108, 110, 164–224; occupations, 39, 47, 57–60, 67–68, 88, 97, 108, 110, 155, 164–224; publishing patterns, 50, 53–54, 61, 88–89; informal recognition, 89–92, 153, 164–224; honors, 26, 75–78, 80, 150, 155–156, 164–224; empiricism, 5, 59–61, 76, 102, 115, 118, 122–140, 143–144, 146–147, 150–151; cosmopolitanism, 5, 76–77, 84, 106–107; cultural nationalism, 1, 70–73, 76–80, 82–84, 94, 96–97, 102–105, 117, 123, 145–147; community of, 1–5, 84–97, 103–107, 112, 114–115, 117–119, 141–144, 146–147, 150–151; professionalization in nineteenth century, 2–3, 148, 150–151

Enlightenment, enlightened governance, 2, 4, 12, 18–20, 36–37, 39, 48, 53, 78, 147–149

French Revolution, 106–107, 112, 119, 148, 209. *See also* central schools

iatrochemistry, 7, 11, 16, 20, 230, 243

libraries, 41, 188

lyceums, 226, 245, 251, 260–261

Masonics, 67, 107

materia medica, 35, 38, 65, 68, 167, 178, 185, 192, 194, 211, 215, 223, 229, 234, 237, 241–242, 251, 255, 257, 263, 265, 268–269

mathematics, mathematicians, 3, 6, 19, 43, 46, 97, 99, 131, 166–168, 179, 182, 188, 192, 201, 204, 212, 218, 246, 253–254, 262, 286

medical-surgical colleges, 10, 21, 31, 33, 35, 38–39, 47, 56, 66, 114, 125, 139, 174–176, 180, 185, 187, 197, 211, 213–214, 226–227, 230–236, 267–268; laboratories, 267

medicine, physicians, 7–8, 14–19, 21–24, 26, 28, 32–42, 49, 54–60, 64–67, 70, 80, 86–88, 108, 110, 130–132, 135, 139, 145, 165–181, 183–205, 208–211, 213–221, 223–224, 226–269, 272–292

mercantilism. *See* cameralism

metallurgy (including assaying and smelting), metallurgists, 10–11, 21, 23–26, 28, 43, 55, 66–67, 71, 78, 80, 135, 164, 170–171, 175, 180–184, 186–188, 192, 194, 196, 199, 201, 203, 212, 222, 234–235, 242, 244, 261, 263, 267, 273–274, 276, 281, 284, 286–291

military education, officers, 20, 22, 198, 211, 233–234, 276–277, 283, 286, 288–289, 291–292

mineralogy, mineralogists, 24–25, 46, 67, 75, 90, 106, 180, 184, 187–188, 194, 196, 200, 203–204, 209, 216, 232, 234, 260–261, 267

mining industry, mining officials, 8, 11, 21–22, 25, 28, 43–45, 49, 53, 58–59, 67, 75, 86–90, 107–110, 126, 130, 135, 164, 170–171, 173, 175, 178–184, 186–188, 190, 192, 194–196, 198, 201, 203–205, 210–214, 218, 220, 222–223, 226, 232, 234–235, 254, 259, 261, 267–268, 272–284, 286–291

mining schools, 31, 43–44, 107, 109, 140, 150, 183, 186, 188, 194, 196, 198, 201, 206, 211, 221–222, 226, 232, 241–242, 261, 263

———, laboratories, 44, 242, 263
moral weeklies, 17–18

natural history, natural historians, 25, 35, 100, 110, 186, 196, 203–204, 209–210, 214–216, 220, 223, 234, 241, 251, 253, 255, 258, 260, 262, 265
natural philosophy, natural philosophers, physics, physicists, 3, 6–8, 15–16, 25–26, 46, 50, 58, 66, 90, 94, 97–98, 100, 107, 110, 112, 118, 125, 130, 135–136, 138, 143, 147, 149, 166–168, 173–174, 182, 188–189, 192, 194, 204, 206, 209–211, 214–216, 223, 233–234, 236–237, 240, 243, 246–247, 251, 253–255, 257–258, 260–261, 264, 268
natural theology, 1, 10, 68–69
nobles, rulers, 7, 13, 16, 20–22, 26, 32, 34, 36–37, 39, 41, 43–49, 67–68, 75–76, 79, 86–88, 90, 106–110, 112, 130–131, 135, 147, 149–150, 164, 166–179, 181–184, 186, 188–189, 191, 193, 195–200, 203–204, 206–210, 212, 214–216, 219–223, 229–236, 238, 240–246, 248–268, 273–292

officers. *See* military education

Paracelsianism, 2, 8, 262
pathology, 35, 167, 175, 178, 200, 215, 241, 256, 261
periodicals, 2, 4, 19, 62, 71, 85, 147, 151
———, *Allgemeine deutsche Bibliothek*, 26–27, 63, 72–75, 79, 84, 93, 100, 105, 121–123
———, *Allgemeine Literatur-Zeitung*, 86, 93–94; its *Intelligenzblatt*, 108, 110, 114, 123, 125–127, 133, 135, 137–143
———, *Annales de Chimie*, 100, 103
———, Baldinger's *Magazin* 68–69, 74, 146
———, Crell's journals, 4–5, 28, 62, 68–81, 84–94, 96, 99–101, 103–105, 108–112, 120, 122, 125–128, 130–134, 136–137, 140–142, 146–147, 199, 220, 271–299
———, *Göttingische Anzeigen*, 68, 72, 83, 102–103
———, Göttling's *Almanach*, 74, 102, 122, 208
———, Gren's *Journal*, 111, 120, 122–123, 125–130, 132–139, 210
———, Hermbstaedt's *Bibliothek*, 101–102, 105, 110, 114–115, 125, 159, 211
———, Scherer's *Allgemeines Journal*, 118, 141–142, 160, 199, 220
———, Trommsdorff's *Journal*, 139, 141, 159, 218
———, Weber's *Physikalischchemisches Magazin*, 74
pharmaceutical schools, 58, 92, 174, 191, 208, 210–211, 218
pharmacy, pharmacists, 4, 7, 11, 14–15, 17, 21–22, 26, 28, 30, 34–36, 38, 49, 54–58, 60, 64–65, 70, 72, 78, 80, 86–89, 92, 110–111, 114, 125–128, 145, 164, 173–174, 178–179, 181, 185, 187, 189–191, 195, 197–198, 200–203, 205, 208–211, 213, 215, 217–221, 223–224, 228–233, 236–239, 241–244, 246, 248–249, 251–252, 254–259, 261–262, 264, 267–268, 272–292
physicians. *See* medicine
physics, physicists. *See* natural philosophy
physiology, 35, 64–65, 108, 167, 175, 198, 209, 214–215, 235, 243, 249
Pietism, 7, 18, 167, 169–173, 176, 179, 182, 197, 200–203
polymathy, polymaths, 6, 40, 64, 198
professors of, academicians for, chemistry, 10, 30–31, 33–45, 47–49, 55–56, 59–60, 64–65, 67, 70, 83, 88, 106, 108–111, 114, 125, 130, 140, 166–169, 173–174, 176–188, 193–198, 202–204, 206–211, 214–223, 226–269, 272–273, 275, 277–278, 281, 285, 291

republic of letters, 2, 6, 23
rulers. *See* nobles

scientific academies and societies, 1–2, 25, 31–32, 35, 45–49, 57–59, 65–66, 70, 75, 77–78, 80–81, 84, 99–101, 115, 147, 149–150, 155–156, 169, 171, 174–176, 180–182, 184, 186, 188, 190, 192–194, 198–199, 201–203, 206, 211, 214, 216–219, 220–225, 227, 230–231, 238, 259–260; laboratories, 47, 176, 180, 206, 230, 238
scientific communities, 1–2, 62, 147–148
scientific revolutions, 3, 97–98, 104, 148
Seven Years' War, 18, 20, 43, 45, 53, 241, 259
smelting. *See* metallurgy
specialization, 2, 40–42, 57, 61, 146–148, 150–151
Stahlianism, 8–11, 13–15, 20, 103–104, 120
state economics. *See* cameralism
subscriptions, subscribers, 3, 21–22, 29, 85–89, 145, 271–299
surgery, surgeons, 22, 33, 35, 56, 64, 165–166, 168–169, 174, 191, 200–201, 205, 214, 228, 246, 248, 252, 267, 274, 276, 281–284, 287, 289

technology, technical chemists, 18–19, 42–43, 29, 57–60, 88, 137, 164, 170–171, 181–184, 190, 193–194, 196–197, 201, 206–207, 210–214, 219, 253, 278
Thirty Years' War, 18, 238

universities, 1–2, 6, 8, 18, 25–26, 43–45, 73, 79, 148–151, 164, 182, 208, 227; laboratories, 30–31, 33–35, 44, 130, 186, 188, 202, 214, 223, 226–228, 236, 240, 242–244, 247, 251, 256–258, 261, 265–266, 269; medical faculties, 21, 30–42, 55–58, 64–68, 70, 165–181, 183–196, 198–204, 209–210, 214–216, 218, 223–224, 226, 228–230, 234, 236–252, 254–259, 261–266, 268–269; philosophical faculties, 42, 64, 167–168, 174, 178–179, 184–185, 192, 198, 202, 204, 208, 210–212, 214, 216–217, 220–221, 223–224, 227, 231, 239–240, 243–244, 247, 249–250, 253, 255, 262
utilitarianism, 2, 18–20, 24, 79, 144, 147, 149–150

women, 29, 64, 80, 96, 170, 172, 175, 179, 186–189, 191, 196–197, 205, 222

# PERSON AND PLACE INDEX

With a few self-explanatory exceptions, this index is restricted to persons and places appearing on two or more pages. Also see the name and geographical indices to Crell's subscription lists, pp. 272–299 above.
Page numbers in italics refer to the Biographical Profiles or Institutional Histories in the appendices.

Abich, R. A., 67, 272
Abildgaard, P. C., 130, 135, 272
Achard, F. C., 44, 47–48, 51–52, 54, 57, 59, 88–92, 146, 153, 180, 194, 198–199, *206–207*, 230–232, 272
Alberti, M., 172, 178, 246
Alembert, J. d', 47–48
Altdorf, 31, 33, 166–167, 177, 226, *228–229*
Altona, 170, 292
Amsterdam, 175, 201, 292
Andreae, J. G. R., 26, 38, 51–52, 55–57, 90, 153, 176, 179, 181, *187*, 197, 203, 205, 272
Annaberg, 164, 190
Ansbach, 166–167, 292

Baader, F., 107, 110–111
Bacon, F., 39
Baier, J. J., 177, 228–229
Baldinger, E. G., 36, 56, 65, 68–69, 72, 74, 79, 146, 157, 159, 190, 273
Bamberg, 33, 226, *229*
Banks, J., 75, 188, 273
Barner, J., 166–167
Basel, 41, 225, 227, *230*
Baumer, J. W., 44–45, 243–244
Bayen, P., 119
Beireis, G. C., 38, 64–67, 70, 79, 199, 209, 249–250, 273
Bergman, T. O., 75, 77, 187, 205, 230, 273
Berlin, 1, 8, 10, 21–25, 28, 31–33, 35, 38–39, 44–49, 52, 55–59, 70, 75, 78, 81, 84, 88, 91, 107, 110, 114–115, 125–128, 130–131, 134–135, 137, 149, 155, 164, 167–170, 173–178, 180–185, 187, 189, 192–199, 203–204, 206, 210–214, 216–217, 219–220, 222–224, 226–227, *230–234*, 241, 293
Bern, 22, 223, 293
Bernburg, 175, 191, 209, 217, 293
Beroldingen, F.C. von, 216, 273
Berthollet, C. L., 99–100, 105, 123, 223
Berzelius, J. J., 217, 219, 223
Best, P., 220, 237
Bindheim, J. J., 79, 86, 274
Bischof, C. G. C., 214, 223
Bischoff, H., 130, 274
Black, J. 65–66, 74–75, 77, 84, 157, 159, 199
Blagden, C., 75, 87
Blankenburg, 58, 181, 187, 235, 293
Blumenbach, J. F., 64, 100–101, 107, 110–111, 158, 216, 274
Boerhaave, H., 8, 156, 173, 175, 181
Börner, F., 10, 22–24, 157
Bohn, J., 51, 153, *165*, 173–174
Bonn, 32, 215, 219, 223, 226–227, *234*
Born, I., 21, 58, 90–91, 106–109, 153, 186, 188, *196–197*
Bourquet, D. L., 127, 135, 231
Boyle, R., 168

Breslau, 185, 212, 216, 223–224, 241, 262
Brugnatelli, L. V., 87, 103, 186, 274
Brunswick, 33, 43–44, 62, 66–68, 70, 88, 139, 146, 180–181, 199–200, 214, 216, 220–221, 226–227, *235–236*, 240, 293
Brussels, 99, 215
Bucholtz, W. H. S., 51–52, 55, 57–58, 72, 74, 90, 119, 153, 187, *191–192*, 208–209, 211, 217–218, 274
Bucholz, C. F., 51, 101, 153, 160–161, 213, *217–218*, 224, 239
Budapest, 31, 33, 52, 195–196, 226, *236*, 267, 293
Büchner, A. E., 187, 239
Bützow, 24, 33, 262
Burghart, G. H., 83, 156
Burghausen, 78, 81

Cajetano, D. M., 7, 170
Carl of Brunswick, 16, 166
Carl August of Saxe-Weimar, 191, 208, 219–220
Carl Wilhelm Ferdinand of Brunswick, 75–76, 200
Cartheuser, F. A., 38, 51–52, 55, 57–58, 76, 153, 157, 176, 178, 190–191, *192*, 193, 199, 204
Cartheuser, J. F., 38, 51, 55, 59, 153, 169, 172, 174, *178*, 192–193, 197, 241
Catharinenburg, 86, 293
Cavendish, H., 98, 275
Chaptal, J. A., 114, 209
Coburg, 192–193, 293
Colmar, 226, *236–237*, 293
Cologne, 31, 220, 226, 234, *237*, 293
Copenhagen, 78, 81, 130, 170, 179, 201, 223, 293
Cothenius, C. A. von, 211, 275
Cramer, C., 168, 238–239
Cramer, J. A., 21, 26, 38, 43, 51, 53–54, 57–58, 67, 153–154, 172, 176–178, *181–182*, 183–187
Crell, L., 1, 4, 26, 38, 51–52, 55, 57, 62–96, 100, 102–105, 117–118, 122, 140–142, 144–148, 153, 158–159, *198–200*, 209, 215, 220, 223, 235, 245, 249, 271
Cullen, W., 65–66, 157, 189

Dalberg, C. von, 39, 46
Danzig, 81, 135, 197, 294
Darmstadt, 7, 167
Dehne, J. C. C., 70, 72, 90–91, 275
Delius, H. F., 111, 240, 275
Dijon, 81, 87, 99, 294
Dippel, J. C., 7, 51, 153, *169–170*, 172
Dolfuss, J. C., 86, 276
Dorpat, 216, 220
Dresden, 33, 164, 171, 182–183, 190, 200–201, 222, 294
Dublin, 81, 87, 294
Düsseldorf, 33, 294
Duisburg, 31–32, 226, *238*
Duising, J. G., 30, 258

Edinburgh, 65–66, 68, 74–75, 77–78, 81, 189, 209
Ehrhart, F., 80, 187, 205, 276
Ehrmann, F. L., 264, 276
Elberfeld, 187, 294
Eller, J. T., 51, 84, 153–154, 166–167, 173–174, *175–176*, 177, 180–181, 194, 203, 231, 255
Elwert, J. K. P., 21, 276
Erfurt, 26, 31, 34, 46, 60, 72, 75, 78, 128, 132, 134, 155, 168, 172, 174, 179, 184–185, 187, 190, 208–210, 213, 217–219, 224, 226, *238–240*, 243–244, 294
Erlangen, 31, 33, 85, 110, 125, 131, 139, 150, 185, 214, 226–227, 235, *240*, 294
Erxleben, J. C. P., 28, 203, 209
Euler, L., 192, 194

Fabricius, P. C., 64, 67
Fabbroni, G. V. M., 105, 118
Ferber, J. J., 90–91, 230, 276
Fiedler, C. W., 217, 276
Florence, 223, 294
Fontana, F., 95, 223
Forster, G., 75, 80, 196, 276
Forster, J. R., 75, 220

## Person and Place Index / 307

Fourcroy, A. F., 99–100, 102, 159, 216, 223
Francke, A. H., 172, 176
Frankfurt am Main, 7, 33, 65, 180, 185, 187, 213, 294
Frankfurt an der Oder, 31, 34, 38, 55, 78, 81, 130, 166, 168, 178, 184, 192–194, 205, 226, *241*, 294
Franklin, B., 100–101, 189
Frederick I of Prussia, 7, 46, 164, 170, 173
Frederick II of Prussia, 16, 20, 46–49, 147, 176–177, 181, 183–184, 195, 198, 206, 230
Frederick William I of Prussia, 46–47
Frederick William II of Prussia, 206
Freiberg, 9, 21, 31, 43–44, 52, 56, 89, 107, 109–110, 140, 150, 171, 173–175, 177, 180, 182–183, 185, 201, 210–211, 221–223, 226, *241–242*, 294
Freiburg im Breisgau, 30–33, 41, 204, 226, *242–243*
Freienwalde, 176–177, 294
Fuchs, J. N., 198, 222
Fulda, 31, 33, 226, *243*

Gahn, J. G., 70, 277
Gaubius, H. D., 50, 55–56, 187, 189, 202
Gehlen, A. F., 51, 142, 153, 160, 198–199, 203–204, 213–214, 218–220, 223, *224,* 260
Gellert, C. E., 21, 43–44, 51, 54, 58–59, 67, 140, 153, 157, 181, *182–183*, 190, 194, 196, 198, 200, 211, 222, 242
Geneva, 81, 126, 206, 294
Geoffroy, C. J., 173, 185
Gerhard, C. A., 21, 38, 51–52, 55, 58–59, 73, 76, 90, 153, 157, 175–176, 178, 180, 184, 192, *193–195*, 198, 206, 211–212, 217, 230–232
Gericke, P., 38, 51, 153, 169, 172, *177–178*, 181, 246, 249
Giessen, 31, 44–45, 58, 169–170, 172, 191–192, 204, 226, *243–244*, 294
Girtanner, C., 50, 90, 94, 103, 110–112, 117, 125–129, 133, 153, 159, 185, 202, *208–209*, 223
Gmelin, J. F., 13, 29, 38–39, 41, 51–53, 55, 57, 59, 72–73, 88–92, 100, 102, 112, 133, 136, 138, 140–141, 146, 148, 153, 158–160, 187–188, 199, *202*, 208–209, 211, 214–216, 221–223, 244–245, 277
Gmelin, L., 202, 248
Gmelin, P. F., 38, 55, 193, 265
Goethe, J. W., 208, 216, 220, 277
Göttingen, 25, 31, 33, 35, 39, 41, 52, 65, 75, 78, 81, 83, 88, 107, 110–111, 125, 128, 130, 150, 155, 175, 178, 184, 187, 199–200, 202–203, 208–209, 211, 214–216, 218, 221, 223, 226, *244–245*, 262, 294
Göttling, J. F. A., 51–52, 55, 57–58, 72, 74, 90–92, 94, 101–102, 122, 137–138, 140, 146, 153, 159–160, 190–191, 202, 205, *207–208*, 217–221, 223, 253, 278
Goslar, 170, 216, 294
Gotha, 166, 294
Gravenhorst, C. J., 67, 278
Graz, 226, *245*
Gren, F. A. C., 51, 88, 90–92, 94, 101–104, 112, 120–137, 140, 142–143, 146, 153, 158, 160, 199, *209–210*, 216, 218, 220, 223, 246–247, 278
Greifswald, 31, 39, 52, 202–204, 226, 241, *246*, 254, 294
Grosse, J., 175, 185
Gruner, C. G., 252, 278
Guyton de Morveau, L. B., 75, 87, 95, 99–101, 158, 223, 271, 278

Hagen, H., 158, 203, 213
Hagen, K. G., 1, 51–52, 55, 90–92, 104, 153, 158, *203–204*, 212–213, 224, 255–256, 278
Hahnemann, C. F. S., 90–92, 158–159, 200
Haidinger, C., 107–109
Halberstadt, 168, 176, 200, 206, 295
Halle, 6, 8, 32–33, 36, 42–43, 56, 58, 81, 88, 107, 110, 126, 167–168, 171–173, 176–181, 185, 187, 192,

Halle (cont.)
209–210, 216, 220, 223–224, 226–227, *246–248*, 249, 268, 295
Haller, A., 16, 41, 45, 55, 68–69, 72, 184
Hamberger, G. E., 175, 184, 252
Hamburg, 72, 135, 164, 181, 200–201, 203–204, 210, 219, 223, 295
Hameln, 1, 53, 88, 107, 122, 199, 205, 215, 295
Hanckewitz, A. G., 173, 175
Hanover, 52, 56, 76, 92, 167, 173, 187, 197, 205, 214, 295
Hardenberg, C. A. von, 214, 278
Hartmann, P. I., 241, 250
Hassenfratz, J. H., 86, 279
Hecker, A. F., 132–133, 135, 137
Heidelberg, 31, 35, 50, 204, 214–215, 226, *248–249*, 253, 295
Heister, L., 64, 177–178, 199
Helmstedt, 1, 31, 38, 52, 58, 62, 64, 66–68, 70, 78–79, 107, 168, 177–178, 181, 198–199, 209, 215–216, 223, 226, 235, 241, *249–250*, 295
Henckel, J. F., 8–11, 14, 16, 21, 43, 51, 56, 153, 166–167, *171–172*, 173–175, 177, 180, 182–183, 185, 241–242
Herborn, 226, *250*
Hermann, J. B., 264, 279
Hermbstaedt, S. F., 51, 88, 90–92, 100–105, 107, 110–115, 117, 125–140, 142, 146–147, 153, 158–159, 191, 194, 198, *210–212*, 213, 217–218, 220, 222, 231, 233–234, 271, 279
Heyer, J. C. H., 51–52, 55, 88, 90, 94, 153, 187, *200*, 279
Heynitz, F. A. von, 21, 44–45, 171
Hildburghausen, 191, 219
Hildebrandt, G. F., 51, 105, 125, 139–140, 153, 202, *214*, 235, 240, 279
Hildesheim, 216, 295
Höpfner, J. G. A., 102, 280
Hoffmann, C. A., 90, 280
Hoffmann, F., 7–8, 43, 51, 56, 153, 166–167, *168–169*, 172–173, 176–178, 180, 185, 187, 230, 246–247
Hoffmann, J. M., 51, 153, *166–167*, 228

Homberg, W., 164
Hopson, C. R., 65, 159
Horix, J. B. von, 26, 34
Hübner, J. H., 6, 14
Humboldt, A. von, 79, 107, 110–112, 115, 130–131, 135, 159, 198, 209, 216, 220

Idria, 43–44, 186, 196, 263
Ilsemann, J. C., 90–92, 280
Ingen-Housz, J., 50, 153, 188, *189*
Ingolstadt, 31–33, 37, 226, *250–251*, 259, 295
Innsbruck, 33, 186, 188, 226, *251–252*

Jacquin, J. F. von, 188, 209, 266
Jacquin, N. J., 21, 38, 50, 55, 58, 89, 106–108, 153, *188–189*, 196, 202, 215, 263, 266
Jars, A. G., 43
Jena, 6, 31, 33, 36, 41, 57, 122, 140, 150, 165–168, 170–171, 174–175, 177–179, 184, 187, 191, 204, 207–208, 219–220, 226–227, *252–253*, 295
Joseph II of Austria, 188, 196
Juch, K. W., 220, 228
Jüngken, J. H., 7
Juncker, J., 51, 153, 157, 167–170, *172*, 176, 178–181, 246
Justi, J. H. G., 29, 267

Kaiserslautern, 44–45, 52–53, 59, 204, 226, 249, *253*, 295
Karsten, D. L. G., 111–112, 126, 142, 280
Karsten, W. J. G., 100, 188, 209, 281
Kassel, 31, 44, 72, 203, 226, *254*, 296
Kastner, C. W. G., 160, 240, 247–248
Kiel, 31, 33, 135, 150, 223, 226, *254–255*, 296
Kielmeyer, C. F., 223, 265
Kirwan, R., 75, 87, 95, 100–102, 114, 137, 281
Klaproth, M. H., 51–52, 55–58, 79, 88–92, 94, 101, 103, 107, 110–112, 114–118, 126–129, 134–137, 142, 146, 153, 159, 180, 194–195, *197–198*, 203–206, 211, 213, 217, 219–222, 224, 231–233, 281
Klopstock, F. G., 26

Klügel, G. S., 107, 110–111, 281
Königsberg, 1, 6, 22, 31, 33, 36, 43, 52, 55–56, 165–166, 172–174, 182–183, 187, 190, 198, 200–201, 203, 213, 224, 226, 232, *256–257*, 296
Krause, R. W., 167, 252
Kunckel, J., 51, 153–154, *164–165*, 167
Kurella, E. G., 36, 231–232, 282

Lampadius, W. A. E., 51, 105, 125, 130, 134, 136, 140, 153, 161, 183, 198, 202, *221–223*, 242
Lange, J. J., 43, 80, 157, 246
Langensalza, 52, 56, 58, 88, 92, 107, 122, 190, 208, 210–211, 296
Laplace, P. S., 99–100
Lasius, G. S. O., 90, 282
Laugier, R. F., 195, 266
Lavoisier, A. L., 3, 75, 95–96, 98–103, 119–120, 220, 223
Lehmann, J. G., 21, 23–24, 51, 55, 57, 83, 94, 153, 157, 175–176, 180–181, *183–184*, 188, 190, 193, 195, 231
Leibniz, G. W., 7–8, 39, 46, 230
Leiden, 43, 50, 55–56, 58, 65, 170, 175, 181, 187–189, 202
Leipzig, 6, 18–19, 22, 31, 33, 36, 43, 52, 55–56, 59, 75, 107, 110, 165–166, 172–174, 177, 179, 181–183, 185, 187, 190, 198–201, 220, 226, 232, *256–257*, 296
Lemery, L., 175, 181
Lemery, N., 7–8
Leonhardi, J. G., 51–52, 55, 90–92, 101–102, 104, 119, 153, 158–159, 190, *200–201*, 282
Leopold II of Austria, 199, 282
Libavius, A., 2, 8
Lichtenberg, G. C., 90, 100, 112, 128, 130, 134, 136, 149, 208–209, 221, 223, 282
Liebig, J., 150, 243
Link, H. F., 103, 107, 110, 114, 153, 159, 194, 198, 202, 211, *216–217*, 220, 223, 247, 262
Linnaeus, C., 181, 195, 209
Lomonosov, M. V., 171, 182
London, 66, 68, 70, 81, 87, 100, 126, 153–156, 168, 173, 175, 187, 189, 196, 199, 202, 209, 223, 296
Louvain, 188–189, 215
Lowitz, J. T., 86, 202, 222, 283
Ludolf, H., 51, 53, 55, 59–60, 153–154, 174, *179–180*, 184, 187, 238–239, 257

Macquer, P. J., 101, 190, 202, 212
Magdeburg, 70, 168, 175, 191, 193, 296
Magnus, G., 211, 217, 223
Mainz, 26, 34, 180, 227, 238, *257–258*, 269
Mangold, C. A., 51, 55, 57, 59, 153–154, 157, 174, 179, *184–185*, 187, 190, 218, 238–239
Mannheim, 33, 46, 78, 81
Marburg, 30, 150, 172, 215, 226, 250, *258–259*, 296
March, C., 246, 254
Marggraf, A. S., 23–24, 38, 47–49, 51, 55–57, 59, 70, 76, 83, 153–154, 169, 171–173, 175–176, 179, *180–181*, 182–186, 190, 192–193, 195, 197, 203, 206, 213, 230–231
Maria Theresa of Austria, 20, 37, 186, 189
Martius, E. W., 107, 110–111, 283
Marum, M. van, 102, 115, 210, 223
Maupertuis, P. L. de, 48
Mayer, J. T., 107, 110–111, 125, 131, 134, 136, 283
Meibom, H., 168, 249
Meidinger, K. von, 107–108, 271
Meier, C. E., 132–133, 135
Meissen, 52, 56, 59, 182, 190, 201
Mencke, J. B., 17, 24
Menn, J. G., 28, 158, 237
Métherie, J. C. de la, 87, 101, 120, 209, 283
Meyer, J. C. F., 38, 51–52, 55, 57, 59, 90–92, 153, 176, 180, *195*, 198, 203, 213, 284
Meyer, J. F., 51, 55, 84, 153, 172, *179*, 187
Michaelis, J. D., 25–26, 40
Mikan, J. G., 107–109, 261
Milan, 87, 296
Minden, 168, 296

Mitscherlich, E., 211, 216–217, 223
Model, J. G., 50, 203
Mönch, C., 72, 258–259
Molitor, N. K., 106, 257
Moll, C. E., 141, 214
Mons, J. B. van, 99, 123, 136–138, 159, 215
Montpellier, 184, 209
Moscow, 86, 254, 296–297
Münster, 226, *259*
Munich, 33, 46, 150, 156, 184, 211, 214, 216, 218–219, 222–224, 227, *259–260*

Napoleon of France, 218, 222
Nebel, D. W., 215, 248
Neumann, C., 10, 38–39, 47, 51, 55–56, 58, 153, 156, 167, *173–174*, 175–176, 180, 185, 203, 230–231
Newton, I., 175
Nicolai, E. A., 204, 252, 285
Nicolai, F., 26, 37, 45, 73, 76, 85, 110, 114, 159
Nietzki, A., 228, 246
Nonne, J. P., 184, 239
Nose, C. W., 69–70, 93, 285
Nuremberg, 7, 165, 185, 228, 297

Offenbach, 209, 297
Olmütz, 226, *260–261*
Orlovius, A. J., 255, 285
Osnabrück, 179, 297

Paracelsus, T. P., 10
Paris, 45–46, 65, 68, 81, 84, 86–87, 97, 99–102, 110, 119–120, 156, 173, 175, 181, 185, 188–189, 202, 209, 214, 216, 220, 223, 254, 297
Pavia, 50, 87, 186, 297
Pein, C. H., 114, 211, 231
Peithner, J. T., 196, 261
Peschier, J., 126–127, 132
Petzold, A. F., 43, 55, 256–257
Pfaff, C. H., 105, 153, 161, 202, *223–224*, 254–255
Philadelphia, 81, 86, 297
Pickel, G., 215, 269, 285
Planer, J. J., 218, 239, 285
Ploucquet, W. G., 193, 286

Poerner, C. W., 51–52, 55–56, 58–59, 153, *190*, 200–201
Pope, A., 40
Pott, J. H., 22, 35, 38–39, 47–48, 51, 55–57, 59, 84, 153–154, 167, 169, 172, 174–175, *176–177*, 179–185, 187, 192–193, 195, 203, 213, 230–231
Prague, 33, 43–44, 107–109, 150, 196, 226, *261–262*, 297
Priestley, J., 95, 101, 208
Procháska, G., 107–109
Pyrmont, 172, 203, 297

Quedlinburg, 175, 178, 181, 190, 197, 297

Réaumur, R. A. F. de, 173, 185
Reden, F. W. von, 135, 212, 286
Rettberg, E. F., 130, 135
Reuss, C. G., 223, 265, 286
Richter, F. A., 101, 246, 286
Richter, J. B., 51, 105, 137–140, 153, 159, 194, 204, *212–213*, 220
Ridiger, A., 55, 190, 257
Riga, 220, 297
Ritter, J. W., 208, 220, 224
Roi, J. P. du, 67, 287
Rolfinck, W., 165–166, 168, 170–171, 174, 252
Rome, 166, 223
Rose, H., 211, 217
Rose, V. Jr., 51, 131, 135, 153, 195, 197–198, 203–204, 211, *213*, 224, 287
Rose, V. Sr., 58–59, 83–84, 157, 197, 213, 232
Rostock, 31, 33, 210, 216, 220, 226–227, 246, 254, *262–263*, 297
Rothe, G., 51, 153, 156, 165, 167, *173*
Rothenburg ob der Tauber, 50, 129, 297
Rousseau, G. L. C., 28, 101, 251, 287
Ruprecht, A. von, 107–109, 263

Sage, B. G., 95, 209
St. Petersburg, 46, 50, 59, 80–81, 86,

130, 156, 167, 171, 173, 177, 181–184, 186, 188, 190, 192, 194, 204, 219–222, 298
Salzwedel, P., 213, 287
Scheele, C. W., 50, 80, 86, 187, 205, 271, 287
Schemnitz, 21, 31, 43–44, 107–109, 186, 188, 196, 226, 261, *263*
Scherer, A. N., 51, 96, 118, 138, 140–142, 153, 160, 199, 208, 216, 219, *220–221*, 237, 247
Scherer, J. A., 106–108, 125, 140, 159, 189, 267
Schiller, J. M., 90, 129, 133, 135, 287
Schlüter, C. A., 51, 54, 153, *170–171*
Schmid, R. J. F., 179, 184
Schnecker, J. E. D., 216, 288
Schöningen, 70, 298
Schrader von Schliestedt, H. B., 67, 181–182
Schreber, J. C. D., 240, 288
Schulpforta, 166, 183
Schulze, J. G. F., 209, 288
Schulze, J. H., 180, 246
Schurer, F. L., 101, 114, 288
Schweigger, J. S. C., 240, 247
Schwerin, 204, 298
Scopoli, G. A., 21, 50, 58, 153, 158, *186–187*, 196, 263, 288
Selle, C. G., 211, 288
Sommer, J. C., 67, 289
Spielmann, J. J., Jr., 185, 289
Spielmann, J. J., Sr., 180, 185
Spielmann, J. R., 38–39, 51–52, 55, 57, 59, 73, 76, 153, 171, 176, 180, 184, *185–186*, 187, 193, 197, 199, 209, 264
Spring, J. P., 251, 259–260
Stahl, G. E., 8–11, 14, 21, 38, 43, 46, 50–51, 83, 96–97, 100, 153, 156, 166, *167–168*, 169, 171–176, 187, 231
Stargard, 218, 298
Stettin, 50, 59, 184, 195, 213, 218, 298
Stockholm, 78, 81, 87, 156, 164, 170, 298
Storr, G. C. C., 90–92, 265, 289
Stralsund, 50, 60, 202, 298

Strasbourg, 31–33, 39, 52, 65, 68, 73, 101, 169, 180, 184–185, 193, 199, 209, 220, 226–227, *263–264*, 298
Stromeyer, F., 199, 202, 218, 245
Stuttgart, 31, 33, 65, 204, 223, 226, *264–265*, 298
Suckow, G. A., 28–29, 44, 51–53, 55, 59, 153, 157–158, *204–205*, 249, 253, 289
Suckow, L. J. D., 157, 204
Suersen, J. F. H., 127, 135–136, 188
Swieten, G. van, 37, 43–44, 181, 186, 188, 263, 266

Teichmeyer, H. F., 51, 153, 156, 165–166, *174–175*, 178–179, 184
Tihavsky, F. von, 108
Tilemann, J., 250, 258
Trebra, F. W. H. von, 90, 290
Trommsdorff, J. B., 51, 97, 105, 125, 132–141, 153, 160–161, 191, 208–210, 213, 217, *218–220*, 239
Trommsdorff, W. B., 72, 209–210, 218, 239
Tschirnhaus, E. W. von, 7
Tübingen, 31, 33, 38, 52, 55, 193, 202, 220, 223, 226, *265–266*, 298

Uppsala, 81, 195, 220, 298
Utrecht, 170, 177

Vienna, 31, 33, 37–38, 43–44, 50, 55, 89, 106–108, 174, 186, 188, 193, 195–197, 199, 202, 215, 223–224, 226, 234, 236, 263, *266–268*, 298
Vogel, R. A., 35, 38–39, 51, 55, 59, 83, 153, 176, 179, 184, *187–188*, 190, 200, 202–203, 209, 218, 244–245
Vogelmann, Jr., 257, 269
Voigt, C. F., 217, 290
Voigt, J. C. W., 90, 290
Volta, A., 186, 223, 290
Voltaire, F. M. A., 79

Watt, J., 66, 100
Weber, J. A., 51–52, 55, 57, 59, 74, 146, 153, 157, 185, *192–193*

Wedel, G. W., 51, 153, *165–166*, 167–168, 171, 174–175, 179, 187, 252
Wegely, J. G., 211, 291
Weigel, B. N., 60, 202, 291
Weigel, C. E., 28, 38–39, 51–52, 55, 59–60, 90–91, 102, 153, 158, 187, *202–203*
Weimar, 52, 58, 72, 142, 191, 208–209, 211, 218–220, 299
Wenzel, C. F., 21–22, 51–52, 55–58, 89, 153, 183, 190, *201*, 242, 271
Werlhof, P. G., 175, 178, 187
Werner, A. G., 90, 222
Wernigerode, 164, 197
Westendorff, J. C., 90, 119, 291
Westrumb, J. F., 1, 51–52, 55–56, 79, 88–92, 94–95, 100, 103–105, 107, 112, 115, 117, 122–137, 140, 142–144, 146, 153–154, 158–159, 197, 199, *205–206*, 215, 291
Wetzlar, 179, 299
Wiegleb, J. C., 26–27, 51–53, 55, 57–58, 60–61, 72–73, 88–92, 94–96, 100, 102–105, 107, 117, 119, 122–123, 138, 140, 144, 146, 148, 153, 157–159, 179, 187, *190–191*, 203, 208, 210–211, 291
Wieland, C. M., 19, 26, 79
Winterl, J. J., 51–53, 55, 153, *195–196*, 236, 291
Wittenberg, 43, 164–165, 169, 173, 182–183, 190, 198, 200, 225, 227, *268*, 299
Wittwer, P. L., 28, 228
Wolfenbüttel, 169, 299
Wolff, C., 18, 45, 171, 177–178
Wolff, F., 1, 107, 110, 114, 131, 135
Wolter, J. A. von, 37, 259
Würzburg, 31–33, 41, 204, 215, 226, 229, 234, *268–269*
Wurzer, F., 105, 153, 159, 161, 188, 199, 202, 205, *215–216*, 234, 258

Zimmermann, E. A. W., 67, 200, 292
Zürich, 223, 299